Finite Elements

ICASE/NASA LaRC Series

Stability of Time Dependent and Spatially Varying Flows
D.L. Dwoyer and M.Y. Hussaini (eds.)

Studies of Vortex Dominated Flows
M.Y. Hussaini and M.D. Salas (eds.)

D.L. Dwoyer M.Y. Hussaini R.G. Voigt
Editors

Finite Elements

Theory and Application

Proceedings of the ICASE Finite Element Theory and Application Workshop Held July 28–30, 1986, in Hampton, Virginia

With 113 Illustrations

Springer-Verlag
New York Berlin Heidelberg
London Paris Tokyo

M.Y. Hussaini
R.G. Voigt
Institute for Computer Applications
in Science and Engineering (ICASE)
ICASE/NASA
NASA Langley Research Center
Hampton, VA 23665
U.S.A.

D.L. Dwoyer
NASA Langley Research Centre
Hampton, VA 23665
U.S.A.

Library of Congress Cataloging-in-Publication Data
ICASE Finite Element Theory and Application Workshop
(1986: Hampton, Va.)
Finite elements.
1. Finite element method—Congresses.
I. Dwoyer, Douglas L. II. Hussaini, M. Yousuff.
III. Voigt, Robert G. IV. Institute for Computer
Applications in Science and Engineering. V. Title.
TA347.F5125 1986 620'.001515353 87-20766

Typeset by Asco Trade Typesetting Ltd., Hong Kong.
Printed and bound by Edwards Brothers Inc., Ann Arbor, Michigan.
Printed in the United States of America.

9 8 7 6 5 4 3 2 1

ISBN 0-387-96610-2 Springer-Verlag New York Berlin Heidelberg
ISBN 3-540-96610-2 Springer-Verlag Berlin Heidelberg New York

Preface

This volume covers the proceedings of the ICASE/LaRC workshop on "Finite Element Theory and Application" held during July 28–30, 1986. The purpose of this workshop was to provide an update on the status of finite element theory, to assess the impact of this theory on practice, and to suggest directions for future research. There were thirteen participants in the workshop. Some of them were leading mathematicians working on the finite element theory, and the rest expert practitioners in the areas of fluid dynamics and structural analysis.

The first six articles in this volume provide a brief review of the theoretical and computational aspects of finite element methods (FEM). The remaining seven articles deal with a variety of applications highlighting the type of results that are possible, and indicating areas which deserve future research.

The first article is by Temam. It provides an introduction and overview of the general finite element methods for the nonspecialist. It also illustrates the power of finite element methods with two specific applications—the free-surface flow/structure interaction problem and the compressible Euler solution to the flow past a finite aspect ratio flat plate at incidence. The second article by Brezzi is again an introduction and overview of mixed finite element methods. It includes a brief discussion of special techniques for solving the discrete problem, as well as some applications to certain basic problems in elasticity and hydrodynamics. The article by Fix treats the singular finite element methods which are of technological importance, particularly in the area of fracture mechanics. It discusses singular solutions for both linear and nonlinear systems. In his paper, Pironneau summarizes the well-known finite element methods for hyperbolic systems of equations. The applications are confined to fluid dynamics. Solution techniques for the algebraic equations resulting from the finite element discretization of elliptic problems is the theme of Nicolaides and Choudhury's article. They survey the recent developments in the area of preconditioned iterative methods and algebraic multigrid methods. Gunzburger's article is a brief survey of the mathematical aspects of finite element methods for the primitive-variable formulation of steady-state incompressible Navier–Stokes equations. Only those mathematical results are discussed which are immediately relevant to practical computations.

All of the remaining articles, with the exception of Park's, deal mainly with the individual researcher's work in the pertinent area. Carlson and Miller discuss their gradient weighted moving finite element method for the solution of nonlinear equations involving steep moving fronts, and include examples of a two-dimensional Burgers' equation and a nonlinear diffusion equation. Morgan, Peraire, and Löhner describe the extension of their adaptive mesh refinement methods for compressible Navier–Stokes equations. They have provided some results for two-dimensional problems which include high Mach number flow past a cylinder and a blunt body at incidence, shock reflection from a flat plate, and shock impingement on a cylinder. Atluri summarizes some recent work on multifield finite element methods (with particular reference to consistency and stability of such methods), singular finite element methods for fracture mechanics, and the field boundary element method for the large deformation problems of Kirchhoff shells. Babuška provides the state of the art on the *p* and *h*-*p* versions of the finite element method with reference to elliptic problems. The article by Park overviews the theoretical and practical aspects of transient analysis methods in structural dynamics, and discusses error and step-size control strategies. It also covers some recent advances in partitioned solution procedures, subcycling techniques, and parallel computations. Oden describes the *h* version of the adaptive finite element method. He presents some representative results from the simulation of supersonic rotor–stator flow fields and the solution of quasistatic problems in viscoplasticity of metals under high temperatures. Last, but not the least, is the work of Thornton and Dechaumphai in the important area of integrated flow-thermal-structural analysis. They report on the recent progress in this field at LaRC, and present some preliminary computational results for the coupled flow, thermal and structural response of a thin panel in a Mach 3 flow.

The editors would like to take the opportunity to thank all the participants in the workshop for their cooperation in making the workshop a success, and the staff of Springer-Verlag for their assistance in bringing out this volume.

Contents

Contributors

Atluri, Satya N.
Center for Computational Mechanics
Georgia Institute of Technology
Atlanta, GA 30332, U.S.A.

Babuška, I.
Institute for Physical Science and Technology
University of Maryland
College Park, MD 20742, U.S.A.

Brezzi, F.
Dipartimento di Meccanica Strutturale
Istituto di Analisi Numerica del Consiglio Nazionale delle Ricerche
27100 Pavia, Italy

Choudhury, Shenaz
Department of Mathematics and Statistics
University of Pittsburgh
Pittsburgh, PA 15260, U.S.A.

Carlson, Neil
Department of Mathematics
University of California
Berkeley, CA 94720, U.S.A.

Dechaumphai, Pramote
Mechanical Engineering and Mechanics Department
Old Dominion University
Norfolk, VA 23508, U.S.A.

Fix, George G.
Department of Mathematics
University of Texas
Arlington, TX 76019, U.S.A.

Gunzburger, Max D.
Department of Mathematics
Carnegie Mellon University
Pittsburgh, PA 15213, U.S.A.

Löhner, R.
Naval Research Laboratory
Washington, DC 20370, U.S.A.

Miller, Keith
Department of Mathematics
University of California
Berkeley, CA 94720, U.S.A.

Morgan, K.
University College of Swansea
Singleton Park
Swansea SA28PP, U.K.

Nicolaides, R.A.
Department of Mathematics
Carnegie Mellon University
Pittsburgh, PA 15213, U.S.A.

Oden, J. Tinsley
Aerospace Engineering and Engineering Mechanics Department
The University of Texas at Austin
Austin, TX 78712, U.S.A.

Park, K.C.
Department of Mechanical Engineering
University of Colorado
Boulder, CO 80309-10019, U.S.A.

Peraire, J.
Naval Research Laboratory
Washington, DC 20370, U.S.A.

Pironneau, O.
INRIA
Rocquencourt
Le Chesnay 78153, France

Temam, Roger
Laboratoire d'Analyse Numerique
Université de Paris-Sud
91405 Orsay, France

Thornton, Earl A.
Mechanical Engineering and Mechanics Department
Old Dominion University
Norfolk, VA 23508, U.S.A.

CHAPTER 1

Survey of the Status of Finite Element Methods for Partial Differential Equations

Roger Temam

1.1. Introduction

It is totally impossible to survey the theory of finite element methods (FEM) within a few pages, and the object of this article is to describe for the non-specialist some very basic ideas and concepts in finite elements approximations, and to discuss some future trends in the theory without any attempt at being exhaustive. Beside this survey part, this article contains in Sections 1.4.3 and 1.5.1 a report on some new results concerning two specific problems, *viz. a free boundary fluid structure interaction problem* and the *Euler equations for inviscid flows*.

There is no agreement about the first appearance of the method. Finite element methods have probably been used for many years for computing and engineering purposes in a more or less explicit form. R. Courant mentions in [10] the approximation of a function in $\mathbb{R}^2$ by continuous piecewise linear functions on a triangulation, and this may be the first appearance in the mathematical literature.

Although it is difficult to track the first appearance of the method, there is no doubt that the first systematic and large scale utilizations of the FEM occurred in the sixties in *solid mechanics* engineering. The period coincides, of course, with the first computers and the early stages of what we now call scientific computing. Probably the reason that FEM immediately became popular among solid mechanics engineers is that, as we recall later, the foundations of the FEM coincide with some very fundamental concepts in solid mechanics. The method has spread with different levels of response in fluid mechanics, in optimization and control theory, and among mathematicians.

Like the solid mechanists, mathematicians (numerical analysts and some more theoretically oriented mathematicians) have been working in FEM because the methods are appropriate for mathematical treatment, and are very close in their fundamental concepts to the ideas and tools that are used in the mathematical treatment of the linear and nonlinear boundary value problems by functional analysis.

The mathematical and engineering literature on FEM for partial differential equations is abundant, and there is no way to survey it here. The questions that we address are the following ones: In Section 1.2.1, we recall the principle of weak formulations, and in Section 1.2.2, we recall the role of domain decomposition in the context of structural mechanics. We return to domain decomposition in Section 1.5.1 as it relates to future developments in the FEM in relation to parallel computation and some possible extensions of the method. Sections 1.3.1 to 1.3.3 are devoted to the description of very typical mathematical results. Section 1.3.1 describes the general mathematical framework and the most common finite elements. Section 1.3.2 provides some convergence and error results, and Section 1.3.3 is an introduction to mixed and hybrid finite elements. Some specific applications (among many others) of the FEM are then described. Section 1.4.1 is related to the Navier–Stokes equations. Section 1.4.2 deals with fluid–structure interaction problems, and Section 1.4.3 deals with the applications of FEM to the solution of the Euler equations. Finally, as indicated, we return in Section 1.5.1 to domain decomposition and the role that this can play in future developments for FEM.

1.2. Foundations of the Finite Element Methods

The FEM lie on two fundamental ideas:

- the *weak formulation* of a boundary value problem;
- the *domain decomposition*, i.e., the decomposition of the domain corresponding to the problem into smaller subdomains, the elements.

As mentioned previously, both ideas are closely related to basic concepts of solid mechanics. The weak formulation of a boundary value problem coincides with the virtual work theorems and energy principles in the statics of solids. Domain decomposition is also an extrapolation of the natural approach in structural mechanics in which large structures consist of smaller substructures that are properly connected or assembled, and the study of the large structure is reduced to that of the elementary structures and their connections.

1.2.1. Weak Formulations

We begin by recalling briefly the weak formulation of some boundary value problems in solid and fluid mechanics. Other examples of weak formulations will appear later (abstract boundary value problems).

1.2.1.1. *Weak Formulations in Solid Mechanics*

Consider a solid body that fills at rest a region Ω of $\mathbb{R}^3$ with boundary Γ. We assume that the body is subjected to volumic forces of density $f = (f_1, f_2, f_3)$ in Ω, and to surface (traction) forces of surface density $F = (F_1, F_2, F_3)$ on some part Γ_1 of Γ, and reaches a new equilibrium position. The unknowns of the problem are:

the field displacements, $u = (u_1, u_2, u_3)$, $u(x)$, $x \in \Omega$, representing the displacement between the position at rest of a particle $x \in \Omega$ and its new equilibrium position $x + u(x)$;

the stress tensors field, $\sigma = (\sigma_{ij})$.

Under the assumption of small displacements, the equilibrium equations read

$$\sum_{j=1}^{3} \frac{\partial \sigma_{ij}}{\partial x_j} + f_i = 0 \quad \text{in } \Omega, \tag{2.1}$$

$$\sum_{j=1}^{3} \sigma_{ij} v_j = F_i \quad \text{on } \Gamma_1, \tag{2.2}$$

where $v = (v_1, v_2, v_3)$ is the unit outward normal on Γ.

Usually, the displacement u is given on the complementary part Γ_0 of Γ_1, $\Gamma_0 = \Gamma \backslash \Gamma_1$,

$$u = U \quad \text{on } \Gamma_0. \tag{2.3}$$

The so-called set of statically admissible stress tensors $\mathscr{S}_{\text{ad}}(f, F)$ is the set of tensor fields σ satisfying (2.1) and (2.2). The set $\mathscr{C}_{\text{ad}}(U)$ is the set of kinematically admissible displacements fields, i.e., the set of u's satisfying (2.3).

The equations (2.1)–(2.3) that hold for any material are supplemented by the *constitutive equations* of the material, which depend on the material and connect stresses and displacements. Without describing these relations, we can already see the weak formulation of the problem. Let σ, u be solution of (2.1)–(2.3) and let v be another kinematically admissible field of displacements, $v \in \mathscr{C}_{\text{ad}}(U)$ (and $w = v - u \in \mathscr{C}^0_{\text{ad}} \equiv \mathscr{C}_{\text{ad}}(0)$). We multiply (2.1) by w_i, add these relations for $i = 1, 2, 3$, integrate over Ω, and use Green's formula and (1.2), (1.3). We obtain

$$\int_\Omega \sigma_{ij} \varepsilon_{ij}(w)\, dx = \int_\Omega f_i w_i \, dx + \int_{\Gamma_i} F_i w_i \, d\Gamma \qquad \text{for all} \quad w \in \mathscr{C}^0_{\text{ad}}, \tag{2.4}$$

where the Einstein summation convention has been used and $\varepsilon(w) = (\varepsilon_{ij}(w))$ is the strain tensor

$$\varepsilon_{ij}(w) = \frac{1}{2}\left(\frac{\partial w_i}{\partial x_j} + \frac{\partial w_j}{\partial x_i}\right).$$

If we remember that $\sigma = \sigma(u)$ because of the constitutive law, we find that (2.4) is the weak formulation for the displacements.[1] For instance, in the simplest case of linear elasticity, we have pointwise

$$\sigma_{ij}(x) = A_{ijkl} \varepsilon_{kl}(u)(x) \qquad \text{for all} \quad x \in \Omega, \tag{2.5}$$

[1] A similar formulation is available for the stresses σ.

where the coefficients A_{ijkl} define a linear positive invertible operator A in the space of symmetric tensors of order two. Hence (2.4) becomes

$$\int_\Omega A_{ijkl}(u)\varepsilon_{kl}(u)\varepsilon_{ij}(w)\,dx = \int_\Omega f_i w_i\,dx + \int_{\Gamma_i} f_i w_i\,d\Gamma \qquad \text{for all} \quad w \in \mathscr{C}^0_{\text{ad}}. \tag{2.6}$$

In linear and nonlinear elasticity, the weak formulation (2.4) (or (2.6)) coincides with the relation given by the virtual work theorem. It also leads to energy principles.

1.2.1.2. *Weak Formulations in Fluid Mechanics*

Weak formulations in fluid mechanics do not have a physical interpretation as natural as in solid mechanics. They have been introduced by J. Leray [16–18] for the study of weak (i.e., nonregular) solutions of the Navier–Stokes equations in an attempt to explain turbulence by the appearance of singularities in the curl vector of the flow. Although we do not yet know whether such singularities arise in space dimension three, there is no doubt that the contribution of J. Leray has been a fundamental step for the mathematical treatment of the Navier–Stokes equations by the modern methods of functional analysis, and also for the numerical treatment of the equations in computational fluid dynamics (CFD).

Consider, for example, the Navier–Stokes equations of an incompressible fluid in the stationary case. The fluid fills a bounded region Ω of $\mathbb{R}^3$ with boundary Γ. In the Eulerian representation of the flow, the unknowns are the velocity field $u = (u_1, u_2, u_3)$ and the pressure field p; $u = u(x)$, $x \in \Omega$, is the velocity of the particle of fluid at x, and $p(x)$ is the pressure at point x. We have the equations

$$-\nu\,\Delta u + (u\cdot\nabla)u + \nabla p = f \quad \text{in } \Omega, \tag{2.7}$$

$$\operatorname{div} u = 0 \quad \text{in } \Omega, \tag{2.8}$$

where $\nu > 0$ is the kinematic viscosity and f represents volumic forces. Equation (2.7) is the equation of conservation of momentum. Equation (2.8) is the incompressibility equation, i.e., the equation of mass conservation. If Γ is materialized and moving with velocity U, then the nonslip condition on Γ is

$$u = U \quad \text{on } \Gamma. \tag{2.9}$$

Let $\mathscr{V}(U)$ be the space of functions satisfying (2.8) and (2.9). Then $u \in \mathscr{V}(U)$, and if v is a test function in $\mathscr{V}(U)$, $w = v - u \in \mathscr{V}(0)$. We take the scalar product of (2.7) with w (pointwise in $\mathbb{R}^3$), integrate over Ω, and use Green's formula. We have

$$-\int_\Omega \Delta u w\,dx = -\int_\Omega \Delta u_i w_i\,dx = \int_\Omega \frac{\partial u_i}{\partial x_j}\frac{\partial w_i}{\partial x_j}\,dx,$$

$$\int_\Omega \operatorname{grad} p\cdot w\,dx = \int_\Gamma p w\cdot\nu\,d\Gamma - \int_\Omega p \operatorname{div} w\,dx = 0.$$

Hence p disappears and we obtain the weak formulation of (1.7)–(1.9):

$$\begin{cases} u \in \mathscr{V}(U) \text{ and for every } w \in \mathscr{V}(0), \\ \nu \sum_{i,j=1}^{3} \int_\Omega \frac{\partial u_i}{\partial x_j} \frac{\partial w_i}{\partial x_j} \, dx + \sum_{i,j=1}^{3} \int_\Omega u_i \frac{\partial u_j}{\partial x_i} w_j \, dx = \sum_{i=1}^{3} \int f_i w_i \, dx. \end{cases} \tag{2.10}$$

It is equivalent to say that u satisfies (2.10) or that u satisfies (2.7)–(2.9). The striking fact in formulation (2.10) is that *the pressure disappears* and we are left with an equation involving u only. Once u is found we know from mathematical results that there exists p that is defined up to an additive constant by (2.7). However, in the practice of numerical computations p is obtained differently, in general, as the Lagrange multiplier of the constraint div $u = 0$ (see Section 1.3.3).

1.2.2. Domain Decomposition

In structural mechanics it is natural to compute a complicated structure by considering the smaller substructures of which it is made. Each substructure is well modeled, its behavior is well understood, and then the mechanical engineers model the interaction (contact laws, etc.) of the different components to obtain the description of the full structure.

As mentioned before, finite elements in solid mechanics have started as an extrapolation of this idea to continuous bodies: the full solid body is decomposed into smaller elements (cf. Figure 2.1); a simplified constitutive law is adopted on each element; and a simplified version of the constitutive law leads to simplified interactions laws between the contiguous elements.

Similarly, the particle and cells methods in fluid mechanics that are very close to the FEM are based on a simplified analysis of the flow in small cells with simplified fluid transfer laws. The generalization and mathematization of the FEM have led to a more systematic view and a more systematic approach.

Beside discretization, there are several other good reasons to decompose a large domain into smaller subdomains. These reasons are also at the heart of future developments in scientific computation and probably in finite elements. We will return on this important question in Section 1.4.3.

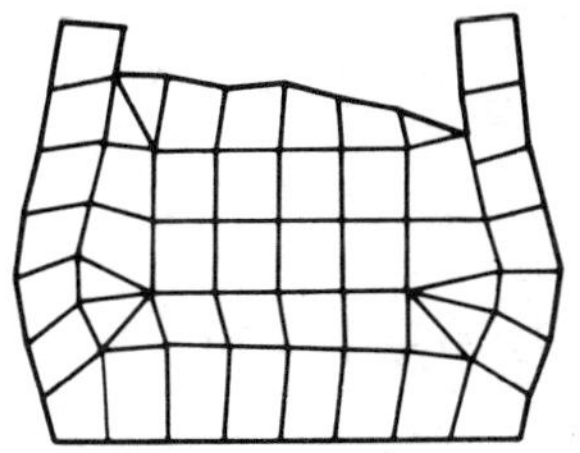

FIGURE 2.1. Discretization grid using triangles and quadrangles.

1.3. Main Methods—Main Mathematical Results

We give an overview of some typical FEM and some typical mathematical results that have been obtained.

1.3.1. The Usual Finite Element Methods

1.3.1.1. *A Model Problem*

We consider as a model problem the following mathematical problem.

We denote by Ω an open bounded domain of $\mathbb{R}^n$, with boundary Γ, and we consider a Laplace equation,

$$-\Delta u + u = f \quad \text{in } \Omega, \tag{3.1}$$

with associated boundary conditions of Dirichlet and Neumann type

$$u = 0 \quad \text{on } \Gamma_0, \tag{3.2}$$

$$\frac{\partial u}{\partial \nu} = 0 \quad \text{on } \Gamma_1, \tag{3.3}$$

where Γ_0, Γ_1 is a partition of Γ. In the two-limit cases $\Gamma_0 = \Gamma$, $\Gamma_1 = \varnothing$ and $\Gamma_0 = \varnothing$, $\Gamma_1 = \Gamma$, we obtain, respectively, the Dirichlet and Neumann problems; the general case is a mixed boundary value problem.

Let V be the space of functions u satisfying (3.2) and possessing a certain level of regularity that we do not specify at the moment. The solution u of (3.1)–(3.3) belongs to V and if v is a test function in V, we multiply (3.1) by v, integrate over Ω, and apply Green's formula. Thanks to (3.3) (and $v = 0$ on Γ_0) we find

$$-\int_\Omega \Delta uv\, dx = \sum_{i=1}^n \int_\Omega \frac{\partial u}{\partial x_i}\frac{\partial v}{\partial x_i}\, dx, \tag{3.4}$$

and thus

$$\begin{cases} u \in V \text{ and} \\ a(u, v) = (f, v) \qquad \text{for all} \quad v \in V, \end{cases} \tag{3.5}$$

where

$$a(u, v) = \sum_{i=1}^n \int_\Omega \frac{\partial u}{\partial x_i}\frac{\partial v}{\partial x_i}\, dx + \int_\Omega uv\, dx, \tag{3.6}$$

and

$$(f, v) = \int_\Omega f(x)v(x)\, dx \tag{3.7}$$

is the scalar product in $L^2(\Omega)$.

Conversely, it can be proved (under suitable regularity assumptions) that if

u satisfies (3.5) then u is the solution of (3.1)–(3.3). Equation (3.1) is derived from (3.5) by appropriate methods using distribution derivatives; (3.2) follows from "$u \in V$," whereas (3.3) is a boundary condition hidden in (3.5). This is a general fact with weak formulations like (3.5): some boundary conditions of the problem are contained in the definition of the space V, and some boundary conditions are contained in equation (3.5).

Let us give a more precise definition of the space V. Roughly speaking, *the space V will be the space of all functions u vanishing on Γ_0 and such that* $a(u, u) < \infty$. More precisely it is easy to see that the expression

$$\{a(u, u)\}^{1/2}$$

is a norm on the space of continuously differentiable functions on $\bar{\Omega}$ that vanish on Γ_0. We define V as the completion of this space for this norm; we obtain the space

$$V = \{v \in H^1(\Omega),\ v|_{\Gamma_0} = 0\}, \tag{3.8}$$

where $H^1(\Omega)$ is the Sobolev space

$$H^1(\Omega) = \left\{v \in L^2(\Omega),\ \frac{\partial v}{\partial x_i} \in L^2(\Omega),\ i = 1, \ldots, n\right\}. \tag{3.9}$$

More generally, $H^m(\Omega)$, the Sobolev space of order m, is the space of functions u square integrable in Ω ($u \in L^2(\Omega)$) such that all derivatives of order $\leq m$ are square integrable also.

1.3.1.2. *Abstract Boundary Value Problem*

The situation in (3.5) is typical of many linear elliptic boundary value problems. The abstract setting is the following one:

We are given a Hilbert space V (norm $\|\cdot\|_V$) and a bilinear form a on $V \times V$ that is *continuous*, i.e.,

$$\begin{cases} \text{There exists } M < \infty \text{ such that} \\ a(u, v) \leq M \|u\|_V \|v\|_V \qquad \text{for all} \quad u, v \in V, \end{cases} \tag{3.10}$$

and *coercive*, i.e.,

$$\begin{cases} \text{There exists } \alpha > 0 \text{ such that} \\ a(u, u) \geq \alpha \|u\|_V^2 \qquad \text{for all} \quad u \in V; \end{cases} \tag{3.11}$$

We are also given a linear continuous form l on V, i.e., an element of the dual V' of V;

and then the problem is

$$\begin{cases} \text{To find } u \in V \text{ such that} \\ a(u, v) = \langle l, v \rangle \qquad \text{for all} \quad v \in V. \end{cases} \tag{3.12}$$

Despite its simplicity, (3.12) is applicable to many interesting boundary value problems in mechanics and physics. The existence and uniqueness of a solution u of (3.12) are classically provided by the Lax–Milgram theorem (see, for instance, [28]).

More generally, nonlinear elliptic boundary value problems can be set in a form similar to (3.12), if we allow V to be a Banach space and a to be nonlinear with respect to its first argument (i.e., a maps $V \times V$ into $\mathbb{R}$ and is linear with respect to its second argument). For instance, it follows readily from (2.10) that the stationary Navier–Stokes equations (2.7)–(2.9) with $U = 0$ can be written in this form. Similarly, consider the problem (3.1)–(3.3) and replace the linear equation

$$-\Delta u + u = f \quad \text{in } \Omega$$

by the nonlinear one

$$-\Delta u + p(u) = f \quad \text{in } \Omega, \tag{3.13}$$

where p is a polynomial of odd degree with a positive leading coefficient. Then (3.13), (3.2), and (3.3) can be set in a form similar to (3.12)

$$V = \{v \in H^1(\Omega) \cap L^{\alpha+1}(\Omega),\ v|_{\Gamma_0} = 0\},$$

$$a(u, v) = \sum_{i=1}^{n} \int_\Omega \frac{\partial u}{\partial x_i}\frac{\partial v}{\partial x_i}\,dx + \int_\Omega p(u)v\,dx,$$

where α is the degree of the polynomial p (see [28]).

In the nonlinear case, there are no general assumptions on a covering all the interesting situations, and we will restrict ourselves to specific examples.

1.3.1.3. *General Form of Finite Element Approximations*

The discretization of the abstract boundary value problem (3.12) consists in choosing:

- a family $(V_h)_{h \in \mathscr{H}}$ of finite-dimensional approximations of V;
- a family $(a_u(u_h, v_h))_{h \in \mathscr{H}}$ of bilinear forms on $V_h \times V_h$ that approximate a.

Roughly speaking, there are two types of discretizations produced by the finite elements:

- the *conforming finite elements* in which the V_h are subspaces of V of higher and higher dimensions as the parameter $h \to 0$;
- the *nonconforming finite elements* in which the V_h are not subspaces of V.

Of course, finite elements have been only used in space dimension $n = 2$ and at a less developed stage when $n = 3$. We consider first the case where Ω is a polygonal set. A basic ingredient of FEM is a triangulation of Ω. By this, we mean a suitable covering of Ω by either:

- a family of triangles;
- a family of rectangles whose sides are parallel to the axes (or more general quadrilateral sets); or
- a combination of triangles and rectangles (or quadrilateral sets).

The triangles or rectangles are the (finite) "elements." The space V_h consists of functions of a given type (usually a polynomial) on each element that are properly connected. The values of the functions of V_h or their derivatives at some particular points of the elements (vertices, midedges, and so on) are the *nodal values* that fully determine the functions in V_h. A natural basis of V_h consists of the *shape functions*: the functions of V_h whose nodal values are 1 for one of them and 0 for all the others. In most cases these functions have a "small" support, and this leads to fairly sparse matrices for the discretized problem. When a function v is defined on Ω (or on an element K), its interpolant on Ω (or K) denoted $r_h v$ (or $r_K v$) is the function of V_h (or the elementary function on K) that has the same nodal values as v.

1.3.1.4. *Conforming Finite Elements* ($n = 2$)

For second-order elliptic boundary value problems, the basic space V is $H^1(\Omega)$ or a product of such spaces or a subspace of such spaces.

The simplest and most common elements used in this case are the P_1 elements on triangles and the Q_1 elements on rectangles. P_1 (resp. P_n) is the set of polynomials of degree ≤ 1 (resp. $\leq m$), whereas Q_1 (resp. Q_m) is the set of polynomials of degree ≤ 1 (resp. $\leq m$) with respect to each variable.

Some other typical elements used for second-order boundary value problems are depicted in Figure 3.1. We will return to the P_1 and Q_1 elements after we briefly describe the elements in Figure 3.1.

Triangles

linear: polynomials of degree ≤ 1 on the triangles; nodal values = values at vertices;

quadratic: polynomials of degree ≤ 2 on the triangles; nodal values = values at vertices and midedges;

cubic: polynomials of degree ≤ 3 on triangles; nodal values = values at vertices, barycenter, and 1/3 points on edges;

reduced cubic: polynomials of degree ≤ 3, vanishing at the barycenter on each triangle; nodal values = values at vertices and 1/3 points on edges.

Rectangles

linear: polynomials of degree ≤ 1 in each variable on rectangles; nodal values = values at vertices;

quadratic: polynomials of degree ≤ 2 in each variable; nodal values = values at vertices, midedges, and center;

cubic: polynomials of degree ≤ 3 in each variable: nodal values = values of function at 16 different points (see Figure 3.1);

reduced quadratic: polynomials of degree ≤ 2 in each variable satisfying a linear relation (on each rectangle); nodal values = values of function at vertices and midedges.

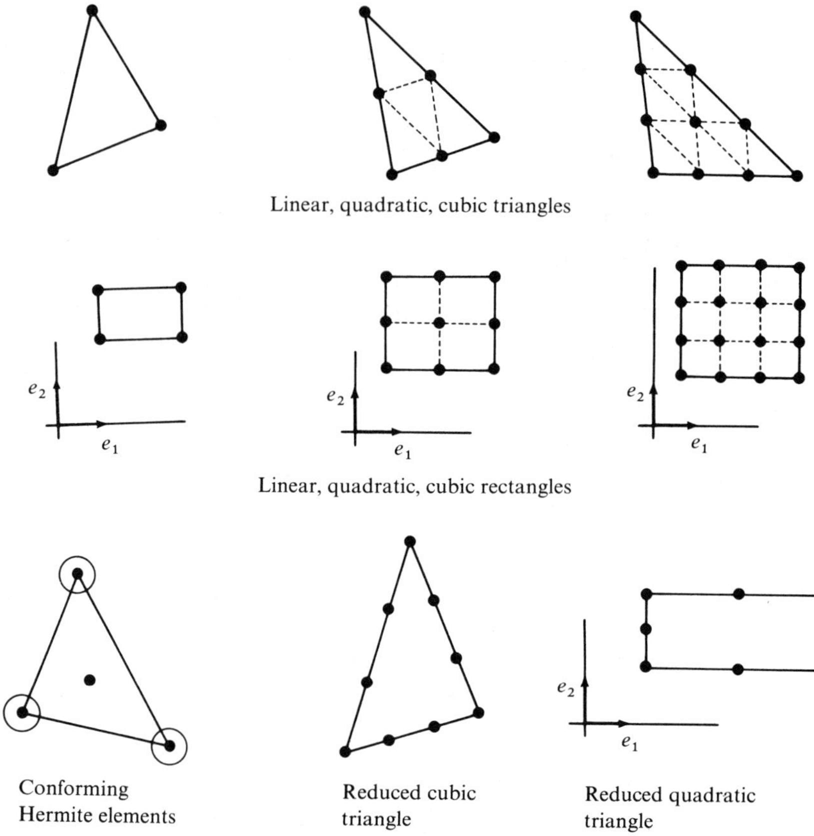

FIGURE 3.1. Conforming finite elements ($n = 2$). (Adapted from Reference [9].)

All functions obtained by these elements are globally C^0 (continuous) except the quadratic Hermite triangle, which produces C^1 approximants (continuously differentiable functions).

More special elements can be found in the literature; see, for instance, the book by P.G. Ciarlet [9] on the mathematical side, and the book by Zienckiewicz [33] and the work of Argyris [1] and others on the engineering or mechanical sides.

The more sophisticated elements produce better (more precise) results that need more computing time and a good expertise in finite elements technology. In a nonspecialized industrial environment, the tendency seems to be the utilization of simple elements of degree one or at most two with a suitable refinment of the mesh.

As mentioned previously, the simplest and most commonly used elements are the P_1 elements on triangles and the Q_1 elements on rectangles with sides parallel to the x and y axes. Let us also mention the quadrilateral elements described hereafter.

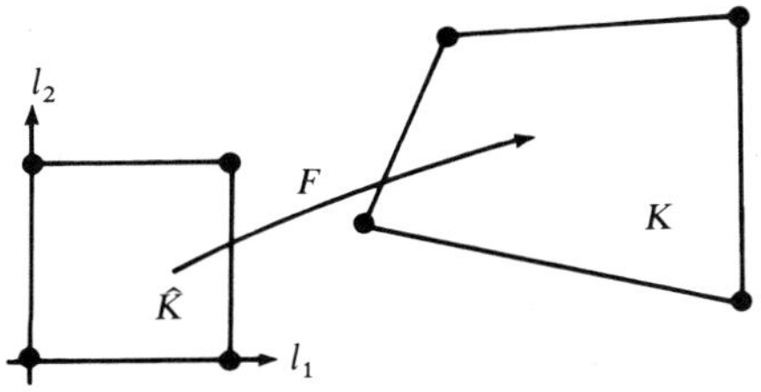

FIGURE 3.2. The mapping F.

Let $\hat{K}$ denote the square $(0, 1) \times (0, 1)$ in the ξ, η plane. We observe that a mapping F with Q_1 components

$$F(\xi, \eta) = \begin{cases} a + b\xi + c\eta + d\xi\eta, \\ a' + b'\xi + c'\eta + d'\xi\eta, \end{cases}$$

can map $\hat{K}$ on any arbitrary quadrilateral K of the x, y plane (cf. Figure 3.2). The image by F of a line in the ξ, η plane is generally a curved line of the x, y plane. However the lines $x =$ constant, $y =$ constant, and in particular, the boundary of $\hat{K}$ are mapped by F onto straight lines of the x, y plane. A natural element on the quadrilateral K is the image by F^{-1} of the Q_1 element on $\hat{K}$:

$$q = q(\xi, \eta) \in Q_1 \to q \circ F^{-1} = q \circ F^{-1}(x, y).$$

In general, these elements *are not polynomials on* K. They are, however, easy to use, and their explicit expression is rarely used.

1.3.1.5. *Conforming Finite Elements* ($n = 3$)

The triangulation is now the covering of Ω ($=$ a polygonal set) by either tetrahedrons or three-dimensional rectangles whose edges are parallel to the axes or combinations of those.

The most common elements are:

linear, quadratic, and cubic tetrahedrons;
linear, quadratic, and cubic three-dimensional rectangles.

The definitions of these elements are the same as previously in the two-dimensional case replacing triangle by tetrahedron and rectangle by three-dimensional rectangle. For the cubic (tetrahedron and three-dimensional rectangle) elements, the nodal values are shown in Figure 3.3. All these elements lead to functions that are globally C^0 (continuous) but not more.

1.3.1.6. *Nonconforming Finite Elements*

As indicated previously, nonconforming finite elements produce approximate function spaces V_h that are not subspaces of V. For instance, the linear nonconforming triangle described later produces, when applied to problem

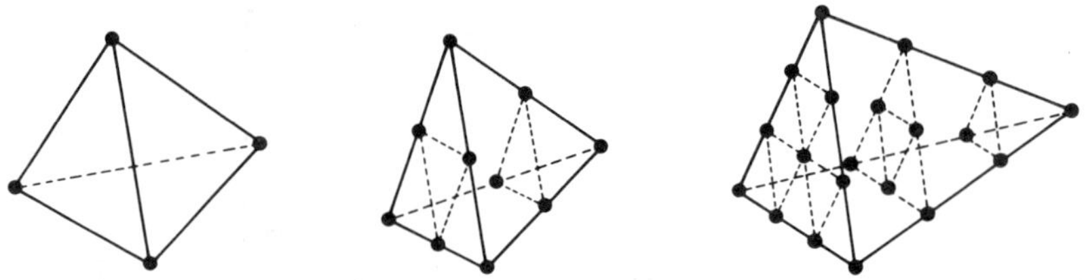

Linear, quadratic, cubic tetrahedrons

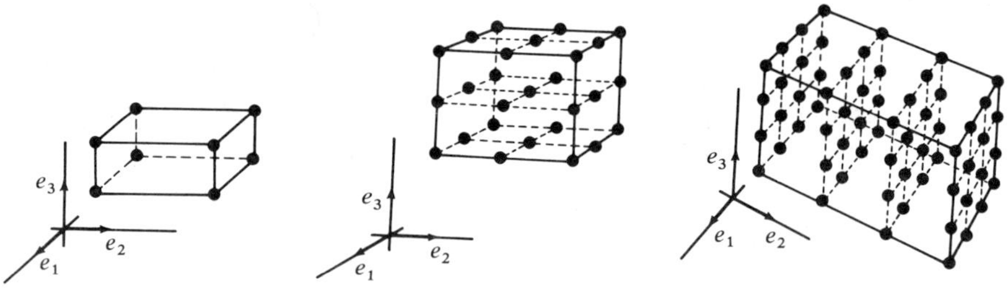

Linear, quadratic, cubic rectangles

FIGURE 3.3. Conforming finite elements ($n = 3$). (Adapted from Reference [9].)

(3.12), approximate functions that are highly discontinuous. Still, it may be useful to use such elements in at least two cases:

fluid flow problems in which, because of the incompressibility condition div $u = 0$, the linear triangle elements cannot be used in a straightforward manner;

higher order problems, such as the biharmonic problem, in which most elements described previously fail to produce C^1 functions, and thus the approximate spaces V_h would not be included in $H^2(\Omega)$ (= the natural space for a biharmonic problem).

Nonconforming Linear Elements: Two-Dimensional Case (Triangles)

Polynomials of degree ≤ 1 on the triangles.
Nodal values = values at midedges.

Nonconforming Linear Elements: Three-Dimensional Case (Tetrahedron)

Polynomials of degree ≤ 1 on the tetrahedrons.
Nodal values = values at barycenter of faces.

The global functions are totally discontinuous with discontinuities along the edges of triangles (or faces of tetrahedrons) except for the barycenters (of edges or faces). The method is nevertheless convergent and efficient, particularly for fluid flows: see the book by F. Thomasset [32], which is fully devoted to the utilization of these elements in two-dimensional flows.

1.3.1.7. *Curved Boundaries*

Curved boundaries can be approximated by polygonal lines. Alternatively, we can use the so-called isoparametric elements: the element is the image by an appropriate (simple) mapping of a triangle or a rectangle, and the function reduces on the element to the composition of a polynomial with that mapping. A similar situation occurred with the Q_1 quadrilateral elements.

1.3.2. Convergence and Error Estimate

Concerning convergence and error estimates the situation is different for linear and nonlinear problems.

1.3.2.1. *Linear Problems*

Two types of results have been derived in relation to error computation and convergence (see, for instance, [9]):

interpolation error;
approximation error.

When V_h is a conforming finite element space and u is a function in V (or usually in a smaller space), we consider the interpolant $r_h u$ of u in V_h (this is the finite element function that assumes the same nodal values as u, whereas $r_K u$ is the interpolant of u on an element K); the interpolation results give an upper bound of the norm of $u - r_h u$ in V and other spaces. The approximation results are of a different nature: when $u \in V$ is a solution of a problem such as (3.5) and $u_h \in V_h$ is a solution of the associated discrete problem, then the error between u and u_h is estimated for various norms. In the optimal cases the error between u and u_h is of the same order as that of the distance of u to V_h.

The general results are too abstract to be presented in detail here; we will just recapitulate the error estimates corresponding to the elements described.

For an element K let ρ_K denote the radius of the smallest ball containing K, let ρ'_K denote the radius of the lagest ball included in K, and let $\sigma_K = \rho_K/\rho'_K$.

The analysis is made under the assumptions that

$$\rho_h = \sup_{K \in \mathscr{T}_h} \rho_K \to 0, \tag{3.14}$$

and

$$\sigma_h = \sup_{K \in \mathscr{T}_h} \sigma_K \quad \text{remains bounded from above.} \tag{3.15}$$

If v is a function in V and $r_h V$ its interpolated function in V_h, we consider the H^m seminorm of $v - r_h v$ on an element K of the triangulation $\mathscr{T}_h$ and on the whole domain Ω:

$$|v - r_h v|_{m,K} = \left\{ \sum_{[\alpha]=m} \int_K |D^\alpha(v - r_h v)|^2 \, dx \right\}^{1/2},$$

$$|v - r_h v|_{m,\Omega} = \left\{ \sum_{[\alpha]=m} \int_\Omega |D^\alpha(v - r_n v)|^2 \, dx \right\}^{1/2},$$

where D^α is a partial derivative of order $[\alpha] = m$ and the sum is extended to all such derivatives.

For the elements described, the interpolation result is the following one:

On an element K assume that the interpolation opertor r_K is such that $r_K p = p$ for each polynomial p of degree $\leq k$, and assume that r_K is linear continuous from $H^{k+1}(K)$ into $H^m(K)$, $0 \leq m \leq k+1$. Then

$$|v - r_K v|_{m,K} \leq c \frac{\rho_K^{k+1}}{(\rho_K')^m} |v|_{k+1,K} \qquad \text{for all} \quad v \in H^{k+1}(K). \tag{3.16}$$

We can also assemble the results on the different elements K of a triangulation $\mathcal{T}_h$ and obtain a similar bound on all of Ω (when Ω is a polygon fully covered by the elements):

$$|v - r_h v|_{m,\Omega} \leq C \rho_h^{k+1-m} \sigma_h^m |v|_{k+1,\Omega} \qquad \text{for all} \quad v \in H^{k+1}(\Omega). \tag{3.17}$$

Finally, in dimensions 2 or 3:

for the linear elements (triangles, rectangles, tetrahedrons, three-dimensional rectangles) if:

$$k = 1, \qquad v \in H^2(\Omega), \quad \text{then} \quad |v - r_h v|_{m,\Omega} = O(\rho_h^{2-m}), \qquad 0 \leq m \leq 2;$$

for the quadratic elements (triangles, rectangles, tetrahedrons, three-dimensional rectangles) if:

$$k = 2, \qquad v \in H^2(\Omega), \quad \text{then} \quad |v - r_h v|_{m,\Omega} = O(\rho_h^{3-m}), \qquad 0 \leq m \leq 3;$$

for the cubic elements (triangles, rectangles, tetrahedrons, three-dimensional rectangles) if:

$$k = 3, \qquad v \in H^4(\Omega), \quad \text{then} \quad |v - r_h v|_{m,\Omega} = O(\rho_h^{4-m}), \qquad 0 \leq m \leq 4.$$

Concerning the approximation error, they are optimal (i.e., the approximation error is of the order of the best interpolation error), for instance, with the preceding elements, for problem (3.1)–(3.5) when $\Gamma_0 = \Gamma$, $\Gamma_1 = \varnothing$ (Dirichlet problem) or $\Gamma_0 = \varnothing$, $\Gamma_1 = \Gamma$ (Neumann problem), and Ω is a polygon fully covered by the elements of the triangulation $\mathcal{T}_h$.

1.3.2.2. *Nonlinear Problems*

For nonlinear problems the situation is more difficult and the results are less complete. Usually convergence results can be proved by using energy-type inequalities and convergence techniques that are appropriate for the kinds of equations considered: see, for instance, [28] for the nonlinear problem (3.13), (3.2), (3.3), and [29] for the Navier–Stokes equations. When compactness methods are used, some involved compactness arguments for finite elements may be necessary: see in R. Temam [29] the proof of convergence of the nonconforming P_1 finite element methods for the Navier–Stokes equations. Also by lack of uniqueness for nonlinear elliptic problems the convergence may be limited to a subsequence or we may assume as usual that we are "close" to the solution.

Error estimates are also more difficult to obtain than in the linear case. They usually assume more regularity on the equation and/or the solution than is necessary for convergence.

1.3.3. Mixed and Hybrid Finite Elements

1.3.3.1. *Minimax Formulation of a Boundary Value Problem*

Consider an abstract boundary value problem of the form (3.12)

$$\begin{cases}\text{To find } u \in V \text{ such that} \\ a(u, v) = \langle l, v \rangle \qquad \text{for all} \quad u \in V.\end{cases} \tag{3.18}$$

When the bilinear form a is furthermore symmetric, then (3.18) is equivalent to a convex minimization problem:

$$\begin{cases}\text{To minimize for } v \in V, \\ J(v) = \frac{1}{2}a(v, v) - \langle l, v \rangle.\end{cases} \tag{3.19}$$

The infimum of J on V is attained at a unique point of V that is called a *solution* (or a *minimizer*) for the variational problem (3.19). In fact, the solution of (3.18) is the same as that of (3.19).

The mixed finite elements are closely related to duality. A natural framework for both questions arises when V is a linear subspace of a Hilbert space X of the form

$$V = \{v \in X, b(v, \phi) = 0 \text{ for all } \phi \in Y\}, \tag{3.20}$$

where Y is another Hilbert space and b is a bilinear continuous form on $X \times Y$. We assume, furthermore, that a is extended as a bilinear continuous form on X and that l is extended as a linear continuous form on X.

In this case, we introduce the *Lagrangean* of the problem (cf. [11]):

$$L(v, \psi) = J(v) + b(v, \psi). \tag{3.21}$$

It is easily verified that

$$\operatorname*{Sup}_{\phi \in Y} L(v, \phi) = \begin{cases} J(v) & \text{if } v \in V, \\ +\infty & \text{if } v \in X \backslash V, \end{cases}$$

and that the minimization problem (for $v \in V$):

$$\operatorname*{Inf}_{v \in V} \left\{ \operatorname*{Sup}_{\psi \in Y} L(v, \psi) \right\} \tag{3.22}$$

has the same solution and the same infimum as (3.19).

Now we can associate with (3.21) the *dual problem* of (3.22), which is a maximization problem in Y

$$\operatorname*{Sup}_{\psi \in Y} \left\{ \operatorname*{Inf}_{v \in V} L(v, \psi) \right\}. \tag{3.23}$$

It is shown in [11] that if L (i.e., here b) satisfies a suitable condition, then

(3.23) has a unique solution denoted ϕ. Furthermore, the pair $\{u, \phi\} \in X \times Y$ is a solution of (3.23) and (3.22) (or (3.18)) if and only if $(\partial L/\partial v)(u, \phi) = (\partial L/\partial \psi)(u, \phi) = 0$, i.e.,

$$\begin{cases} a(u, v) + b(v, \phi) = \langle l, v \rangle & \text{for all} \quad v \in X, \\ b(u, \psi) = 0 & \text{for all} \quad \psi \in Y. \end{cases} \tag{3.24}$$

The initial problem (3.18), (3.19) is written in X as a constrained minimization problem

$$\begin{cases} \text{To minimize } J(v) \text{ for } v \in X, \text{ subject to the constraint} \\ b(v, \psi) = 0 \qquad \text{for all} \quad \psi \in Y. \end{cases} \tag{3.25}$$

The preceding framework associates with the initial problem (3.18), (3.19), (3.25) an element ϕ of X that is the *Lagrange multiplier* for the constrained optimization problem (3.25).

The necessary condition on b that guarantees the existence of ϕ is the *inf–sup condition*, introduced independently by Babuska [2] and Brezzi [5]:

$$\begin{cases} \text{There exists } \beta > 0 \text{ such that} \\ \displaystyle \operatorname*{Inf}_{\psi \in Y} \operatorname*{Sup}_{v \in X} \frac{b(v, \psi)}{\|v\|_X \|\psi\|_Y} \geq \beta. \end{cases} \tag{3.26}$$

Equivalently, (5.8) means that the linear operator B from X into Y' defined by

$$\langle Bv, \psi \rangle = b(v, \psi) \qquad \text{for all} \quad v \in x \quad \text{and for all} \quad \psi \in Y' \tag{3.27}$$

is an isomorphism from the orthogonal of V in X onto Y' or that the adjoint B' of B that maps X' into Y is an isomorphism from X onto the polar set V^0 of V

$$V^0 = \{\theta \in X', \langle \theta, v \rangle = 0 \text{ for all } v \in V\}.$$

The reader is referred for more details to the article of Brezzi in Chapter 2 of this volume. Note that the form (3.24) of the problem can be studied independently of the corresponding Lagrangean and variational problems and is suitable for several types of generalizations:

given a linear continuous form χ on Y, we can replace the second equation (3.24) by

$$b(u, \psi) = \langle \chi, \psi \rangle \qquad \text{for all} \quad \psi \in Y;$$

more importantly, the form a may be nonlinear with respect to its first argument u, and this corresponds to considering nonlinear partial differential equations, in particular the Navier–Stokes equations (discussed later) or monotone operators (see [24]).

See, also in, [11] a different point of view for duality that includes (3.24) as a particular case.

Remark 3.1. Let us also mention here the *penalization* of (3.24) that leads to consideration of the following problems:

$$\begin{cases} \text{To find } u^\varepsilon \in X,\ \phi^\varepsilon \in Y \text{ such that} \\ a(u^\varepsilon, v) + b(v, \phi^\varepsilon) = \langle l, v\rangle \quad \text{for all} \quad v \in X, \\ -\varepsilon c(\rho^\varepsilon, \psi) + b(u^\varepsilon, \psi) = 0 \quad \text{for all} \quad \psi \in Y, \end{cases} \tag{3.28}$$

where c is a bilinear continuous coercive symmetric form on Y and $\varepsilon > 0$ is a fixed positive parameter that is intended to tend to 0. A solution u^ε, ϕ^ε of (3.25) exists for every $\varepsilon > 0$, and u^ε, ϕ^ε converges to the solution u, ϕ of (3.26) when $\varepsilon \to 0$; see M. Bercovier [3].

1.3.3.2. *Examples*

Stokes Equations. The Stokes equations provide one of the most typical examples in which the framework (3.24) is suitable. The Stokes problem is the problem (2.7)–(2.10) when $U = 0$ and the nonlinear term $(u \cdot \nabla)u$ is dropped. In this case ($\Omega \subset \mathbb{R}^n$, $n = 3$ or more generally $n \neq 3$):

$$X = H_0^1(\Omega)^n = \left\{v \in L^2(\Omega)^n, \frac{\partial v}{\partial x_i} \in L^2(\Omega)^n, \text{ for all } i, v = 0 \text{ on } \Gamma\right\},$$

$$V = \{v \in X, \operatorname{div} v = 0\},$$

$$Y = \left\{\phi \in L^2(\Omega), \int_\Omega \phi(x)\,dx = 0\right\},$$

$$a(u, v) = \nu \sum_{i,j=1}^n \int_\Omega \frac{\partial u_i}{\partial x_j}\frac{\partial v_i}{\partial x_j}\,dx,$$

$$b(v, \phi) = -\int_\Omega (\operatorname{div} v)\phi\,dx.$$

It can be shown that (3.24) is equivalent to the Stokes problem

$$\begin{cases} -\nu\Delta u + \operatorname{grad} p = f & \text{in } \Omega, \\ \operatorname{div} u = 0 & \text{in } \Omega, \\ u = 0 & \text{on } \Gamma. \end{cases} \tag{3.29}$$

The operator div is a surjection from V onto Y (see, for instance, [29]), and it follows immediately that the Babuška–Brezzi condition (3.26) is satisfied.

As indicated, we can set the Navier–Stokes equations in the framework (3.24) with a replaced by a nonlinear form

$$a(u, v) = a_0(u, v) + a_1(u, v),$$

$$a_0(u, v) = \nu \sum_{i,j=1}^n \int_\Omega \frac{\partial u_i}{\partial x_j}\frac{\partial v_i}{\partial x_j}\,dx,$$

$$a_1(u, v) = \sum_{i,j=1}^n \int_\Omega u_i \frac{\partial u_j}{\partial x_i} v_j\,dx.$$

In that case, cf. [6].

Dirichlet Problem. The framework (3.24) applies to (3.5) and provides the dual of this problem (see [11]). For simplicity, we restrict ourselves to the case where $\Gamma_0 = \Gamma$, $\Gamma_1 = \varnothing$; i.e., we consider the Dirichlet problem in Ω. We set

$$X = H_0^1(\Omega) \times L^2(\Omega)^n, \qquad Y = L^2(\Omega)^n,$$
$$V = \{u = \{u_0, u_1\} \in X, u_1 = \text{grad } u_0\},$$

and for $u = \{u_0, u_1\}$, $v = \{v_0, v_1\} \in X$, and $\psi \in Y$:

$$a(u, v) = \int_\Omega u_0 \cdot v_0 \, dx + \int_\Omega u_1 \cdot v_1 \, dx,$$

$$\langle l, v \rangle = \int_\Omega f \cdot v_0 \, dx,$$

$$b(v, \psi) = \int_\Omega (v_1 - \text{grad } v_0) \cdot \psi \, dx.$$

We identify (3.24) with (3.5). Condition (3.26) is trivially satisfied. The dual (3.23) reads

$$\begin{cases} \text{To maximize} \quad -\dfrac{1}{2}\displaystyle\int_\Omega |\psi|^2 \, dx, \\ \text{for} \quad \psi \in L^2(\Omega)^n, \quad \text{div } \psi + f = 0 \quad \text{in } \Omega. \end{cases} \tag{3.30}$$

Biharmonic Problem. We consider the problem

$$\Delta^2 u = f \quad \text{in } \Omega, \tag{3.31}$$

$$u = 0, \qquad \frac{\partial u}{\partial \nu} = 0 \quad \text{on } \Gamma. \tag{3.32}$$

It is set in the form (3.24) with

$$V = H_0^2(\Omega)$$
$$= \left\{ v \in L^2(\Omega), \frac{\partial v}{\partial x_i}, \frac{\partial^2 v}{\partial x_i \partial x_j} \in L^2(\Omega) \text{ for all } i, j, \text{ and } v = 0, \frac{\partial v}{\partial \nu} = 0 \text{ on } \Gamma \right\},$$

$$a(u, v) = \int_\Omega \Delta u \, \Delta v \, dx,$$

$$\langle l, v \rangle = \int_\Omega f v \, dx.$$

Then we set it in the form (3.24) with

$$X = H_0^2(\Omega) \times L^2(\Omega), \qquad Y = L^2(\Omega),$$
$$V = \{u = \{u_0, u_1\} \in X, u_1 = \Delta u_0\},$$

and for $u = \{u_0, u_1\}, v = \{v_0, v_1\} \in X$, and $\psi \in Y$

$$a(u, v) = \int_\Omega u_1 v_1 \, dx, \qquad \langle l, v \rangle = \int_\Omega f v_0 \, dx,$$

$$b(v, \psi) = \int_\Omega (\Delta v_0 - v_1) \cdot \psi \, dx.$$

We identify (3.24) with (3.5). Condition (3.26) is trivially satisfied. The dual (3.23) reads

$$\begin{cases} \text{To maximize} \quad -\dfrac{1}{2} \displaystyle\int_\Omega |\psi|^2 \, dx \\ \text{for} \quad \psi \in L^2(\Omega), \quad \Delta\psi = f \quad \text{in } \Omega. \end{cases} \tag{3.33}$$

Problems involving the biharmonic appear in elasticity and in fluid mechanics for the treatment of the Stokes problem by utilization of a stream function.

1.3.3.3. *Mixed and Hybrid Elements*

Once we have reduced problem (3.18) to (3.24), we are naturally led to approximate this last problem; i.e.,

to find X_h, Y_h, a_h, b_h, c_h that approximate X, Y, a, b. c;
solve for each h a discrete problem similar to (3.24)

$$\begin{cases} \text{To find } u_h \in X_h, \ \phi_h \in Y_h \text{ such that} \\ a(u_h, v_h) + b_h(v_h, \phi_h) = \langle l_h, v_h \rangle \qquad \text{for all} \quad v_h \in X_h, \\ \qquad b_h(u_h, \psi_h) = 0 \qquad \text{for all} \quad \psi_h \in Y. \end{cases} \tag{3.34}$$

Finite element methods appear naturally in the construction of the spaces X_h and Y_h. We have more flexibility than in an ordinary FEM since we can combine various finite elements for X_h and for Y_h. The major difficulty arises in the verification of the condition (3.26), which leads sometimes to delicate algebraic questions. A thorough investigation of the inf–sup condition for various finite elements related to the Navier–Stokes equations can be found in [21]. In some cases, the number β in (3.26), corresponding to the discrete case, depends on h and tends to 0 as $h \to 0$. In other cases, the inf–sup condition does not hold in the discrete case and we can make it true by reducing the space Y_h in order to suppress the kernel of the discrete analog B'_h of B'; in practice, this amounts to a filtering procedure. The most famous example is the classical checkerboard instability for a Stokes problem corresponding to the $Q_1 - P_0$ element: X_h is a Q_1 approximation of $H_0^1(\Omega)^2$ ($n = 2$) and Y_h is a P_0 approximation of $L^2(\Omega)$; the filtering procedure is standard in this case; see also the analysis of the inf–sup condition in [4], who show with a counterexample that the best value of β in this (discrete) case is of the form ch.

We will not develop further this question here since it is the object of [6] and of other articles in this volume.

1.4. Specific Applications

Finite element methods have been the object of many applications in mechanics and physics. We now describe some specific applications.

1.4.1. Navier–Stokes Equations

The notations being the same as in Section 1.2.1 we consider the time-dependent Navier–Stokes equations for a viscous incompressible flow in a domain Ω

$$\frac{\partial u}{\partial t} - \nu \, \Delta u + (u \cdot \nabla)u + \nabla p = f \quad \text{in } \Omega \times (0, T), \tag{4.1}$$

$$\operatorname{div} u = 0 \quad \text{in } \Omega \times (0, T), \tag{4.2}$$

$$u = U \quad \text{on } \Gamma \times (0, T). \tag{4.3}$$

The unknowns are the velocity vector $u = u(x, t)$ and the pressure $p = p(x, t)$; the volumic forces f and the boundary velocity U (which may both depend on time) are given.

We recall that from the mathematical point of view the initial value problem for the Navier–Stokes equations, i.e., (4.1)–(4.3) supplemented by an initial condition

$$u(x, 0) = u_0(x), \qquad x \in \Omega, \quad u_0 \text{ given}, \tag{4.4}$$

is well set in space dimension two ($\Omega \subset \mathbb{R}^2$). However, we do not yet know whether the same result is true in space dimension three ($\Omega \subset \mathbb{R}^3$), i.e., we do not know whether the curl vector remains bounded or may become infinite even if the data are regular; see, for instance, [29].

The interval of time that we consider may be finite or infinite. Finite intervals of time occur naturally in the study of transient phenomena, whereas "infinite" intervals of time appear in the study of permanent regimes. For instance, if f and U are independent of time, then in some cases the solution u, p of (4.1)–(4.4) converges as $t \to \infty$ to a stationary solution, i.e., a solution of (2.7)–(2.9). A sufficient condition for this to occur is that the Reynolds number R_e is sufficiently small

$$R_e = \frac{U_* L_*}{\nu},$$

where U_* is a typical velocity of the flow and L_* a typical length of Ω.[2] If R_e

[2] Typical velocities are provided by ν and f in the form $L_*^{\alpha_1} \nu^{\alpha_2} \operatorname{norm}(U)$, $L_*^{\beta_1} \nu^{\beta_2} \operatorname{norm}(f)$, where appropriate norms are considered and the α_i, β_i are such that the corresponding expressions have the dimension of a velocity. The sum of these two velocities is an appropriate definition of U_*.

is large, the convergence to a stationary solution is not guaranteed anymore. On the basis of experimental observations related to turbulence, we actually expect that $u(\cdot, t)$, $p(\cdot, t)$ *no longer converge to time-independent solutions even if the data* f, U *are independent of time*. From the numerical point of view this will be the source of new difficulties that have not yet been explored and will not be addressed here [30]. Actually, the computing power that is presently available leaves us at the threshold of the occurrence of nonstationary phenomena at least in space dimension two.

Up to now most of the numerical computations on the *full* Navier–Stokes equations dealt with stationary phenomena (and transient phenomena). At this level a major difficulty for the numerical solution of (4.1)–(4.4) (or (2.7)–(2.9)) is the handling of the free divergence conditions that introduce complicated algebraic conditions in the discrete problem if it is not treated correctly. A certain number of methods related to or independent of the finite elements have been proposed to overcome the difficulties associated with the condition $\operatorname{div} u = 0$.

1.4.1.1. *Utilization of the Penalty Method*

The penalty method, which is due to R. Courant [10] in the context of constrained optimization, has been applied to the Navier–Stokes equations in [25, 26]. The idea that stems from the variational form of the Stokes problem (see Section 1.3.3) is to treat the condition $\operatorname{div} u = 0$ as a constraint and to "penalize" it, i.e., to replace (4.1), (4.2) by

$$\frac{\partial u_\varepsilon}{\partial t} - \nu\,\Delta u_\varepsilon + (u_\varepsilon \cdot \nabla)u_\varepsilon - \frac{1}{\varepsilon}\nabla(\nabla \cdot u_\varepsilon) = f \quad \text{in } \Omega \times (0, T), \tag{4.5}$$

where $\varepsilon > 0$ is a small parameter that is intended to tend to 0. It can be proved [26] that the solution of (4.3)–(4.5) converges to that of (4.1)–(4.4) when $\varepsilon \to 0$. A full asymptotic expansion of u^ε, p^ε in terms of ε can even be obtained in the simpler case of Stokes flows [29].

The penalty method has been applied in several ways by many authors to the finite element approximations of the Navier–Stokes equations, in particular with the mixed finite elements (see Remark 3.1).

1.4.1.2. *Utilization of Algorithms: Artificial Time Dependence*

The utilization of artificial time dependence applies to the solution of the stationary problem (2.7)–(2.9), or to the solution of the stationary problems arising from time discretization of (4.1)–(4.4). In these cases, an artificial evolution problem is introduced whose stationary solution is also the solution of the stationary Navier–Stokes equation.

For instance, we can consider the artificial evolution problem

$$\begin{cases} -\nu\Delta u + (u \cdot \nabla)u + \operatorname{grad} p = f, \\[1ex] \qquad\qquad \dfrac{\partial p}{\partial t} + \alpha \operatorname{div} u = 0, \end{cases} \tag{4.6}$$

($\alpha > 0$), or the equations of slightly compressible fluids

$$\begin{cases} \dfrac{\partial u}{\partial t} - \nu \, \Delta u + (u \cdot \nabla) u + \operatorname{grad} p = f, \\ \\ \qquad\qquad \dfrac{\partial p}{\partial t} + \alpha \operatorname{div} u = 0. \end{cases} \tag{4.7}$$

In both cases the condition div $u = 0$ is not imposed at all times and follows simply from the properties of (4.6) and (4.7) for large t. Consequently, ordinary finite elements, i.e., finite elements not containing the condition div $u = 0$, are used.

1.4.1.3. *Utilization of the Projection Method*

The projection method introduced by A.J. Chorin [8] and R. Temam [27] is connected to the fractional step method. It consists in solving the time evolution of (4.1) without (4.2) and then, more or less frequently, imposing (4.2) by projecting the velocity obtained on the free divergence vector fields.

The time discretization (time mesh $= \Delta t$) when $U = 0$ is given by

$$\begin{cases} \dfrac{\tilde{u}^m - u^{m-1}}{\Delta t} - \nu \, \Delta \tilde{u}^m + (\tilde{u}^m \cdot \nabla) \tilde{u}^m = f^m \quad \text{in } \Omega, \\ \tilde{u}^m = 0 \quad \text{on } \Gamma, \end{cases} \tag{4.8}$$

and $u^m =$ projection of $\tilde{u}^m$, which amounts to saying that

$$\begin{cases} u^m = \tilde{u}^m - \operatorname{grad} q^m \quad \text{in } \Omega, \\ \operatorname{div} u^m = 0 \quad \text{in } \Omega, \\ u^m \cdot \mathbf{v} \ (= \text{normal component of } u^m \text{ on } \Gamma) = 0 \quad \text{on } \Gamma. \end{cases} \tag{4.9}$$

Alternatively, setting $q^m = \Delta t \pi^m$ we can rewrite (4.9) as

$$\begin{cases} \dfrac{u^m - \tilde{u}^m}{\Delta t} + \operatorname{grad} \pi^m = 0 \quad \text{in } \Omega, \\ \operatorname{div} u^m = 0 \quad \text{in } \Omega, \\ u^m \cdot \mathbf{v} = 0 \quad \text{on } \Gamma, \end{cases} \tag{4.10}$$

and π^m provides an approximation for the pressure. It is, however, a poor approximation near the boundary since it satisfies the following nonphysical boundary condition that we infer from (4.10):

$$\frac{\partial \pi^m}{\partial \mathbf{v}} = 0 \quad \text{on } \Gamma. \tag{4.11}$$

See, however, in [35], [36] an improvement of the projection method that yields a better approximation of the pressure near the boundary.

In order to determine u^m, it is actually necessary to compute q^m, π^m, or at least their gradient; π^m is a solution of the Neumann problem, which consists of (4.11) and

$$\Delta \pi^m = \frac{1}{\Delta t} \operatorname{div} \tilde{u}^m. \tag{4.12}$$

It seems better, for a more accurate determination of the pressure that avoids the undesirable boundary layer resulting from (6.11), to consider q^m, π^m as auxiliary functions and to compute the approximation p^m of the pressure by using the boundary value problem

$$p = \psi(u), \qquad p^m = \psi(u^m), \tag{4.13}$$

that we deduce directly from the Navier–Stokes equations; see [29]. At this point we can either use a Dirichlet or a Neumann boundary condition for p [15].

Many other forms of (4.8), (4.9) can also be considered: we can split the operators differently, leaving, for example, some viscosity in (4.9); we can use an explicit scheme in (4.8); or we can solve for the evolution (4.8) for several steps and only perform the projection (4.9) periodically.

In all cases when the projection method is used we need a space of free divergence vector functions so that the projection (4.9) can be performed in a satisfactory manner.

1.4.1.4. *Utilization of Free Divergence Finite Element Spaces*

The simplest element, the piecewise linear (P_1) function on triangles, cannot be used since the condition $\operatorname{div} u = 0$ imposed on each triangle leads to too many algebraic relations, and the spaces of discrete divergence free P_1 vector functions may be reduced to the function 0. We can either impose the condition $\operatorname{div} u = 0$ "less often," or go to more complicated elements such as the non-conforming P_1 element (F. Thomasset [32]) or $P_2, Q_1, Q_2, \ldots$, elements.

1.4.2. Fluid Structure Interactions

In many industrial fields of interest, including aerospace, engineers are confronted with fluids (water, oil, kerosene, gases, and so on) interacting with structures (tanks, containers, obstacles and so forth) along a more or less extended area.

In some cases deformations of the structure may be fairly important and even affect the motion of the fluid; engineers then have to solve problems including a coupling between fluid displacements and elastic deformations of the structure.

We describe here the interaction between a free surface fluid and the structure, assumed to be elastic, that contains it, in an external force field. We follow J. Mathieu [13] and J. Mathieu *et al.* [20], who computed the transient simulation of such a process with the BACCHUS code.

1.4.2.1. *The Arbitrary Lagrange–Euler (ALE) Description*

We consider a moving domain $\Omega^F(t)$ deforming with velocity $w = w(x, t)$, and filled with an incompressible fluid of (constant) density ρ^F that obeys the Navier–Stokes equations. We also consider a second domain $\Omega^S(t)$ made of elastic material and limiting Ω^F. The fluid is limited by a free surface S and the contact surface $\Pi = \Pi(t)$ with Ω^S.

The equations are

$$\begin{cases} \rho^F \left\{ \dfrac{\delta}{\delta t} v^F + ((v^F - w)\nabla) v^F \right\} = \operatorname{div} \sigma^F + \rho^F f^F \quad \text{in } \Omega^F, \\ \operatorname{div} v^F = 0, \\ \rho^S \dfrac{\partial v^S}{\partial t} = \operatorname{div} \sigma^S + \rho^S f^S \quad \text{in } \Omega^S, \\ \dfrac{d}{dt}(\rho^S |J|) = 0, \end{cases} \tag{4.14}$$

where

$$\frac{\delta \rho}{\delta t} = \frac{\partial \rho}{\partial t} + (w \cdot \nabla)\rho = \text{convection derivative associated with the vector field } w;$$

v^i = velocity field in Ω^i;

σ^i = stress tensor in Ω^i;

ρ^i = density in Ω^i (constant in Ω^F);

f = external forces;

J = Jacobian of the mapping $\Omega^S(0) \to \Omega^S(t)$.

The ALE description is determined by the actual velocity field w, which is defined as follows:

$w = 0$ in an internal Eulerian region;

$w = v^F$ on the free surface $S(t)$;

$w = v^S$ on the wet part of the wall $\Pi(t)$.

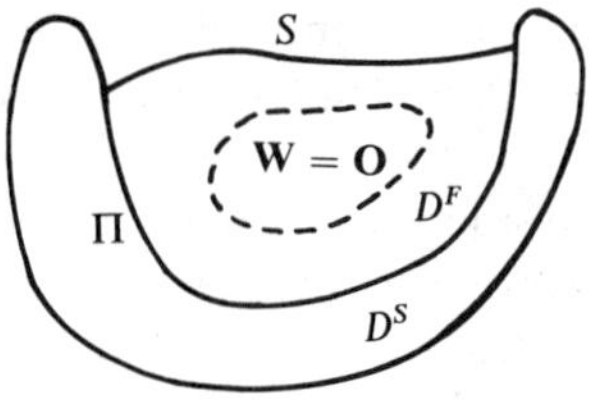

FIGURE 4.1. Fluid–solid interactions.

The *constitutive laws* are

$$\sigma^F = \nu[\nabla v^F + (\nabla v^F)^t] + p\mathbf{I}$$

for the fluid, ν = the kinematic viscosity, p = the hydrostatic pressure, and

$$\sigma^S(t) = \frac{\rho^S(t)}{\rho^S(0)} F(t)\tau(t)F(t)^t$$

for the solid where τ is the standard second Piola–Kirchhoff tensor, $F(t)$ is the gradient of the mapping $\Omega^S(0) \to \Omega^S(t)$, and $\tau(0) = \sigma^S(0) = 0$.

The *boundary conditions* are as follows:

on the structure, displacements are given on some part $\Gamma_u(t)$ of $\partial\Omega^S(t)\backslash\Pi(t)$ and normal stresses are given along the remaining part of $\partial\Omega^S(t)\backslash\Pi(t)$;
on the wet part of the wall, $\Pi(t)$, we have no cavitation,

$$v^F \cdot \mathbf{v} = v^S \cdot \mathbf{v},$$

$\mathbf{v}$ is the normal on Π, and we have a partial slip condition of the fluid along the wall

$$\sigma^F_{tg} = -\beta(v^F_{tg} - v^S_{tg}),$$

and finally, the normal stresses are continuous

$$(\sigma^F - \sigma^S)\cdot \mathbf{v} = 0.$$

The *discretization* of the problem is made with a finite element discretization in space, providing an easy handling of the complex geometric configurations that are caused by large displacements. The elements are of degree one for the velocities (= P_1 for triangles, Q_1 for rectangles, Q_1-transported Q_1 for quadrilaterals), and piecewise constants for the hydrostatic pressure.

A one-step explicit finite difference scheme is used in time. This scheme is subject to the usual stability conditions limiting the time step. The criteria taken into account are:

free surface wave and viscous wave stability in the fluid;
acoustic wave stability in the structure.

Because of the more drastic limitation due to the stability criterion in the structure, a subcycling procedure is used for the structure calculation.

A *mesh adaptation* procedure is necessary. At each fluid-calculation step, the free surface has to be repositioned. The displacements of the free surface, and of the interface between the fluid and the structure, induce a modification of the mesh in the mixed Lagrange–Euler region and possibly a modification of the Euler region (when its boundary intersects the free surface or the wall) or a degeneracy of some element. Thus a rezoning is automatically performed during the calculation. The rezoning is also necessary on $\pi(t)$ since any node belonging to $\partial\Omega^F(t) \cap \partial\Omega^S(t)$ should be at the same time a vertex of some element of $\Omega^S(t)$ *and* of some element of $\Omega^F(t)$.

A sample two-dimensional calculation is shown in Figure 4.2.

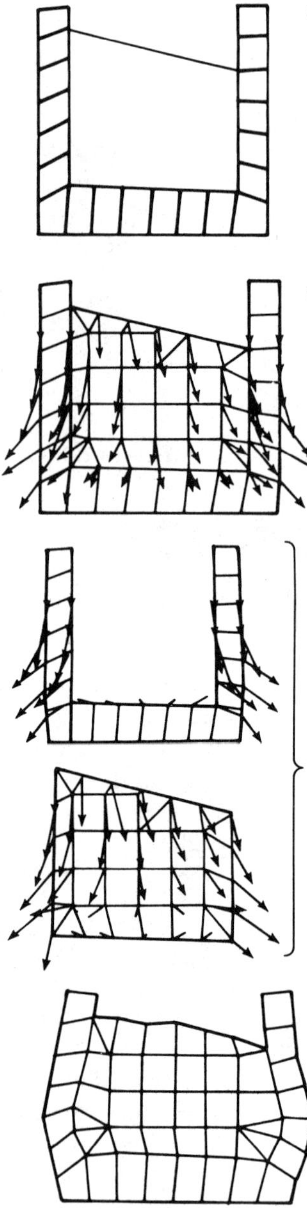

t = 0 — D^s and S are given.

The internal mesh will be automatically designed.

t = 0.2 — The bottom of the tank is in a compressive phasis.

Notice the behavior of the velocity fields on Π: double-valued with normal component continuity.

t = 0.3 — The bottom of the tank has entered an expansive phasis.

t = 0.5 — The tank has reached large enough deformations.

The free-surface tends to stabilize towards the horizontal. Notice the modification of the mesh in D^F.

FIGURE 4.2. Successive deformations.

1.4.3. Three-Dimensional Euler Equations

The finite element method has been applied to the computation of the solution of the Euler equations. We describe here the latest results in [7], which concern the computation of steady vortex flows past a flat plate at a high angle of attack.

1.4.3.1. *Description of the Problem*

The purpose of the computations in [7] is to investigate the developments of vortices at the tip of the plate and their propagation after the trailing edge. In the computations, the plate has no thickness and we expect a strong vortex structure to develop at the tip of the plate; it is not possible to use potential equations, and the full Euler equations are necessary. For the computations, the plate is embedded in a three-dimensional rectangular domain as shown in Figure 4.3; the aspect ratio is 0.5 and the angle of incidence is $\alpha = 15^\circ$ or $\alpha = 30^\circ$. Figure 4.3 shows only half of the plate since symmetry with respect to the y variable is assumed.

The incoming flow is given by $\vec{q}_\infty = (u_\infty, v_\infty, w_\infty) = (q_\alpha \cos\alpha, 0, q_\alpha \sin\alpha)$, α as previously, and M_∞ the Mach number at infinity, $M_\infty^2 = q_\infty^2/a_\infty^2$, $a_\infty^2 = (\gamma - 1)(H - (q_\infty^2/2))$. In these computations, the flow is subsonic, $M_\infty = 0.7$.

Euler Equations. The Euler equations are written in conservative form

$$\frac{\partial \rho u}{\partial x} + \frac{\partial \rho v}{\partial y} + \frac{\partial \rho w}{\partial z} = 0, \qquad \text{conservation of mass,}$$

$$\begin{aligned} &\frac{\partial \rho u^2 + p}{\partial x} + \frac{\partial \rho uv}{\partial y} + \frac{\partial \rho uw}{\partial z} = 0, \\ &\frac{\partial \rho uv}{\partial x} + \frac{\partial \rho v^2 + p}{\partial y} + \frac{\partial \rho vw}{\partial z} = 0, \qquad \text{conservation of momentum,} \\ &\frac{\partial \rho uw}{\partial x} + \frac{\partial \rho vw}{\partial y} + \frac{\partial \rho w^2 + p}{\partial z} = 0, \end{aligned} \tag{4.15}$$

$$\frac{\gamma p}{(\gamma - 1)\rho} + \frac{q^2}{2} = H, \qquad \text{Bernoulli's equation,} \tag{4.16}$$

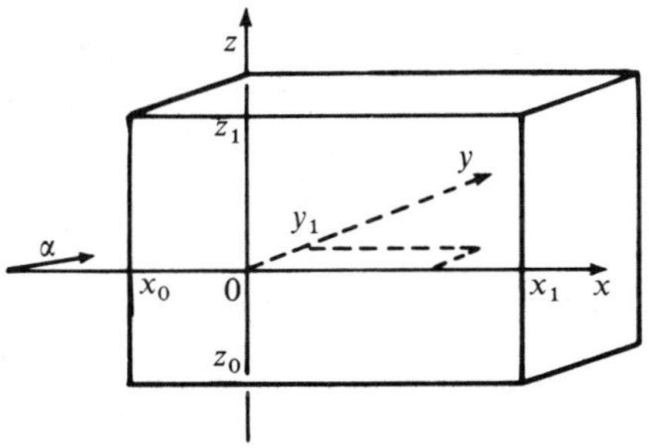

FIGURE 4.3. The flat plate and the corresponding coordinates system.

where $\rho, \vec{q} = (u, v, w), p$ represent, respectively, the density, the velocity vector, and the pressure; H is the total enthalpy; and γ is the ratio of specific heats.

1.4.3.1.1. Boundary Conditions

The boundary conditions on the plate are easy; the tangency condition on the plate (plane $z = 0$) reads $w = 0$, and the symmetry condition is $v = 0$ on $y = 0$. On the contrary, the boundary conditions at the limits of the domain are not easy, especially downstream, where it should allow the vortex to go through the exit plane.

The flow variables have been set to the freestream values at the incoming boundaries ($x = x_0, z = z_0$), and no condition has been imposed at the exit boundaries ($x = x_1, z = z_1$), so that the values there are computed from the variational formulation. As a result of the utilization of a least squares method, this simply amounts to requiring that the first-order equations be satisfied at those boundaries (see [7] for a discussion of the boundary conditions). At $y = y_1$ the far field condition $v = 0$ is used.

1.4.3.2. *The Numerical Method and the Numerical Results*

The equations are solved iteratively as follows:

A fixed point algorithm is based on equation (4.16), computing the density when u, v, w, p are known.

The nonlinear system (4.15) provides u, v, w, p, using the value of ρ from the previous step of the iteration. This system is linearized by Newton's method with linearized variables $\tilde{u}, \tilde{v}, \tilde{w}, \tilde{p}$.

The system for $\tilde{u}, \tilde{v}, \tilde{w}, \tilde{p}$ is discretized by an appropriate Q_1 finite element method (discretization of the conservative variables $\rho u, \rho v, \rho w, \ldots$), and the system is solved by a least squares method.

Figures 4.4 and 4.5 show a sample of the discretization grid. Finally, Figures 4.6–4.8 show the cross-flow velocities at 70, 90, and 110 percent of the plate.

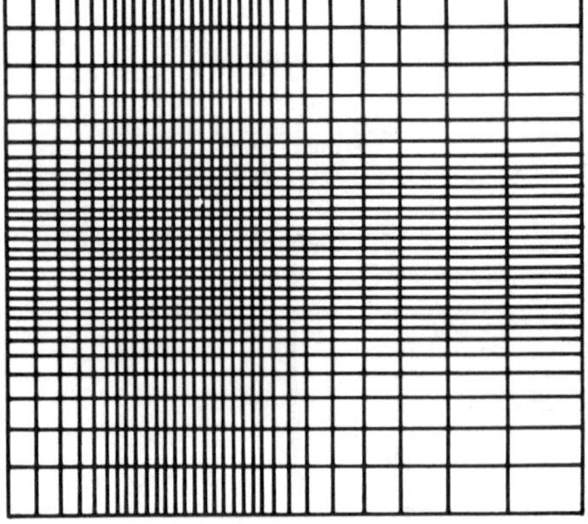

FIGURE 4.4. Discretization grid.

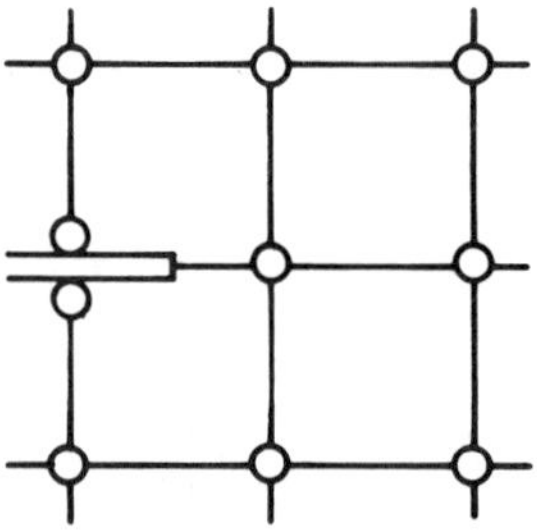

FIGURE 4.5. Discretization grid near the tip of the flat plate.

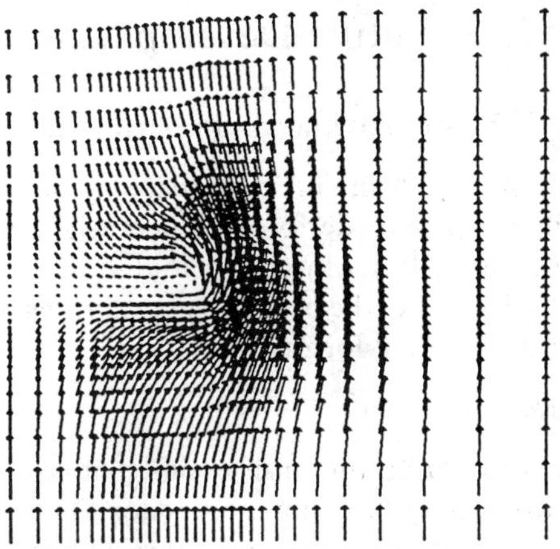

FIGURE 4.6. Cross flow velocities at 70% of the plate.

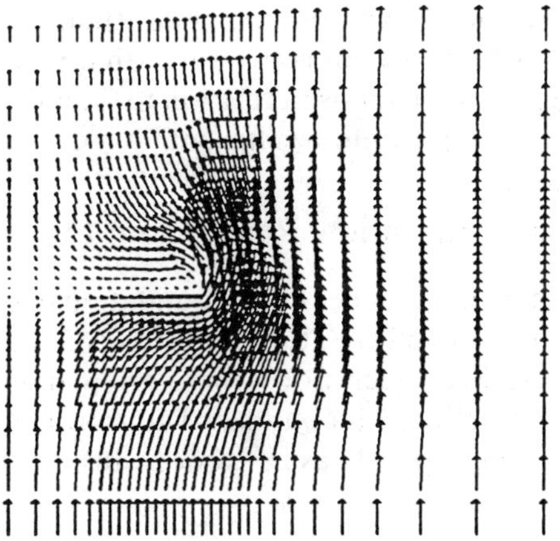

FIGURE 4.7. Cross flow velocities at 90% of the plate.

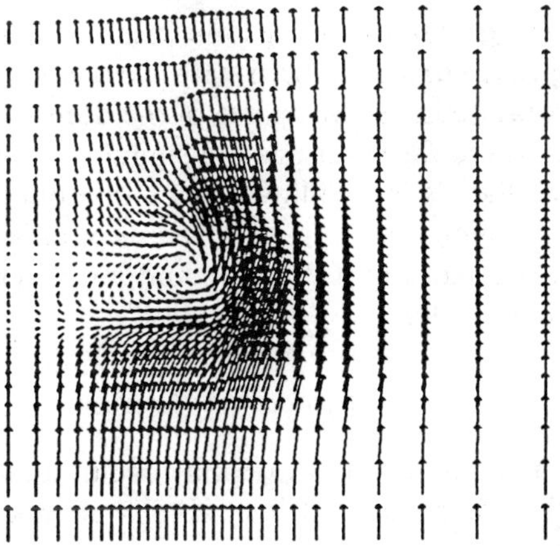

FIGURE 4.8. Cross flow velocities at 110% of the plate.

1.5. Future Developments

The advantages of FEM are well known. In particular:

accurate and automating fitting of complicated geometries;
lowering (dividing by 2) the order of the differential operators that appear, as a result of the utilization of weak formulations;
avoiding the discretization of the boundary conditions of the Neumann type that disappear in the weak formulations.

The inconveniences are:

need of a triangulation program for domains that must be written once for all, but is a discouraging preliminary step for the nonexpert;
computational costs that in no way can compete with the achievements of multigrid and spectral methods.

However, the multigrid and spectral methods attain their optimal efficiency for rectangular domains. It is then conceivable that a combination of FEM and multigrid and spectral methods in relation to domain decomposition and parallel computation can prove to be efficient.

1.5.1. Domain Decomposition: Remarks on Future Developments

The domain decomposition that lies at the foundation of the FEM can appear to be, in a different form, directly related to future developments in the method. Besides the geometrical considerations, there are many other good reasons for decomposing the solution of a boundary value problem in a large domain into the solution of similar problems in subdomains. Let us mention some of them:

1.5.1.1. *Adaptive Meshes*

Adaptive meshes may be suitable for geometrical or analytical reasons, for instance, the fitting of a curved domain with elements of a different nature that are more appropriate in various parts of the domain. The refinement of the mesh in regions where the solutions are singular or nearly so (boundary layers, shocks, front flames, etc.) allows extra computational effort to be concentrated in such regions. Thus finite elements of different types and sizes may be used in various parts of a domain, and a different treatment of the different regions can be useful.

1.5.1.2. *Physical Motivations*

Physical phenomena of a different nature may occur in different regions, and these regions should not be treated in a similar manner. For instance, in aeronautics:

a turbulence model is necessary near the airfoil;
the Navier–Stokes equations with viscous effects and without turbulence are necessary at a certain distance but not at too great a distance;
the Euler equations (i.e., no viscous effect) are sufficient far from the airfoil.

Similar situations occur in combustion or in solid mechanics and lead naturally to the decomposition of the whole domain into smaller ones.

1.5.1.3. *Parallel Computation*

The technology of computers will apparently move more and more toward parallel computation. The decomposition of a domain into subdomains on which the computations are made simultaneously is appropriate for parallel computations. The difficulty is then to synthesize the information from all the subdomains periodically to ensure a correct interaction of the subdomains.

At the present time work is being done on domain decomposition. If this technique is satisfactorily mastered, then we can consider, as has already been done (see, for instance, [13, 23, 19]), combining the advantages of FEM and spectral and multigrid methods by using domain decomposition, as suggested in Figure 5.1. This has already been done, but the goal is to match the performance of the fast methods as closely as possible.

1.5.1.4. *Multigrid Methods*

After this work was completed, we have seen emerge the multigrid methods for finite elements approximation. These methods allow for a faster convergence of the FEM: see, among other references, [37], [38], [39].

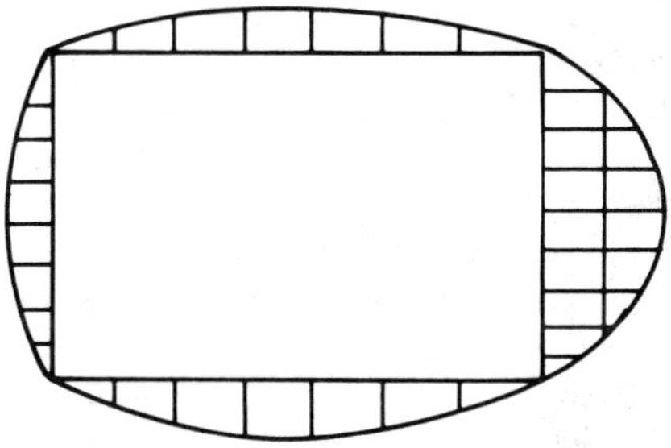

FIGURE 5.1. Example of subdomains decomposition (for finite elements/spectral method discretization).

1.5.2. Concluding Remark

The FEM cannot pretend to be the best method in all situations, but it has proved to be an efficient and performant method in many cases and can hope to be the object of future interesting developments. R. Feynman says in his book [12] that he was able to solve some problems that other people could not solve just because he had some tools in his box that others did not have. It would be too bad not to have the FEM tool in his box of numerical methods.

Acknowledgment. This work was supported under the National Aeronautics and Space Administration, under NASA Contract No. NAS1-18107, while the author was in residence at the Institute for Computer Applications in Science and Engineering (ICASE), Hampton, VA 23665-5225, U.S.A.

References

1. J.H. Argyris, "Energy theorems and structural analysis, I: General theory," *Aircraft Engrg.*, **26** (1954), 347–356, 383–387, 394; **27** (1955), 42–58, 80–94, 125–134.
2. I. Babuška, "The finite element method with Lagrangian multipliers," *Numer. Math.* **20** (1973), 179–192.
3. M. Bercovier, "Perturbation of mixed variational problems: Application to mixed finite element methods," *RAIRO Anal. Numér.*, **12** (1978), 211–236.
4. J. Boland and R. Nicolaides, "Stable and semistable low order finite elements for viscous flows," Storrs, University of Connecticut, Preprint, 1984.
5. F. Brezzi, "On the existence, uniqueness and approximation of saddle-point problems arising from Lagrangian multipliers," *RAIRO Anal. Numér.*, **8** (1974), 129–151.
6. F. Brezzi, "Finite element approximations of the von Karmán equations," *RAIRO Anal. Numér.*, **12**, No. 4 (1978), 303–312.
7. C.H. Bruneau, J.J. Chattot, J. Laminie, and R. Temam, "Computation of vortex flows past a flat plate at high angle of attack," in *Proceedings of the 10th International Conference on Numerical Methods in Fluid Mechanics* (*Peking* 1986), Lecture Notes in Physics, Vol. 264, Springer-Verlag, New York, 1987.
8. A.J. Chorin, "A numerical method for solving incompressible viscous flow problems," *J. Comput. Phys.* **2** (1967), 12–26.
9. P.G. Ciarlet, *The Finite Element Method for Elliptic Problems*, North-Holland, Amsterdam, 1978.
10. R. Courant, "Variational methods for the solution of problems of equilibrium and vibrations," *Bull. Amer. Math. Soc.*, **49** (1943), 1–23.
11. I. Ekeland and R. Temam, *Convex Analysis and Variational Problems*, North-Holland, Amsterdam, 1976.
12. R. Feynman, *Surely You're Joking, Mr. Feynman*, Norton, New York, 1985.
13. B. Gest, "Methode de resolution rapide des systèmes linéaires," Thesis, Laboratoire d'Analyse Numérique, Université Paris-Sud, Orsay, 1985.
14. V. Girault and P.A. Raviart, *Finite Element Methods for the Navier-Stokes Equations*, Springer Series in Computational Mathematics, Vol. 5, Springer-Verlag, New York, 1986.

15. P.M. Gresho and R.L. Sani, "On pressure boundary conditions for the incompressible Navier–Stokes equations," University of Colorado, Boulder, Preprint, 1986.
16. J. Leray, "Etude de diverses équations integrales non linéaires et de quelques problèmes que pose l'hydrodynamique," *J. Math. Pures Appl.*, **12** (1933), 1–82.
17. J. Leray, "Essais sur les mouvements plans d'un liquide visqueux que limite des parois," *J. Math. Pures Appl.*, **13** (1934), 331–418.
18. J. Leray, "Essai sur le mouvement d'un liquide emplissant l'espace," *Acta Math.*, **53** (1934), 193–248.
19. C. Lesartre, Thesis in preparation, Laboratoire d'Analyse Numérique, Université Paris-Sud, Orsay, 1988.
20. J. Mathieu, P. Ravier, J. Boujot, P. Gendre, and M. Hittinger, "Interaction between structure and free surface fluid with large displacements by finite elements" in *Proceedings of the* 10*th International Conference on Numerical Methods in Fluid Mechanics* (*Peking* 1986), Lecture Notes in Physics, Vol. 264, Springer-Verlag, New York, 1987.
21. J.T. Oden and O.P. Jacquotte, "Stability of some mixed finite element methods for Stokesian flows," *Comput. Methods Appl. Mech. Engrg.* **43** (1984), 231–248.
22. J. Periaux, Lecture at the First International Conference on Industrial and Applied Mathematics, Paris, June 1987.
23. B. Ravier, "Méthode de décomposition des domaine," Thesis, Laboratoire d'Analyse Numerique, Université Paris-Sud, Orsay, 1986.
24. B. Scheurer, "Existence et approximation de points selles pour certains problèmes non linéaires," *RAIRO Anal. Numér.*, **11** (1977), 369–400.
25. R. Temam, "Sur l'approximation des solutions des équations de Navier–Stokes," *C.R. Acad. Sci. Paris, Série A*, **262** (1966), 219–221.
26. R. Temam, "Une méthode d'approximation de la solution des équations de Navier–Stokes," *Bull. Soc. Math. France*, **98** (1968), 115–152.
27. R. Temam, "Sur l'approximation de la solution des équations de Navier–Stokes par la méthode des pas fractionnaires II," *Arch. Rational Mech. Anal.*, **33** (1969), 377–385.
28. R. Temam, *Numerical Analysis*, Reidel, Dordrecht, Holland, 1973.
29. R. Temam, *Navier–Stokes Equations*, 3rd ed., North-Holland, Amsterdam, 1984.
30. R. Temam, "Chaos in Fluids," Lecture at the SIAM National Meeting, Boston, July 1986.
31. R. Temam, Article in preparation.
32. F. Thomasset, *Implementation of Finite Element Methods for Navier–Stokes Equations*, Springer Series in Computational Physics, Springer-Verlag, New York, 1981.
33. O.C. Zienckiewicz, *The Finite Element Method in Engineering Sciences*, McGraw-Hill, London. 1971.
34. D.W. Dwoyer and R. Voigt (eds.), *Proceedings of the* 11*th ICNMFD*, *Williamsburg*, 1988. To be published by Springer-Verlag, New York.
35. J. Shen, Article in [34].
36. J. Shen and R. Temam, Article in preparation.
37. C.H. Bruneau, C. Jouron, and Zhang Lin Bo, Article in [34].
38. D. Mauripolis, Thesis, Princeton University, 1987.
39. Zhang Lin Bo, Thesis, Université Paris-Sud, Orsay, 1987.

CHAPTER 2

A Survey of Mixed Finite Element Methods

F. Brezzi

2.1. Introduction

The aim of this paper is to present an introductory survey of mixed finite element methods. We shall deal first with the so-called mixed formulation of some problems arising in elasticity and hydrodynamics. Then we shall analyze the difficulties connected with the choice of appropriate finite element discretizations for a mixed formulation. Finally, we shall discuss some special techniques that are often helpful for solving the discretized problem.

The notation of Ciarlet [17] is followed throughout.

2.2. Mixed Formulations

A precise and satisfactory definition of *mixed method* (or of *mixed formulation*) does not exist. The term originated in the engineering literature (Herrmann [33, 34]; Hellan [32]) in connection with the elasticity theory to denote methods based on the Hellinger–Reissner principle, in which both displacements and stresses were approximated simultaneously. Even among mathematicians, such as Glowinski [29], Babuška [6], Crouzeix and Raviart [18], and Johnson [36], whose work can now be considered as pioneering, the term *mixed* was used only by Johnson in the context of plate bending problems. The term is now used in a much wider sense and has become rather vague. Here we will live with such vagueness, and we shall not try a new unsatisfactory definition. Instead, we present a few cases: linear elliptic problems, the Stokes equations for incompressible fluids, and linear elasticity problems. The first case will be dealt with in more detail because it is formally much simpler, whereas only a few essential points will be stressed for the other two cases.

EXAMPLE 2.1 (Linear Elliptic Operators). The use of mixed formulations for linear elliptic operators is rather recent and, as we shall see, is recommended only in some special cases. However, its presentation is very simple, and this makes it an ideal example. Consider the model problem:

$$\operatorname{div}(A(\mathbf{x})\,\mathbf{grad}\,u) = f \quad \text{in } D \subset \mathbb{R}^d, \tag{2.1}$$

$$(A(\mathbf{x})\,\mathbf{grad}\,u)\cdot\mathbf{n} = g_1 \quad \text{on } \Gamma_{\text{Neu}}, \tag{2.2}$$

$$u = g_0 \quad \text{on } \Gamma_{\text{Dir}}, \tag{2.3}$$

where

(i) $\Gamma_{\text{Dir}} \cup \Gamma_{\text{Neu}} = \Gamma = \partial D$ is a splitting of ∂D, the boundary of the domain D;
(ii) $A(\mathbf{x})$ is a smooth function on $\bar{D}$, with $A(\mathbf{x}) \geq \alpha > 0$ for every $\mathbf{x}$ in D;
(iii) $\mathbf{n}$ is the unit outward normal to ∂D; and
(iv) f, g_1, g_0 are given smooth functions in D and on $\Gamma_{\text{Neu}}, \Gamma_{\text{Dir}}$, respectively.

Introducing, for $g = g_0$ or $g = 0$, the manifold

$$H_g^1(D) = \{v | v \in H^1(D);\ v = g \text{ on } \Gamma_{\text{Dir}}\},$$

we can write the variational formulation of (2.1)–(2.3) as follows:

Find $u \in H_{g_0}^1$ such that

$$\int_D A(\mathbf{x})\,\mathbf{grad}\,u\cdot\mathbf{grad}\,v\,d\mathbf{x} = -\int_D fv\,d\mathbf{x} + \int_{\Gamma_{\text{Neu}}} g_1 v\,d\Gamma, \qquad \forall v \in H_0^1(D).$$

In order to derive the mixed formulation of (2.1)–(2.3) we introduce the variable

$$\mathbf{p} = A(\mathbf{x})\,\mathbf{grad}\,u \quad \text{in } D, \tag{2.4}$$

so that (2.1) and (2.2) become, respectively,

$$\operatorname{div}\mathbf{p} = f \quad \text{in } D, \tag{2.5}$$

$$\mathbf{p}\cdot\mathbf{n} = g_1 \quad \text{on } \Gamma_{\text{Neu}}. \tag{2.6}$$

The formulation (2.3)–(2.6) is often called the *mixed formulation* of (2.1)–(2.3). Two reasonable variational formulations for (2.3)–(2.6) are now possible. The first one is

Find $u \in H_{g_0}^1(D)$ and $\mathbf{p} \in (L^2(D))^d$ such that

$$\int_D (A(\mathbf{x}))^{-1}\mathbf{p}\cdot\mathbf{q}\,dx - \int_D \mathbf{q}\cdot\mathbf{grad}\,u\,d\mathbf{x} = 0, \qquad \forall\mathbf{q} \in (L^2(D))^d, \tag{2.7}$$

$$-\int_D \mathbf{p}\cdot\mathbf{grad}\,v\,d\mathbf{x} = \int_D fv\,d\mathbf{v} - \int_{\Gamma_{\text{Neu}}} g_1 v\,d\Gamma, \qquad \forall v \in H_0^1(D). \tag{2.8}$$

In order to introduce the second variational formulation of (2.3)–(2.6) we define, for $g = g_1$ or $g = 0$, the manifold

$$H_g(\text{div}; D) = \{q | q \in (L^2(D))^d;\ \operatorname{div}\mathbf{q} \in L^2(D);\ \mathbf{q}\cdot\mathbf{n} = g \text{ on } \Gamma_{\text{Neu}}\}.$$

The second variational formulation of (2.3)–(2.6) is now

Find $u \in L^2(D)$ and $\mathbf{p} \in H_{g_1}(\text{div}; D)$ such that

$$\int_D (A(\mathbf{x}))^{-1}\mathbf{p}\cdot\mathbf{q}\,d\mathbf{x} + \int_D u \,\text{div}\,\mathbf{q}\,d\mathbf{x} = \int_{\Gamma_{\text{Dir}}} g_0\mathbf{q}\cdot\mathbf{n}\,d\Gamma, \qquad \forall \mathbf{q} \in H_0(\text{div}; D), \tag{2.9}$$

$$\int_D v \,\text{div}\,\mathbf{p}\,d\mathbf{x} = \int_D fv\,d\mathbf{x}, \qquad \forall \mathbf{v} \in \mathbf{L}^2(D). \tag{2.10}$$

The difference between (2.7)–(2.8) and (2.9)–(2.10) is clearly a simple integration by parts (or, if you prefer, a Green's formula). However, it must be pointed out that the *regularity* required a priori for u and $\mathbf{p}$ is somehow interchanged. This implies that, in discretizing (2.7)–(2.8), we have to use a continuous "u" and can also use a discontinuous "$\mathbf{p}$", whereas in discretizing (2.9)–(2.10) we can use a discontinuous "u" but must also use a "$\mathbf{p}$" with divergence in $L^2(D)$ (and hence $\mathbf{p}\cdot\mathbf{n}$ is continuous at the interfaces). Note also the inversion in the treatment of the boundary conditions.

It is questionable whether (2.7)–(2.8) should be called a mixed formulation for problem (2.1)–(2.3). On the other hand, it seems acceptable to call the formulation (2.9)–(2.10) "mixed". In general, the original formulation (2.1)–(2.3) is to be preferred. It is simpler, it uses just one variable, and many robust methods are based on this approximation. However, in some applications, the "auxiliary" unknown $\mathbf{p}$ defined in (2.4) is actually the more relevant physical variable and/or is the only information that has to be transferred into other equations that are coupled with (2.1)–(2.3). In such cases, the use of a mixed formulation might be preferred, as far as it provides (as it often does) a better accuracy for $\mathbf{p}$. In general, the formulation (2.9)–(2.10) is then used, since it deals with a smoother vector field $\mathbf{p}$. It is often said that the crucial feature in the mixed approach is that it averages $(A(\mathbf{x}))^{-1}$ instead of $A(\mathbf{x})$. This is surely a better thing to do, at least in one dimension, being connected with the homogenization theory; see, for instance, [7]. However, dramatic improvements have been obtained by using (2.9)–(2.10) with a constant $A(\mathbf{x})$; see, for instance, [40]. The *true* reason (if any) for the better behavior of the mixed formulations than the classical ones is still not completely understood. Practical experiences suggest the use of a mixed formulation for "badly behaved" problems, in which the variable $\mathbf{p}(\mathbf{x})$ is expected to be "smoother" than the variable $u(\mathbf{x})$, but clearly this is not the whole story.

EXAMPLE 2.2 (Incompressible Fluids). The Stokes equations for incompressible fluids are of the type

$$-\Delta\mathbf{u} + \mathbf{grad}\,p = \mathbf{f} \quad \text{in } D \subset \mathbb{R}^d, \tag{2.11}$$

$$\text{div}\,\mathbf{u} = 0 \quad \text{in } D. \tag{2.12}$$

Various kinds of boundary conditions can be used in connection with (2.11),

(2.12). For the sake of simplicity, we shall consider only the (physically uninteresting) Dirichlet boundary conditions

$$\mathbf{u} = 0 \quad \text{on } \Gamma = \partial D. \tag{2.13}$$

The natural variational formulation of (2.11)–(2.13) is:

Find $\mathbf{u} \in (H_0^1(D))^d$ and $p \in L^2(D)$ such that

$$\int_D \underline{\underline{\mathbf{grad}}}\,\mathbf{u} : \underline{\underline{\mathbf{grad}}}\,\mathbf{v}\,d\mathbf{x} - \int_D p \operatorname{div} \mathbf{v}\,d\mathbf{x} = \int_D \mathbf{f}\cdot\mathbf{v}\,d\mathbf{x}, \qquad \forall \mathbf{v} \in (H_0^1(D))^d, \tag{2.14}$$

$$\int_D q \operatorname{div} \mathbf{u}\,d\mathbf{x} = 0, \qquad \forall q \in L^2(D). \tag{2.15}$$

The formulation (2.14)–(2.15) had been used for years before the term "mixed method" came into use; however, it is recognized now that (2.14)–(2.15) behaves like a mixed formulation as far as the difficulties in finding good approximations are concerned. We shall also see that (2.14)–(2.15) easily falls into the same abstract framework that is commonly used for mixed methods. Hence, we are somehow allowed to consider (2.14)–(2.15) as a mixed formulation.

EXAMPLE 2.3 (Linear Elasticity Problems). For a vector-valued function $\mathbf{v}(\mathbf{x})$ we define $\underline{\underline{\varepsilon}}(\mathbf{v})$ by

$$\varepsilon_{ij} = \frac{1}{2}\left(\frac{\partial v_i}{\partial x_j} + \frac{\partial v_j}{\partial x_i}\right) \qquad (i, j = 1, \ldots, d). \tag{2.16}$$

The linear elasticity equations are now

$$\underline{\underline{\sigma}} = E : \underline{\underline{\varepsilon}}(\mathbf{u}) \quad \left(\text{i.e., } \sigma_{ij} = \sum_{r=1}^{d} \sum_{s=1}^{d} E_{ijrs}\varepsilon_{rs}(\mathbf{u})\right), \tag{2.17}$$

$$\mathbf{div}\,\underline{\underline{\sigma}} = \mathbf{f} \quad \text{in } D. \tag{2.18}$$

Substituting (2.16), (2.17) into (2.18) gives a second-order linear elliptic system in the unknowns $\mathbf{u}$. Clearly, E is the elasticity tensor and is assumed here to have constant coefficients (and nice "ellipticity" properties). Its inverse (compliance tensor) will be denoted by C. Hence

$$\underline{\underline{\tau}} = E : \underline{\underline{\varepsilon}}(v) \Leftrightarrow \underline{\underline{\varepsilon}}(v) = C : \underline{\underline{\tau}}. \tag{2.19}$$

We assume again the simple boundary conditions:

$$\mathbf{u} = 0 \quad \text{on } \partial D. \tag{2.20}$$

This, of course, is strongly unrealistic: usually we have $\mathbf{u} = \tilde{\mathbf{u}}$ given on Γ_{Dir} and $\underline{\underline{\sigma}} \cdot \mathbf{n} = \mathbf{t}$ given on Γ_{Neu}. However, the proper way of dealing with realistic boundary conditions coincides with the one used in Example 2.1: we chose then to give more details (with simpler notations) while discussing the linear elliptic problem.

We can notice here that the splitting of the problem in more than one unknown is natural for solid physical reasons. This is probably why the first mixed formulations were used in elasticity theory. We shall present here only one mixed formulation, which is similar to the formulation (2.9)–(2.10) for a single elliptic equation. We set

$$H(\mathbf{div}; D) = \{\underline{\underline{\tau}} | \underline{\underline{\tau}} \in (L^2(D))^{d^2};\ \tau_{ij} = \tau_{ji},\ \forall i, j;\ \mathbf{div}\ \underline{\underline{\tau}} \in (L^2(D))^d\},$$

and we consider the problem:

Find $\mathbf{u} \in (L^2(D))^d$ and $\boldsymbol{\sigma} \in H(\mathbf{div}; D)$ such that

$$\int_D (C:\underline{\underline{\sigma}}):\underline{\underline{\tau}}\, dx + \int_D \mathbf{u}\cdot\mathbf{div}\ \underline{\underline{\tau}}\, d\mathbf{x} = 0, \qquad \forall \boldsymbol{\tau} \in H(\mathbf{div}; D), \tag{2.21}$$

$$\int_D \mathbf{v}\cdot\mathbf{div}\ \underline{\underline{\sigma}}\, d\mathbf{x} = \int_D \mathbf{f}\cdot\mathbf{v}\, d\mathbf{x}, \qquad \forall \mathbf{v} \in (L^2(D))^d. \tag{2.22}$$

We can see that (2.21)–(2.22) practically coincides with the variational formulation of the Hellinger–Reissner principle. The use of this principle in the framework of finite elements can be traced back to the pioneering work of Herrmann [33, 34] and Hellan [32]. The interest in using the stress field $\underline{\underline{\boldsymbol{\sigma}}}$ as an independent variable is questionable in as simple a case as the present one, but its use is clear in more general and more complicated problems involving nonlinearities, plasticity, and so on.

We shall now state an abstract existence theorem that is a simplified version of a more general result proved in [11].

Theorem 1. *Let Ξ and Ψ be real Hilbert spaces, $a(\xi_1, \xi_2)$ a bilinear form on $\Xi \times \Xi$, and $b(\xi, \psi)$ a bilinear form on $\Xi \times \Psi$. Set*

$$K = \{\xi | \xi \in \Xi,\ b(\xi, \psi) = 0,\ \forall \psi \in \Psi\},$$

and assume that

$$\exists \alpha > 0 \quad \text{such that} \quad a(\xi, \xi) \geq \alpha \|\xi\|^2_{\Xi}, \qquad \forall \xi \in K, \tag{2.23}$$

$$\exists \beta > 0 \quad \text{such that} \quad \sup_{\xi \in \Xi - \{0\}} \frac{b(\xi, \psi)}{\|\xi\|_{\Xi}} \geq \beta \|\psi\|_{\Psi}, \qquad \forall \psi \in \Psi. \tag{2.24}$$

Then for every $l_1 \in \Xi'$ and $l_2 \in \Psi'$ there exists a unique solution $(\bar{\xi}, \bar{\psi})$ of the problem

$$a(\bar{\xi}, \xi) + b(\xi, \bar{\psi}) = \langle l_1, \xi \rangle, \qquad \forall \xi \in \Xi, \tag{2.25}$$

$$b(\bar{\xi}, \psi) = \langle l_2, \psi \rangle, \qquad \forall \psi \in \Psi. \tag{2.26}$$

Remark. Actually a stronger result is proved in [11]. Namely, {problem (2.25), (2.26) has a unique solution for every $l_1 \in \Xi'$ and $l_2 \in \Psi'$} *if and only if* {(2.24) holds and the bilinear form $a(\xi_1, \xi_2)$, restricted to K, is nonsingular (in the sense that it induces an ismorphism from K on K')}. Clearly, if we assume that $a(\xi_1, \xi_2)$ is symmetric and positive semidefinite, then (2.23) and (2.24) are *necessary and sufficient* for the existence and uniqueness of the solution of (2.25)–(2.26).

Remark. It is clear that, if $a(\xi_1, \xi_2)$ is symmetric, the solution $(\bar{\xi}, \bar{\psi})$ of (2.25)–(2.26) minimizes the functional

$$J(\xi) = \tfrac{1}{2}a(\xi, \xi) - \langle l_1, \xi \rangle, \tag{2.27}$$

on the subspace of Ξ

$$K(l_2) = \{\xi | b(\xi, \psi) = \langle l_2, \psi \rangle, \forall \psi \in \Psi\}, \tag{2.28}$$

and the formulation (2.25)–(2.26) corresponds to the introduction in (2.27)–(2.28) of the Lagrange multiplier $\bar{\psi}$.

2.3. Discretizing a Mixed Formulation

Let us deal first with the abstract framework (2.25)–(2.26). Assume that we are given two sequences $\{\Xi_h\}_{h>0}$ and $\{\Psi_h\}_{h>0}$ of subspaces of Ξ and Ψ, respectively. We set

$$K_h = \{\xi_h | \xi_h \in \Xi_h, b(\xi_h, \psi_h) = 0, \forall \psi_h \in \Psi_h\}. \tag{3.1}$$

We have the following approximation theorem [11].

Theorem 2. *Assume that*

$$\exists \alpha_h > 0 \quad \textit{such that} \quad a(\xi, \xi) \geq \alpha_h \|\xi\|_\Xi^2, \qquad \forall \xi \in K_h, \tag{3.2}$$

$$\exists \beta_h > 0 \quad \textit{such that} \quad \sup_{\xi \in \Xi_h - \{0\}} \frac{b(\xi, \psi)}{\|\xi\|_\Xi} \geq \beta_h \|\psi\|_\Psi, \qquad \forall \psi \in \Psi_h. \tag{3.3}$$

Then for every $l_1 \in \Xi'$ and $l_2 \in \psi'$ and for every $h > 0$ the discrete problem

$$a(\bar{\xi}_h, \xi) + b(\xi, \bar{\psi}_h) = \langle l_1, \xi \rangle, \qquad \forall \xi \in \Xi, \tag{3.4}$$

$$b(\bar{\xi}_h, \psi) = \langle l_2, \psi \rangle, \qquad \forall \psi \in \Psi_h, \tag{3.5}$$

has a unique solution. Moreover, there exists a constant $\gamma_h(\alpha_h, \beta_h) > 0$ such that

$$\|\bar{\xi} - \bar{\xi}_h\|_\Xi + \|\bar{\psi} - \bar{\psi}_h\|_\Psi \leq \gamma_h \left(\inf_{\xi_h \in \Xi_h} \|\bar{\xi} - \xi_h\|_\Xi + \inf_{\psi_h \in \Psi_h} \|\bar{\psi} - \psi_h\|_\Psi \right). \tag{3.6}$$

The dependence of γ_h on α_h and β_h can be easily traced (see [11]). Clearly, if (3.2) and (3.3) hold with constants $\bar{\alpha}$, $\bar{\beta}$ independent of h, then (3.6) holds with a constant $\bar{\gamma}$ independent of h. More general versions of Theorem 2 (and also of Theorem 1) can be found, for instance, in [21] or in [9].

We are now going to see the implications of Theorem 2 in the examples of the previous section.

EXAMPLE 3.1 (Discretizations of the Mixed Formulations for Linear Elliptic Operators). Many examples of successful discretizations of (2.9)–(2.10) are known. The first ones were introduced by Raviart and Thomas [44] and then elaborated and extended to more general cases by Nedelec [41]. Other families of possible discretizations were introduced years later by Brezzi *et al.* [15], and then elaborated and extended in several more recent papers (see, e.g., [13, 14, 42]). All of them share a very helpful property, the so-called *commuting diagram property*, whose importance was first fully recognized in [19]. Let us discuss it in a particular case: the BDM (Brezzi–Douglas–Marini) element of degree two for two-dimensional problems ($D \subset \mathbb{R}^2$). Let $\mathcal{T}_h$ be a regular sequence of decompositions of D into triangles. We assume for the sake of simplicity that $\Gamma_{\text{Neu}} = \varnothing$ in (2.2) and $A(\mathbf{x}) = 1$. As a discretization of $H(\text{div}; D)$ and $L^2(D)$, respectively, we take

$$\Xi_h = \{\mathbf{q} | \mathbf{q} \in H(\text{div}; D); \mathbf{q}_{|T} \in (P_2)^2, \forall T \in \mathcal{T}_h\}, \tag{3.7}$$

$$\Psi_h = \{v | v \in L^2(D); v_{|T} \in P_1, \forall T \in \mathcal{T}_h\}. \tag{3.8}$$

Here and in the following $P_k(S)$ (or simply P_k) will denote the set of polynomials of degree $\leq k$ on the set S. We consider now the discretized problem

Find $\mathbf{p}_h \in \Xi_h$ and $u_h \in \Psi_h$ such that

$$\int_D \mathbf{p}_h \cdot \mathbf{q}\, d\mathbf{x} + \int_D u_h \,\text{div}\, \mathbf{q}\, d\mathbf{x} = \int_{\partial D} g_0 \mathbf{q} \cdot \mathbf{n}\, d\Gamma, \qquad \forall \mathbf{q} \in \Xi_h, \tag{3.9}$$

$$\int_D v \,\text{div}\, \mathbf{p}_h\, d\mathbf{x} = \int_D fv\, d\mathbf{x}, \qquad \forall v \in \Psi_h. \tag{3.10}$$

We now define an operator M_h from $(H^1(D))^2$ into Ξ_h by

$$\begin{cases} \text{(i)} \quad \displaystyle\int_e (\mathbf{q} - M_h \mathbf{q}) \cdot \mathbf{n} p_2\, ds = 0, & \forall e, \text{ edge in } \mathcal{T}_h, \forall p_2 \in P_2(e); \\ \text{(ii)} \quad \displaystyle\int_T (\mathbf{q} - M_h \mathbf{q})\, d\mathbf{x} = 0, & \forall T, \text{ triangle in } \mathcal{T}_h; \\ \text{(iii)} \quad \displaystyle\int_T (\mathbf{q} - M_h \mathbf{q}) \cdot \mathbf{rot}\, b_T\, d\mathbf{x} = 0, & \forall T, \text{ triangle in } \mathcal{T}_h; \end{cases} \tag{3.11}$$

where $b_T := \lambda_1 \lambda_2 \lambda_3$ is the cubic vanishing on ∂T, and $\mathbf{rot}\, \phi := (-\partial\phi/\partial x_1, \partial\phi/\partial x_2)$. We also define an operator $\mathcal{P}_h$ from $L^2(D)$ into Ψ_h by

$$\int_T (v - \mathcal{P}_h v) p_1\, d\mathbf{x} = 0, \qquad \forall T, \text{ triangle in } \mathcal{T}_h, \forall p_1 \in P_1(T). \tag{3.12}$$

Let us now check that $\text{div}\, M_h \mathbf{q} = \mathcal{P}_h \,\text{div}\, \mathbf{q}$ for all $\mathbf{q} \in (H^1(D))^2$. Actually, for all $v_h \in \Psi_h$, we have

$$\int_T v_h \operatorname{div} M_h\mathbf{q}\, d\mathbf{x} = \int_{\partial T} v_h(M_h\mathbf{q}\cdot\mathbf{n})\, ds - \int_T \mathbf{grad}\, v_h \cdot M_h\mathbf{q}\, d\mathbf{x}$$

$$= \int_{\partial T} v_h\mathbf{q}\cdot\mathbf{n}\, ds - \int_T \mathbf{grad}\, v_h\cdot\mathbf{q}\, d\mathbf{x} = \int_T v_h \operatorname{div}\mathbf{q}\, d\mathbf{x}$$

$$= \int_T v_h \mathscr{P}_h \operatorname{div}\mathbf{q}\, d\mathbf{x}. \tag{3.13}$$

It is also easy to check that the divergence operator is linear, continuous, and surjective from $(H^1(D))^2$ onto $L^2(D)$. This can be summarized in the following diagram:

$$\begin{array}{ccccc} (H^1(D))^2 & \xrightarrow{\operatorname{div}} & L^2(D) & \longrightarrow & 0 \\ {\scriptstyle M_h}\downarrow & & \downarrow{\scriptstyle \mathscr{P}_h} & & \\ \Xi_h & \xrightarrow{\operatorname{div}} & \Psi_h & \longrightarrow & 0. \end{array} \tag{3.14}$$

It is easy to check that (3.14) implies, in particular, (3.2) and (3.3), but it is much more powerful than that. For instance, it implies

$$\|\mathbf{p} - \mathbf{p}_h\|_{(L^2(D))^2} \le \gamma_1 \|\mathbf{p} - M_h\mathbf{p}\|_{(L^2(D))^2}, \tag{3.15}$$

$$\|u - u_h\|_{L^2(D)} \le \gamma_2(\|\mathbf{p} - M_h\mathbf{p}\|_{(L^2(D))^2} + \|u - \mathscr{P}_h u\|_{L^2(D)}), \tag{3.16}$$

with γ_1, γ_2 independent of h (whenever $\mathbf{p} \in (H^1(D))^2$). In particular, with the choice (3.7)–(3.8) this yields

$$\|\mathbf{p} - \mathbf{p}_h\|_{(L^2(D))^2} \le \gamma_1 h^3 \|\mathbf{p}\|_{(H^3(D))^2}, \tag{3.17}$$

$$\|u - u_h\|_{L^2(D)} \le \gamma_2 h^2(\|u\|_{H^2(D)} + \|\mathbf{p}\|_{(H^2(D))^2}). \tag{3.18}$$

Note that (3.17) does not follow from the abstract error estimate (3.6).

The commuting diagram has other nice properties. For instance, it allows a simple proof of error estimates in dual norms, as in [20] or in [15]. Error estimates in L^∞ norms are also available: see, for instance, [45, 46] and [26, 27].

The most popular scheme for (2.9)–(2.10), that is, the "lowest order Raviart–Thomas," can be obtained by using, instead of (3.7), (3.8),

$$\begin{cases} \Xi_h = \{\mathbf{q} \mid \mathbf{q} \in H(\operatorname{div}; D);\ \mathbf{q}_{|T} \in (P_1(T))^2, \forall T;\ \mathbf{q}\cdot\mathbf{n}_{|e} \in P_0(e), \forall e\}, \\ \Psi_h = \{v \mid v \in L^2(D);\ v_{|T} \in P_0, \forall T \in \mathscr{T}_h\}. \end{cases} \tag{3.19}$$

Accordingly, we then use $P_0(e)$ instead of $P_2(e)$ in (3.11)(i) and drop (3.11)(ii) and (iii); similiarly, we use $P_0(T)$ instead of $P_1(T)$ in (3.12). It follows immediately that (3.13) still holds, and then (3.14) also holds. Clearly, only an $O(h)$ rate can now be achieved in both (3.17) and (3.18).

EXAMPLE 3.2 (Discretizations of the Stokes Equations). Life is much harder when we go from (2.9)–(2.10) to (2.14)–(2.15). The only positive aspect is that now the bilinear form $a(\mathbf{u}, \mathbf{v})$ is such that (2.23) actually holds in the whole $(H_0^1(D))^2$ (our present Ξ) so that (3.2) also holds regardless of the choice of the discretization. This might partially excuse all the "Stokes-thinking" people who consider (3.3) as *the* condition for mixed methods. If we try to discretize even the easy (2.9)–(2.10) with a scheme that does not satisfy (3.14), we will see that (3.2) can cause great difficulty.

However, coming back to Stokes, it is true that the only condition to be satisfied by the discretization is (3.3), which now reads

$$\exists \beta_h > 0, \qquad \sup_{v \in \Xi_h - \{0\}} \frac{\int_D q \operatorname{div} \mathbf{v} \, d\mathbf{x}}{\|\mathbf{v}\|_1} \geq \beta_h \|q\|_{L^2(D)/\mathbb{R}}, \qquad \forall q \in \Psi_h, \tag{3.20}$$

with, if possible, β_h independent of h. A sufficient condition is the following so-called Fortin's trick [22]: we have to find a linear operator M_h from $(H^1(D))^2$ into Ξ_h such that

$$\|M_h \mathbf{v}\|_1 \leq c_1 \|\mathbf{v}\|_1, \qquad \forall \mathbf{v} \in (H_0^1(D))^2, \tag{3.21}$$

$$\int_D q_h \operatorname{div}(\mathbf{v} - M_h \mathbf{v}) \, d\mathbf{x} = 0, \qquad \forall \mathbf{v} \in (H_0^1(D))^2, \quad \forall q_h \in \Psi_h. \tag{3.22}$$

We consider one example. Let $\mathscr{T}_h$ be a decomposition of D into rectangles, R, with sides parallel to the axes (the use of isoparametric elements is obviously also allowed, but more complicated to describe), and choose

$$\Xi_h = \{\mathbf{v} | \mathbf{v} \in (H_0^1(D))^2; \mathbf{v}_{|R} \in (Q_2(R))^2, \forall R \in \mathscr{T}_h\}, \tag{3.23}$$

$$\Psi_h = \left\{ q | q \in L^2(D); \int_D q \, d\mathbf{x} = 0; q_{|R} \in P_1(R), \forall R \in \mathscr{T}_h \right\}. \tag{3.24}$$

In (3.23), $Q_2(R)$ means the set of polynomials of degree ≤ 2 in each variable. Let us see how to construct the operator M_h at least for a smooth $\mathbf{v}$. To deal with a general $\mathbf{v}$ in $(H_0^1(D))^2$ is technically more complicated, but the philosophy is the same. In each R we set

$$M_h \mathbf{v} = \mathbf{v} \qquad \text{at the vertices (eight conditions);} \tag{3.25}$$

$$\int_e (M_h \mathbf{v} - \mathbf{v}) \, ds = 0 \qquad \text{on each edge (eight conditions);} \tag{3.26}$$

$$\int_R \operatorname{div}(M_h \mathbf{v} - \mathbf{v}) x_i \, d\mathbf{x} = 0, \qquad i = 1, 2 \text{ (two conditions).} \tag{3.27}$$

We have a total of 18 conditions (note that the dimension of Q_2 is 9). It is easy to check that they are independent. Let us check (3.22); that is, let us check that

$$\int_R \operatorname{div}(M_h \mathbf{v} - \mathbf{v}) p_1 \, d\mathbf{x} = 0, \qquad \forall p_1 \in P_1(R). \tag{3.28}$$

Clearly, (3.27) implies that (3.28) holds for $p_1 = x_1$ and $p_1 = x_2$. We need only to check $p_1 \equiv 1$:

$$\int_R \operatorname{div}(M_h \mathbf{v} - \mathbf{v})\, d\mathbf{x} = \int_{\partial R} (M_h \mathbf{v} - \mathbf{v}) \cdot \mathbf{n}\, ds = 0, \tag{3.29}$$

due to (3.26). We can now apply (3.6) and obtain

$$\|\mathbf{u} - \mathbf{u}_h\|_1 + \|p - p_h\|_{L^2(D)/\mathbb{R}} \leq ch^2(\|\mathbf{u}\|_3 + \|p\|_2). \tag{3.30}$$

There are many other known choices available for obtaining a discretization of (2.14)–(2.15) that satisfies (3.20). An almost complete list of them can be found in Brezzi and Fortin[1] together with the references. In particular, Scott and Vogelius [47] proved that, under minor restrictions on the decomposition of D into triangles, we can always use a continuous velocity field of local degree k and a discontinuous pressure field of local degree $k - 1$, provided $k \geq 4$. For the low degrees, a special headache is provided by the use of bilinear velocities and constant pressures: its convergence has been proved in a variety of cases (see [38, 43, 48]), but not yet in the general case. In any event, a filtering of the pressure field is always required to eliminate the checkerboard modes. General strategies for constructing discretizations that fulfill (3.20) are given in [10, 16]. Modifications of the discrete equations that allow violation of (3.3) have received considerable attention in recent times. Roughly speaking, in the special case of equations (2.14), (2.15), they consist in substituting (2.15) with the "perturbed" equation [16]

$$\int_D q \operatorname{div} \mathbf{u}\, d\mathbf{x} + h^2 \int_D \mathbf{grad}\, p \cdot \mathbf{grad}\, q\, d\mathbf{x} = 0, \qquad \forall q \in H^1(D) \tag{3.31}$$

provided that we use a continuous pressure field in the finite element discretization. A modification of (3.31) of the form

$$\int_D q \operatorname{div} \mathbf{u}\, dx = \sum_{T \in \mathcal{T}_h} \alpha h_T^2 \int_T (\Delta \mathbf{u} - \mathbf{grad}\, p + \mathbf{f}) \cdot \mathbf{grad}\, q\, d\mathbf{x} \tag{3.32}$$

(α = small parameter to be adjusted; h_T = diameter of T) has been introduced by Hughes *et al.* [35]. Note that (3.31) is simpler but has a *consistency error* of order h^2 that is not present in (3.32). For more details and additional results in this direction, see [12].

For the use of more general boundary conditions, the basic reference is Verfürth [49]; see also the references contained therein.

Additional references for the Stokes and Navier–Stokes equations can be found in Glowinski and Pironneau [30], Glowinski [31], Girault and Raviart [28], and Brezzi and Fortin.[2]

---

[1] F. Brezzi and M. Fortin, book in preparation.

[2] F. Brezzi and M. Fortin, book in preparation.

EXAMPLE 3.3 (Discretizations of Linear Elasticity Problems). It is difficult, in general, to find convenient finite element discretizations for equations (2.21)–(2.22). We shall briefly indicate here four possible ways to proceed. The first possibility is to try to construct spaces that verify the commuting diagram property (as in (3.14)). This has been possible, up to now, only by means of *composite elements*: that is, each element is split into subelements and we use trial functions that are polynomials in each subelement (plus suitable continuity requirements from one subelement to the other). Examples of this approach can be found in Johnson and Mercier [37] or in Arnold *et al.* [4]. A second possibility is to give up the symmetry condition that appears in the definition of $H(\mathbf{div}; D)$, and to enforce it a posteriori by means of a Lagrange multiplier. After discretization we deal then with stress fields having only a *weak symmetry*. This idea was first used by Fraeijs de Veubeke [23] and then modified and analyzed by Amara and Thomas [1] and Arnold *et al.* [3]. A third possibility is to change the "auxiliary function" and use a *different, nonsymmetric tensor field* instead of $\underline{\underline{\boldsymbol{\sigma}}}$. This will, in general, produce some trouble at Γ_{Neu} (if $\Gamma_{\text{Neu}} \neq \emptyset$) that can be treated with the introduction of an additional Lagrange multiplier on Γ_{Neu}. We refer to Arnold and Falk [5] for more details on this approach. Finally, a fourth possibility is the addition of a stabilizing term in the style of (3.31) or (3.32). For this we refer to Franca *et al.* [25].

It should be pointed out that additional difficulties arise when dealing with nearly incompressible materials. In these cases (2.23) ceases to hold (in the limit) in the whole space but still holds for divergence-free tensor fields. This implies that (3.2) must also be checked if the discretization is such that $K_h \not\subset K$. Additional references for the preceding (and many other) applications can be found in Brezzi and Fortin.[3]

2.4. Numerical Methods for Solving the Discretized Problem

The major difficulty that arises in solving a linear problem such as (3.4), (3.5) is that the associated matrix

$$\begin{pmatrix} A & B \\ B^{\mathrm{T}} & 0 \end{pmatrix} \tag{4.1}$$

is indefinite. There are many ways of overcoming this difficulty, mostly using some particular feature of the problem under consideration in order to rewrite it in a different form. Here we shall briefly sketch two of them: one that is mostly used in Examples 2.1 and 2.3 and the other used in Example 2.2.

The first technique, which is very old (see [24]) starts from the following simple observation. If the space Ξ_h is made of functions that are completely

[3] F. Brezzi and M. Fortin, book in preparation.

discontinuous from one element to the other, then the most natural choice of basis functions for Ξ_h will produce a matrix A in (4.1) that is block-diagonal. Then the inverse matrix A^{-1} can be easily computed explicitly. Solving (3.4) element by element for $\bar{\xi}_h$ and substituting into (3.5) (static condensation) leaves us with the final matrix $B^{\mathrm{T}}A^{-1}B$ and the only unknown $\bar{\psi}_h$. Note that (3.2) and (3.3) will imply that $B^{\mathrm{T}}A^{-1}B$ is symmetric and positive definite (if $a(\xi_1, \xi_2)$ is symmetric). Now if instead the functions in Ξ_h have some continuity properties from one element to the other (as in Example 3.1 we had $\mathbf{p}_h \cdot \mathbf{n}$ continuous at the interfaces), this cannot be done. However, we can choose to work in a larger space, say $\tilde{\Xi}_h$, made of discontinuous functions, and then require the continuity by means of a Lagrange multiplier. Let us see the procedure in the particular case of Example 3.1. We set

$$\tilde{\Xi}_h = \{\mathbf{q} | \mathbf{q} \in (L^2(D))^2;\ \mathbf{q}_{|T} \in (P_2)^2,\ \forall T \in \mathscr{T}_h\}, \tag{4.2}$$

$$\Lambda_h = \{\mu | \mu_{|e} \in P_2(e),\ \forall e, \text{ internal edge in } \mathscr{T}_h,\ \mu = 0 \text{ on } \partial D\}, \tag{4.3}$$

$$c(\mathbf{q}, \mu) = \sum_{T \in \mathscr{T}_h} \int_{\partial T} (\mathbf{q} \cdot \mathbf{n}) \mu \, ds. \tag{4.4}$$

Clearly, if $\mathbf{q} \in \tilde{\Xi}_h$, then

$$\mathbf{q} \in \Xi_h \Leftrightarrow c(\mathbf{q}, \mu) = 0, \qquad \forall \mu \in \Lambda_h. \tag{4.5}$$

It is not difficult to check that the new problem:

Find $\tilde{\mathbf{p}}_h \in \tilde{\Xi}_h$, $\tilde{u}_h \in \Psi_h$, $\lambda_h \in \Lambda_h$, such that

$$\int_D \tilde{\mathbf{p}}_h \cdot \mathbf{q} \, d\mathbf{x} + \sum_T \int_T \tilde{u}_h \operatorname{div} \mathbf{q} \, d\mathbf{x} = \int_{\partial D} g_0 \mathbf{q} \cdot \mathbf{n} \, ds + c(\mathbf{q}, \lambda_h), \qquad \forall \mathbf{q} \in \tilde{\Xi}_h, \tag{4.6}$$

$$\sum_T \int_T v \operatorname{div} \tilde{\mathbf{p}}_h \, d\mathbf{x} = \int_D f v \, d\mathbf{x}, \qquad \forall v \in \Psi_h, \tag{4.7}$$

$$c(\tilde{\mathbf{p}}_h, \mu) = 0, \qquad \forall \mu \in \Lambda_h, \tag{4.8}$$

has a unique solution, and that $\tilde{\mathbf{p}}_h = \mathbf{p}_h$, $\tilde{u}_h = u_h$. Now both the unknowns $\tilde{\mathbf{p}}_h$ and $\tilde{u}_h$ are a priori discontinuous, and they can be eliminated, at the element level, by static condensation. The final matrix, in the unknown λ_h, will be symmetric and positive definite. It is clear that λ_h itself should be an approximation of u at the interfaces, and it has been used as such by engineers. However, it was only rather recently that it was proved mathematically that λ_h converges to u, and in general with a *better* order of convergence than u_h itself (see [2]). For instance, in the present case, once λ_h is known we can construct, element by element, a $u_h^* \in P_3(T)$ such that

$$\begin{cases} \displaystyle\int_e (u_h^* - \lambda_h) p_2 \, ds = 0, & \forall e \text{ edge of } T,\ \forall p_2 \in P_2(e), \\ \displaystyle\int_T (u_h^* - u_h) \, d\mathbf{x} = 0, & \end{cases} \tag{4.9}$$

and show that

$$\|u - u_h^*\|_{L^2(D)} \leq O(h^4) \tag{4.10}$$

instead of (3.18) (for the proof of (4.10), see [15]). A similar result (but only with $O(h^2)$ in (4.10)) can also be achieved with the lowest order Raviart–Thomas element described at the end of Example 3.1. However, the best way to compute the solution for this last element is to solve with the so-called P_1-nonconforming method and then use the postprocessing of Marini [39]. For additional results on the convergence of λ_h to u see Brezzi *et al.* [15], Brezzi *et al.* [14], and Gastaldi and Nochetto [27].

This same idea (disconnect Ξ_h and use a Lagrange multiplier to force back the continuity) can be used for elasticity problems and in many other cases. However, it has not been possible, so far, to use it, for instance, for the Stokes equations (and more generally when continuity at the *vertices* is required in Ξ_h). Then we can use the following other trick, that was first analyzed by Bercovier [8]. If the space Ψ_h is made of discontinuous functions (as it was the case in our Example 3.2), then we can perturb equation (3.5) into

$$b(\bar{\xi}_h, \psi) = \varepsilon(\bar{\psi}_h, \psi)_\Psi + \langle l_2, \psi\rangle, \qquad \forall\psi \in \Psi_h. \tag{4.11}$$

The corresponding matrix (for (3.4), (4.11)) becomes, roughly,

$$\begin{pmatrix} A & B \\ B^{\mathrm{T}} & -\varepsilon I \end{pmatrix}. \tag{4.12}$$

Now the discontinuity in Ψ_h allows us to eliminate $\bar{\psi}_h$ at the element level. We obtain in that way a matrix $A + \varepsilon^{-1}BB^{\mathrm{T}}$. If (3.2) and (3.3) are satisfied, this new matrix will be symmetric and positive definite (always if $a(\xi_1, \xi_2)$ is symmetric). Moreover, calling $(\xi_h^\varepsilon, \psi_h^\varepsilon)$ the solution of (3.4) and (4.11), we have

$$\|\bar{\xi}_h - \xi_h^\varepsilon\|_\Xi + \|\bar{\psi}_h - \psi_h^\varepsilon\|_\Psi = O(\varepsilon). \tag{4.13}$$

The method can also be applied when Ψ_h consists of continuous functions, provided that some lumping procedure is used to compute the inner product in (4.11). However, in such cases, we get for $\varepsilon^{-1}BB^{\mathrm{T}}$ a bandwidth that is generally larger than that of A, and this is often a considerable drawback.

A different attempt to reduce (3.4), (3.5) to a single equation in the case of the Stokes problem can be found in Bramble and Pasciak.[4]

Acknowledgments. This research was supported under NASA Contract No. NAS1-18107 while the author was in residence at the Institute for Computer Applications in Science and Engineering (ICASE), NASA Langley Research Center, Hampton, VA 23665, U.S.A.

---

[4] J.H. Bramble, and J. Pasciak "A preconditioning technique for indefinite systems resulting from mixed approximations of elliptic problems" to appear in *Math. Comput.*

References

1. M. Amara and J.M. Thomas, "Equilibrium finite elements for the linear elastic problem," *Numer. Math.*, **33** (1979), 367–383.
2. D.N. Arnold and F. Brezzi, "Mixed and nonconforming finite element methods: Implementation, postprocessing, and error estimates," *RAIRO M_2AN*, **19** (1985), 7–32.
3. D.N. Arnold, F. Brezzi, and J. Douglas, Jr., "PEERS: A new mixed finite element for plane elasticity," *Japan J. Appl. Math.*, **1** (1984), 347–367.
4. D.N. Arnold, J. Douglas, Jr., and C.P. Gupta, "A family of higher order finite element methods for plane elasticity," *Numer. Math.* **45** (1984), 1–22.
5. D.N. Arnold and R.S. Falk, "A new mixed formulation for elasticity," *Numer. Math.*, to appear.
6. I. Babuška, "The finite element method with Lagrangian multipliers," *Numer. Math.*, **20** (1973), 179–192.
7. I. Babuška and J.E. Osborn, "Numerical treatment of eigenvalue problems for differential equations with discontinuous coefficients," *Math. Comp.*, **32** (1978), 991–1023.
8. M. Bercovier, These de Doctorat d'Etat, Université de Rouen, 1976.
9. C. Bernardi, C. Canuto, and Y. Maday, "Generalized inf–sup condition for Chebyshev approximation of the Navier–Stokes equations," submitted to *SIAM J. Numer. Anal.*
10. J. Boland and R. Nicolaides, "Stability of finite elements under divergence constraints," *SIAM J. Numer. Anal.*, **20**, No. 4 (1983), 722–731.
11. F. Brezzi, "On the existence, uniqueness, and approximation of saddle-point problems arising from Lagrangian multipliers," *RAIRO Anal. Numér.*, **8–32** (1974), 129–151.
12. F. Brezzi and J. Douglas, Jr., "Stabilized mixed methods for the Stokes problem," *Numer. Math.*, to appear.
13. F. Brezzi, J. Douglas, Jr., R. Duran, and M. Fortin, "Mixed finite elements for second-order elliptic problems in three variables," *Numer. Math.*, **51** (1987), 237–250.
14. F. Brezzi, J. Douglas, Jr., M. Fortin, and L.D. Marini, "Efficient rectangular mixed finite elements in two and three space variables," *RAIRO M_2AN*, **21**, No. 4 (1987), 581–604.
15. F. Brezzi, J. Douglas, Jr., and L.D. Marini, "Two families of mixed finite elements for second-order elliptic problems," *Numer. Math.*, **47** (1985), 217–235.
16. F. Brezzi and J. Pitkäranta, "On the stabilization of finite element approximations of the Stokes equations," *Efficient Solutions of Elliptic Systems* (*Kiel*, 1984), Braunschweig, Vieweg, 1984.
17. P.G. Ciarlet, *The Finite Element Method for Elliptic Problems*, North-Holland, Amsterdam, 1978.
18. M. Crouzeix and P.A. Raviart, "Conforming and non-conforming finite element methods for solving the stationary Stokes equations," *RAIRO Anal. Numér.*, **R3** (1973) 33–76.
19. J. Douglas, Jr. and J.E. Roberts, "Mixed finite element methods for second-order elliptic problems," *Mat. Apl. Comput.*, **1** (1982), 91–103.
20. J. Douglas, Jr. and J.E. Roberts, "Global estimates for mixed methods for second-order elliptic equations," *Math. Comp.*, **44** (1985), 39–52.

21. R.S. Falk and J.E. Osborn, "Error estimates for mixed methods," *RAIRO Anal. Numér.*, **14** (1980), 309–324.
22. M. Fortin, "An analysis of the convergence of mixed finite element methods," *RAIRO Anal. Numér.*, **11** (1977), 341–354.
23. B.X. Fraeijs de Veubeke, "Stress function approach," in *World Congress on the Finite Element Method in Structural Mechanics* (*Bournemouth*, 1975.).
24. B.X. Fraeijs de Veubeke, "Displacement and equilibrium models in the finite element method," in *Stress Analysis* (O.C. Zienkiewicz and G. Hollister, eds.), Wiley, New York, 1965.
25. L.P. Franca, T.J.R. Hughes, A.F.D. Loula, and I. Miranda, "A new family of stable elements for nearly incompressible elasticity: A mixed Petrov–Galerkin finite element method," *Numer. Math.*, to appear.
26. L. Gastaldi and R.H. Nochetto, "Optimal L^∞-error estimates for nonconforming and mixed finite element methods of lowest order," *Numer. Math.*, **50** (1987), 587–611.
27. L. Gastaldi and R.H. Nochetto, "On L^∞-accuracy of mixed finite element methods for second-order elliptic problems," *Mat. Apl. Comput.*, to appear.
28. V. Girault and P.A. Raviart, "Finite element approximation of the Navier–Stokes equations," Lecture Notes in Mathematics, Vol. 749, Springer-Verlag, New York, 1979.
29. R. Glowinski, "Approximations externes par éléments finis d'ordre un et deux du probleme de Dirichlet pour Δ^2," in *Topics in Numerical Analysis*, Vol. I (J.J.H. Miller, ed.), Academic Press, London, 1973, pp. 123–171.
30. R. Glowinski and O. Pironneau, "Numerical methods for the first biharmonic equation and for the two-dimensional Stokes problem," *SIAM Rev.*, **21**, No. 2 (1979), 167–212.
31. R. Glowinski, *Numerical Methods for Nonlinear Variational Problems*, Springer Series in Comparative Physics, Springer-Verlag, New York, 1984.
32. K. Hellan, "Analysis of elastic plates in flexure by a simplified finite element method," *Acta Polytech. Scand. Math. Comput. Sci. Ser.*, **46** (1967).
33. L.R. Herrmann, "Finite element bending analysis for plates," *J. Engrg. Mech. Div. ASCE EMS*, **93** (1967), 49–83.
34. L.R. Herrmann, "A bending analysis for plates," in *Proceedings of the Conference on Matrix Methods in Structural Mechanics*, AFFDL-TR-66-88, pp. 577–604.
35. T.J.R. Hughes, L.P. Franca, M. Balestra, "A new finite element formulation for computational fluid dynamics, *V*: Circumventing the Babuska–Brezzi condition: A stable Petrov–Galerkin formulation of the Stokes problem accommodating equal order interpolations," *Comput. Meth. Appl. Mech. Engrg.*, **59** (1986), 85–101.
36. C. Johnson, "On the convergence of some mixed finite element methods for plate bending problems," *Numer. Math.*, **21** (1973), 43–62.
37. C. Johnson and B. Mercier, "Some equilibrium finite element methods for two-dimensional elasticity problems," *Numer. Math.*, **30** (1978), 103–116.
38. C. Johnson and J. Pitkäranta, "Analysis of some mixed finite element methods related to reduced integration," *Math. Comp.*, **38** (1982), 375–400.
39. L.D. Marini, "An inexpensive method for the evaluation of the solution of the lowest order Raviart–Thomas mixed method," *SIAM J. Numer. Anal.*, **22**, No. 3 (1985), 493–496.
40. L.D. Marini and A. Savini, "Accurate computation of electric field in reverse-biased semiconductor devices: A mixed finite element approach," *COMPEL*, **3**, No. 3 (1984), 123–135.

41. J.C. Nedelec, "Mixed finite elements in $\mathbb{R}^3$," *Numer. Math.*, **35** (1980), 315–341.
42. J.C. Nedelec, "A new family of mixed finite elements in $\mathbb{R}^3$," *Numer. Math.*, **50** (1986), 57–82.
43. J. Pitkäranta and R. Stenberg, "Error bounds for the approximation of the Stokes problem using bilinear/constant elements on irregular quadrilateral meshes," in MAFELAP V (Brunel, 1984, J. Whiteman, ed.), Academic Press, London, 1985, pp. 325–334.
44. P.A. Raviart and J.M. Thomas, "A mixed finite element method for second-order elliptic problems," in *Mathematical Aspects of Finite Element Methods* (*Rome*, 1975), Lecture Notes in Mathematics, Vol. 606, Springer-Verlag, New York, 1975, pp. 292–315.
45. R. Scholz, "L^∞-convergence of saddle-point approximations for second-order problems," *RAIRO Anal. Numér.*, **11**, No. 2 (1977), 209–216.
46. R. Scholz, "Optimal L^∞-estimates for a mixed finite element method for elliptic and parabolic problems," *Calcolo*, **20** (1983), 355–377.
47. L.R. Scott and M. Vogelius, "Norm estimates for a maximal right inverse of the divergence operator in spaces of piecewise polynomials," *RAIRO* M_2AN, **19** (1985), 111–143.
48. R. Stenberg, "Analysis of mixed finite element methods for the Stokes problem: A unified approach," *Math. Comp.*, **42** (1984), 9–23.
49. R. Verfürth, "Finite element approximation of stationary Navier–Stokes equations with slip boundary conditions," Ruhr Universitätat Bochum, B.75, June 1986.

CHAPTER 3

Singular Finite Element Methods

George J. Fix

3.1. Introduction

The key property of elliptic systems is that their solution tends to be as smooth as the data and other factors permit. This condition strikingly constrasts with hyperbolic systems, in which singular behavior (e.g., shocks) can arise even if all inputs are smooth.

To illustrate this, consider the following model problem defined in a bounded region Ω of $\mathbb{R}^n$:

$$\operatorname{div}[a \operatorname{grad} \phi] = f \quad \text{in } \Omega, \tag{1.1}$$

$$B[\phi] = g \quad \text{on } \partial\Omega. \tag{1.2}$$

The boundary operator B has thc form

$$B[\phi] = \alpha \frac{\partial \phi}{\partial \mathbf{v}} + \beta\phi, \tag{1.3}$$

where $\mathbf{v}$ is the outer normal to $\partial\Omega$. Thus the inputs to this system are the coefficients a, α, β; the data f, g; and the region Ω. If all of these are smooth, then the same is true of the solution ϕ.

This means that singular behavior can arise only in the cases where the inputs are irregular. The following are examples of technical importance.

CASE 1 (Irregular Boundaries). The study of fracture and crack propagation involves elliptic systems of various orders depending on the type of problem being modeled [1], [2]. Torsion problems are second-order (like (1.1)), whereas plates involve fourth-order equations, and shell problems are even of higher order [3, 4]. Nevertheless, the region Ω in question generally contains a slit as in Figure 1.1 with the point P being the crack tip. In the linear, second-order case ($a = 1$ in (1.1)) it can be shown that with Dirichlet boundary condition ($\alpha = 0$, $\beta = 1$ in (1.2)) the solution ϕ behaves like

$$\phi = \sigma r^{1/2} \sin\left(\frac{\theta}{2}\right) + o(r) \tag{1.4}$$

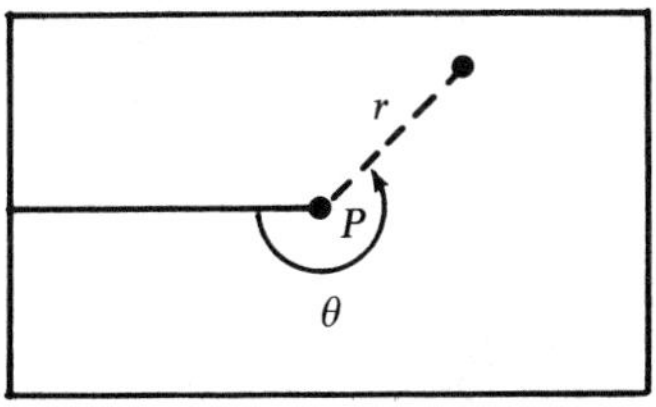

FIGURE 1.1. The slit region Ω: (r, θ) are polar coordinates at P.

near the crack tip ($r \to 0$). Thus the gradient is singular at the crack tip

$$|\nabla\phi| = O(r^{-1/2}) \quad \text{as } r \to 0. \tag{1.5}$$

The coefficient σ of the singularity is of great technical importance and is called the *stress intensity factor*. Knowing it permits estimates on crack behavior via energy release rates [5, 6].

Whereas singular behavior at corners in solid mechanics is of great technical significance, in fluid mechanics corner singularities are often irrelevant artifacts. For example, linear potential flow over a flat plate involves the same equations and geometry as discussed previously [7]. In this case, the gradient $\nabla\phi$ is the velocity field and hence (1.5) predicts a square root singularity at P. This does not occur in real flows and is an artifact of the linearization. A proper nonlinear analysis shows that the velocity singularity is a good deal milder. This and other nonlinear effects are discussed in Section 3.3.

CASE 2 (Discontinuous Boundary Operators). Singularities can arise when the boundary conditons (1.2) abruptly change type or, what is the same, when the coefficients α, β in (1.3) are discontinuous. An example is given in Figure 1.2. Observe that in this case the Neumann condition

$$\frac{\partial\phi}{\partial v} = 0$$

is a symmetry condition, and hence this case is exactly the one represented in Figure 1.1. In particular, the singularity is described by (1.4). Most of the

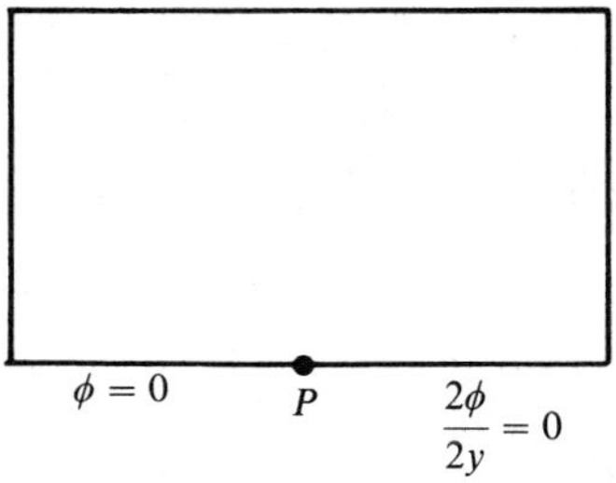

FIGURE 1.2. Discontinuous boundary conditions.

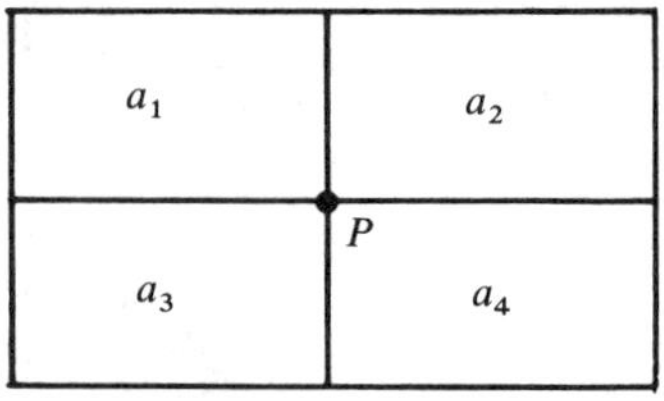

FIGURE 1.3. The region Ω for the a priori bound (2.22).

applications in fluid and solid mechanics involving discontinuous boundary operators arise in this manner.

CASE 3 (Discontinuous Coefficients). Discontinuities in the coefficient a in (1.1) can also generate singularities in the solution ϕ. Cases of technical importance include diffusion problems in regions Ω consisting of different materials [8, 9]. A typical example is shown in Figure 1.3. Using polar coordinates (r, θ) at the point P, it can be shown that the singularity behaves like

$$\phi = \sigma r^{\lambda} \Phi(\theta), \tag{1.6}$$

plus higher order terms in r for a suitable function Φ. The exponent depends on the (constant) values a_j of a in each of the regions Ω_j ($j = 1, \ldots, 4$). In general, these singularities can be far more serious than the square root singularity in (1.4). In particular, $\lambda > 0$ can be made arbitrarily small for appropriate choices of $a_1, \ldots, a_4$ [9].

CASE 4 (Nonsmooth Data). Cases where the boundary data g in (1.1), (1.2) is not smooth arise in solid mechanics when bodies are subject to random loads [10]. To date, these have been best treated by using the techniques of stochastic differential equations since singular behavior of the solution ϕ tends to be distributed over the entire region Ω. These techniques are not treated in this paper, and the reader is referred to [11] for more details.

The presence of singularities, such as described previously, tend to affect the rates of convergence for both finite difference as well as finite element schemes significantly [12]. Except for Case 4, two approaches have been used to treat the singular behavior. One, generally called the *singular element method* because of the way it is used in finite element schemes, attempts to incorporate the singular behavior into the approximation. The other approach uses grid refinement.

To put this paper into proper perspective, it should be clearly noted that, in general, grid refinement is the better approach. This is particularly the case when adaptive strategies can be incorporated into the approximation [13–15]. Singular elements tend to be useful only in very special circumstances, some of which are cited later. However, the methodology used to derive singular elements still remains quite relevant. The point is that a priori knowledge of the nature of the singularity can be of practical benefit even if this information is used only indirectly. This is certainly true of grid refinement as well as h, p, and $h - p$ versions of the finite element method [15, 16]. Because of this, these methodologies will receive the primary emphasis in this paper.

Sections 3.2 and 3.3 are devoted to the types of singular behavior that can arise in elliptic systems. Section 3.3 considers nonlinear effects, and this material is apparently new. The final section concentrates on practical issues associated with singular elements along with selected numerical results.

3.2. Singular Behavior in Linear Systems

For simplicity, attention will be confined to problems with corner singularities. Interface problems (Case 3 in Section 3.1) are treated with similar techniques.

In the linear case the starting point is the construction of explicit solutions using separation of variables for constant coefficient problems and simple geometries. Then, using well-developed techniques from the theory of partial differential equation (e.g., modifiers and frozen coefficients iterations), we can analyze problems with variable coefficients and rather broad classes of regions Ω [17, 18].

Apparently, the first researcher to realize that the form of singularities can be obtained by a local separation of variables was L. Williams [19]. To describe the results in this classic paper, consider the sectorial region shown in Figure 2.1, letting (r, θ) denote polar coordinates; then (1.1) (with $a = 1$) becomes

$$\frac{\partial^2 \phi}{\partial r^2} + \frac{1}{r}\frac{\partial \phi}{\partial r} + \frac{1}{r^2}\frac{\partial^2 \phi}{\partial \theta^2} = 0. \tag{2.1}$$

Assuming a solution of the form

$$\phi = r^{\lambda}\Phi(\theta), \tag{2.2}$$

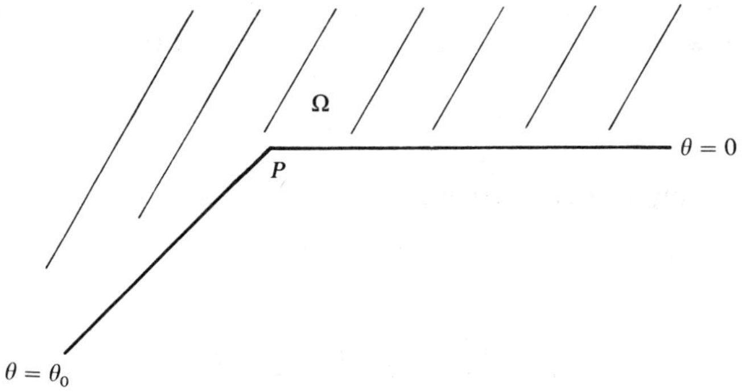

FIGURE 2.1. The sector Ω.

we are led to an eigenvalue problem for the exponent λ and the function Φ:

$$\frac{d^2\Phi}{d\theta^2} + \lambda^2\Phi = 0. \tag{2.3}$$

The type of singularity obtained thus depends on the boundary conditions. Dirichlet conditions

$$\Phi(0) = \Phi(\theta_0) = 0 \tag{2.4}$$

yield

$$r^{\pm\alpha} \sin \sigma\theta, \qquad \sin \sigma\theta_0 = 0. \tag{2.5}$$

The relevant solution is the one with the smallest positive index σ. This gives

$$\phi = r^{\pi/\theta_0} \sin\left(\frac{\pi\theta}{\theta_0}\right) \tag{2.6}$$

as the dominant singular term. Observe that $\nabla\phi$ is finite at the corner point P, if and only if $0 < \theta_0 < \pi$; i.e., re-entrant corners ($\theta_0 > \pi$) yield unbounded gradients. In the case of a crack shown in Figure 1.1, we have $\theta_0 = 2\pi$, and hence (2.6) reduces to the square root singularity given by (1.4).

It is important to note that the order of the differential operator exerts an important influence on the type of singularities that are obtained. For example, consider the fourth-order equation

$$\Delta^2\phi = 0 \tag{2.7}$$

defined in the section shown in Figure 2.1. Assuming a solution of the form

$$\phi = r^{\lambda+1}\Phi(\theta), \tag{2.8}$$

we obtain

$$\Phi(\theta) = a_1 \sin(\lambda + 1)\theta + a_2 \cos(\lambda + 1)\theta + b_1 \sin(\lambda - 1)\theta + b_2 \cos(\lambda - 1)\theta. \tag{2.9}$$

The constants a_j, b_j depend on the boundary conditions to be imposed on $\theta = 0$ and $\theta = \theta_0$. For example, along a clamped radial edge we have

$$\omega = 0, \qquad \frac{1}{r}\frac{\partial\omega}{\partial\theta} = 0; \tag{2.10}$$

whereas a simply supported edge gives

$$\omega = 0, \qquad \sigma_\theta = 0, \tag{2.11}$$

where

$$\sigma_\theta = \frac{1}{r^2}\frac{\partial^2\omega}{\partial\theta^2} + \frac{1}{r}\frac{\partial\omega}{\partial r} + \nu\frac{\partial^2\omega}{\partial r^2}, \tag{2.12}$$

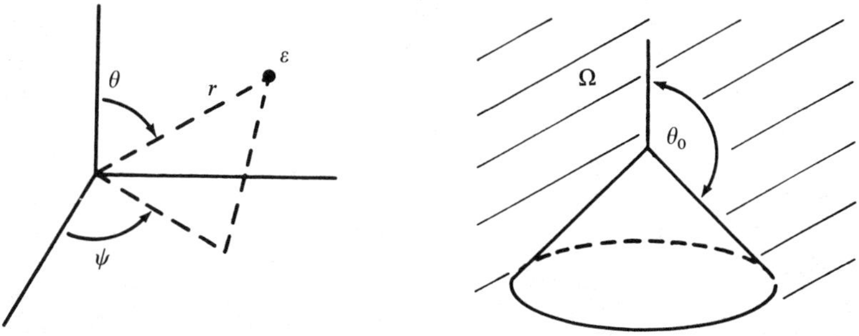

FIGURE 2.2. Spherical coordinates and the conical region Ω.

ν being the Poisson ratio. Substituting into either of the boundary conditions gives a nonlinear equation for the exponent λ. Except for special cases this equation must be solved numerically; i.e., explicit formulas for the exponents are not known. Nevertheless, a number of qualitative features of solutions to these equations are known (along with specific numerical values in technically important cases [19]).

An important point is that solutions λ can be complex. This implies that oscillatory behavior in the radial directions can occur. This is particularly relevant for cases where the stress intensity factors are approximated by using only nodal values of the solution ϕ.

In the case of a crack (Figure 1.2) it can be shown that $\lambda = (1/2)$. Thus the stress σ_θ given by (2.12), which is the analog of the gradient for the fourth-order case, displays a square root singularity.

Another issue of technical importance concerns the effect of dimension. The direct generalization of Williams's work to three dimensions involves the conical region shown in Figure 2.2. Letting (r, θ, ψ) denote spherical coordinates, we seek a solution of the form

$$\phi = r^\lambda \Phi(\theta, \psi). \tag{2.13}$$

In the second-order case, we are again led to an eigenvalue problem, except that now for the Laplace–Beltrami operator on the boundary surface of Ω

$$\frac{1}{\sin^2\theta}\frac{\partial^2\Phi}{\partial\psi^2} + \frac{1}{\sin\theta}\frac{\partial}{\partial\theta}\left(\sin\theta\frac{\partial\Phi}{\partial\theta}\right) + \lambda(\lambda+1)\Phi = 0. \tag{2.14}$$

A further separation of variables gives

$$\Phi(\theta, \psi) = \sin(m\psi + \alpha)P(\theta), \tag{2.15}$$

where P satisfies the Legendre equation

$$\frac{d}{d\mu}\left[(1-\mu^2)\frac{dP}{d\mu}\right] + \left[\lambda(\lambda+1) - \frac{m^2}{1-\mu^2}\right]P = 0 \tag{2.16}$$

with $\mu = \cos\theta$. In the case of Dirichlet boundary conditions, solutions are obtained by requiring.

$$P(\mu_0) = 0, \qquad \mu_0 = \cos\theta_0. \tag{2.17}$$

Properties of these solutions have been studied for the case $0 < \theta_0 < \pi$ [20]. The dominant singularity is independent of ψ (i.e., $m = 0$, $\alpha = (\pi/2)$ in (2.15)). In this case (2.16) can be solved with Legendre functions $P = P_\lambda$, and (2.17) reduces to a nonlinear equation for λ. Many of the characteristics of the two-dimensional case reappear here. For example, $|\nabla\phi|$ is finite at P only in the case of convex region $(0 < \theta_0 < (\pi/2))$.

Interestingly, this analysis runs into trouble in the case of a crack $\theta_0 = \pi$.[1] Here, (2.17) does not have solutions with $m = 0$ since the only solutions to (2.16) with $m = 0$ that are finite on $-1 \leq \mu \leq 1$ are Legendre polynomials $P_n(\lambda = n)$ that satisfy $|P_n(\pm 1)| = 1$. The net result is the question of the completeness of the functions in which the singular behavior is described. Nevertheless, if we were to accept the limiting behavior as $\theta \to \pi$ as valid, then the apparently ubiquitous square root singularity reemerges.

The open question of completeness of the functions used to describe singular behavior in three dimensions is also present in regions other than cones. Thus a rigorous account of how solutions behave near planar slices in Ω, a case of considerable technical importance, still cannot be given.

Elementary solutions such as those described previously can be used to obtain results for rather general regions Ω and for problems with variable coefficients. Typically this approach does not yield explicit behavior, but rather shows that the solutions lie in appropriate Sobolev spaces. With most grid refinement techniques, this information is almost as useful as knowing the explicit behavior.

To fix ideas, consider the second-order case (1.1), (1.2), in which a is a smooth function of the spatial variables and $\alpha = 0$, $\beta = 1$ (Dirichlet boundary conditions). Let Ω be the region shown in Figure 2.3, which has a corner at P but is otherwise smooth. Note that the edges leading to P do not necessarily have to be straight. For this region we introduce the following weighted Sobolev spaces in terms of the radial distance r to P. Indeed, let

$$\|\omega\|_{0,\beta} = \left\{\int_\Omega r^\beta |\omega|^2\right\}^{1/2} \tag{2.18}$$

be a weighted L_2 norm for the given exponent $\beta \in \mathbb{R}$. The higher order spaces

[1] The case $\theta_0 = \pi$ is exactly the one in which the region Ω fails to satisfy an exterior cone condition [17, 18]. Most of the results from the theory of partial differential equations, as well as embedding theorems for Sobolev spaces, are not valid in the absence of this condition. In two dimensions, this appears to be only a technical point limiting the analytical techniques used, whereas in three dimensions it seems to be a fundamental issue.

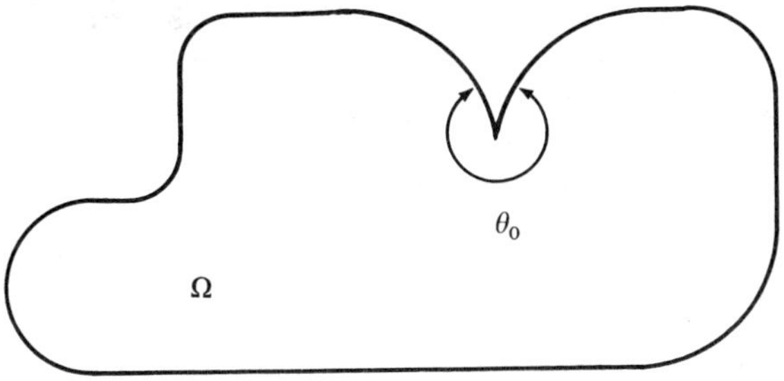

FIGURE 2.3.

involve derivatives. Thus

$$\|\omega\|_{1,\beta} = \left\{ \int_\Omega r^\beta [|\nabla\omega|^2 + \omega^2] \right\}^{1/2}, \tag{2.19}$$

with analogous definitions for $\|\omega\|_{t,\beta}$ $(t > 1)$. A priori bounds on the solution ϕ were first obtained in a fundamental paper by Kondrat'ev [17] (see also [18]). A typical result in the case of homogeneous boundary conditions ($g = 0$ in (1.2)) shows that if

$$t + 1 - \left(\frac{\pi}{\theta_0}\right) < \beta \leq t + 1, \tag{2.20}$$

and

$$\|f\|_{t,\beta} < \infty, \tag{2.21}$$

then there is a constant $0 < C < \infty$, depending only on t, β, Ω, and a, for which

$$\|\phi\|_{t+2,\beta} \leq C \|f\|_{t,\beta}. \tag{2.22}$$

The analysis requires that Ω satisfy an exterior cone condition, and hence the interior angle θ_0 is restricted to $0 < \theta_0 < 2\pi$. This inequality can be verified directly in the cases where the explicit singular behavior is known.

The practical use of a priori bounds like (2.22), as well as that of the explicit singular behavior (when known), will be discussed in the last section.

3.3. Singular Behavior in Nonlinear Systems

In this section, we shall study the effects of nonlinearities on singular behavior. In this setting, the basic paradigm underlying the analysis in the linear case, namely separation of variables, is not available; hence different techniques are needed.

To fix ideas, we consider (1.1) with Dirichlet boundary conditions ($\alpha = 0$,

$\beta = 1$) in the sector shown in Figure 2.1. We assume the coefficient depends on the gradient $\nabla\phi$ as

$$a = a(|\nabla\phi|^2). \tag{3.1}$$

For the problem to be elliptic, we need

$$\frac{d}{dt}[a(t^2)t] > 0 \tag{3.2}$$

uniformly throughout Ω.

An early approach to this problem was given by Tolksdorf [21]. He looked for solutions of the form

$$\phi = r^\lambda \Phi(\theta), \tag{3.3}$$

and showed that λ, Φ are related through a nonlinear eigenvalue problem. A key parameter in the latter is the number

$$q = \lim_{t\to\infty}\left[\frac{t[da(t)/dt]}{a(t)}\right]. \tag{3.4}$$

As will be seen, this number determines the nonlinear effects on the singularity. Observe that $q = 0$ in the linear case, and for this system to be elliptic it is necessary that

$$q > -\tfrac{1}{2}. \tag{3.5}$$

Because of the nonlinearities, it is not possible to get explicit solutions to the nonlinear eigenvalue problem. However, using maximum principle arguments with comparison functions, Tolksdorf was able to obtain the following bounds on the minimal exponent λ in (3.3):

$$\lambda_1(\theta_0) \le \lambda \le \lambda_2(\theta), \tag{3.6}$$

where

$$\lambda_1(\theta) = \frac{q\alpha_0 + \sqrt{(2q+1)(4q+1) + q^2\alpha_0^2}}{\alpha_0(4q+1)}, \tag{3.7}$$

and

$$\lambda_2(\theta) = \frac{q\alpha_0 + \sqrt{(2q+1) + q^2\alpha_0^2}}{\alpha_0(2q+1)} \tag{3.8}$$

with $\theta_0 = \alpha_0\pi$.

Unfortunately, as will be shown later, the inequalities in (3.6) are not strict except in the linear case $q = 0$. Nevertheless, the results do contain useful information. For example, $0 < \lambda < 1$ for re-entrant corners ($\pi < \theta_0 < 2\pi$), and hence $|\nabla\phi|$ is singular in these cases. Moreover, this is the only property that survives from the linear case, and the nonlinear terms definitely affect the asymptotic behavior as $r \to 0$. It is noteworthy that this approach can also be used in three and higher dimensions.

Sharp results can be obtained for the case of two dimensions. The starting point is a fundamental paper by L. Lehman [22]. This work was limited to the linear case; however, the function-theoretic approach used did not require the explicit construction of solutions via separation of variables. The exact form of the singularity was obtained, but in an indirect manner using appropriate transformations and analytic continuation. This is why the results are limited to planar regions; however, they do point the way to an approach for the nonlinear case via a hodograph transformation.

In particular, we let u, v denote the components of the gradient $\nabla\phi$ and rewrite (1.1) as

$$(a + 2a^{\nabla}u^2)\frac{\partial u}{\partial x} + (2a^{\nabla}uv)\frac{\partial u}{\partial y} + (2a^{\nabla}uv)\frac{\partial v}{\partial x} + (a + 2a^{\nabla}v^2)\frac{\partial v}{\partial y} = 0, \quad (3.9)$$

$$\frac{\partial u}{\partial y} - \frac{\partial v}{\partial x} = 0, \quad (3.10)$$

where $a^{\nabla} = da/dt$. The hodograph transformation takes the form

$$x = x(u, v), \qquad y = y(u, v). \quad (3.11)$$

At points where the Jacobian of the transformation

$$J = \frac{\partial u}{\partial x}\frac{\partial v}{\partial y} - \frac{\partial u}{\partial y}\frac{\partial v}{\partial x} = \frac{\partial^2 \phi}{\partial x^2}\frac{\partial^2 \phi}{\partial y^2} - \frac{\partial^2 \phi}{\partial x\, \partial y} \quad (3.12)$$

is nonzero, we have

$$(a + 2a^{\nabla}u^2)\frac{\partial y}{\partial v} - (2a^{\nabla}uv - (2a^{\nabla}uv)\frac{\partial y}{\partial u} + (a + 2a^{\nabla}v^2)\frac{\partial x}{\partial t} = 0, \quad (3.13)$$

$$\frac{\partial x}{\partial v} - \frac{\partial y}{\partial u} = 0. \quad (3.14)$$

Because of the last relation, a potential ψ can be introduced so that

$$\frac{\partial \psi}{\partial u} = x, \qquad \frac{\partial \psi}{\partial v} = y. \quad (3.15)$$

This gives

$$(a + 2a^{\nabla}u^2)\frac{\partial^2 \psi}{\partial v^2} - 4a^{\nabla}uv\frac{\partial^2 \psi}{\partial u\, \partial v} + (a + 2a^{\nabla}v^2)\frac{\partial^2 \psi}{\partial u^2} = 0. \quad (3.16)$$

Thus the hodograph takes the original nonlinear problem into a linear one. Observe that ellipticity of the equation is not affected by the transformation.

Hodograph transformations have been used in a number of problems [7]. The technical difficulty has typically been in deriving boundary conditions in the hodograph plane. In fact, the traditional approach in fluid dynamics is to let the boundary conditions so obtained actually define the problem that is being solved [7, Ch. XX]. Here no such problems arise, as can be anticipated

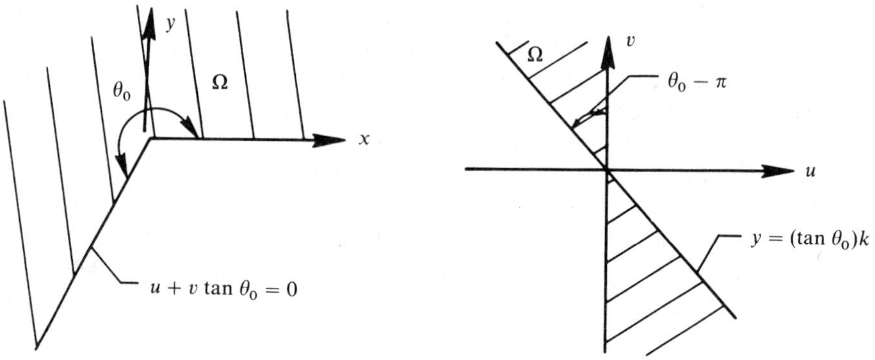

FIGURE 3.1. The hodograph.

from a careful study of Lehman's work. The cases of a re-entrant ($\theta_0 > \pi$) or convex ($\theta_0 < \pi$) corner are different, so for brevity we consider only the former since it is the most important.

The boundary transformations under the hodograph are given in Figure 3.1. In particular, the sector Ω: $0 \leq \theta \leq \theta_0$ is mapped into a double cone Ω with angle $\theta_0 - \pi$.

Introducing polar coordinates,

$$u = \rho \cos \omega, \qquad v = \rho \sin \omega. \tag{3.17}$$

It can be shown that (3.16) reduces to

$$\left(\frac{a}{a + 2a^\nabla \rho^2}\right)\frac{\partial^2 \psi}{\partial p_2} + \frac{1}{\rho}\frac{\partial \psi}{\partial \rho} + \frac{1}{\rho^2}\frac{\partial^2 \psi}{\partial \omega^2} = 0. \tag{3.18}$$

Moreover, Dirichlet boundary conditions $\psi = 0$ on $\partial\hat{\Omega}$ can be used.

Tolksdorf's work shows that $\nabla\phi$ is singular at the corner point $r = 0$. Thus in the hodograph plane it is the behavior at $\rho = \infty$ that is relevant. Since the coefficient in (3.18) is smooth, Kondret'ev's work [17] shows that we can freeze it in the neighborhood of the singularity, i.e., replace it with

$$\lim_{t\to\infty}\left(\frac{a}{a + 2a^\nabla \rho^2}\right) = \frac{1}{1 + 2q}, \tag{3.19}$$

where q is the index (3.4) appearing in Tolksdorf's work. The resulting problem

$$\left(\frac{1}{1 + 2q}\right)\frac{\partial^2 \psi}{\partial \rho^2} + \frac{1}{\rho}\frac{\partial \psi}{\partial \rho} + \frac{\partial^2 \psi}{\partial \theta^2} = 0 \tag{3.20}$$

can now be solved by separation of variables. In particular, letting

$$\psi = \rho^\sigma \sin\left(\frac{\omega\pi}{\theta_0 - \pi}\right), \tag{3.21}$$

we obtain

$$\sigma = \pm\sigma_0, \qquad \sigma_0 = q + \sqrt{q^2 + \frac{2q + 1}{(\alpha_0 - 1)^2}}, \tag{3.22}$$

where $\theta_0 = \alpha_0 \pi$. Since it is the behavior at ∞ that is relevant, we take $\sigma = -\sigma_0$. Expressing the results in terms of the radial coordinate r in the physical plane, we have

$$|\nabla\psi| \sim r. \tag{3.23}$$

Hence

$$\nabla\phi \sim r^{\sigma_0/(\sigma_0+1)-1}, \tag{3.24}$$

or, what is the same, the exact exponent λ in (3.3) is given by

$$\lambda = \frac{\sigma_0}{1+\sigma_0}, \qquad \sigma_0 = q^2 + \sqrt{q^2 + \frac{2q+1}{(\alpha_0 - 1)^2}}. \tag{3.25}$$

Observe that in terms of the lower bound λ_1 and upper bound λ_2 cited previously we have

$$\lambda_1 < \lambda < \lambda_2, \tag{3.26}$$

except in the linear case ($q = 0$) where

$$\lambda_1 = \lambda = \lambda_2 = \frac{1}{\alpha_0}.$$

Consider now the case of a slit where $\alpha_0 = 2$. The exponent λ reduces to

$$\lambda = \frac{2q+1}{2q+2}. \tag{3.26}$$

Observe that if q is negative, then

$$\lambda < \tfrac{1}{2}; \tag{3.27}$$

i.e., the singularity is stronger than that occurring in the associated linear problem. The limiting case is $q = -\frac{1}{2}$, and the problem is no longer elliptic for smaller values of q. That this type of analysis fails is to be anticipated. Indeed, the equations will be hyperbolic in such cases, and the presence of real characteristics means that the type of singularity obtained depends on information in the far field away from the singular point.

If, on the other hand, $q > 0$ we have a situation in which the nonlinear terms smooth the singularity; i.e.,

$$\lambda > \tfrac{1}{2}. \tag{3.28}$$

A case of interest is compressible potential flow [7], in which

$$a(t)^{\delta-1} = M_\infty^2\left[C_0^2 - \left(\frac{\delta-1}{2}\right)t\right]. \tag{3.29}$$

It is readily seen that

$$q = \frac{1}{\delta - 1} \tag{3.30}$$

in this case, and so

$$\lambda = \frac{1 + \delta}{2\delta}. \tag{3.31}$$

For dry air $\delta = 1.405$ and λ is approximately 0.9. As far as numerical approximations are concerned, this is a far less serious singularity than $\lambda = 0.5$ predicted for linear flows.

3.4. Practical Considerations

To fix ideas, consider the planar region shown in Figure 2.3 and the linear second-order case.

As far as (nonadaptive) mesh refinement is concerned, the starting point is the identification of a region Ω_s about the singular point where mesh refinement is used. An important practical consideration is that with local mesh refinement the number of points used in Ω_s need only grow like $O(|\ln h|^2)$, where h is the far field mesh spacing. The key to this is the existence of an a priori bound like (2.22). For simplicity, we take the case $t = 0$ so

$$1 - \frac{\pi}{\theta_0} < \beta < 1. \tag{4.1}$$

The grid is as follows. For triangles adjacent to P the diameter δ satisfies

$$\delta \leq (\text{constant}) h^{(1/1-\beta)}. \tag{4.2}$$

This is increased by a constant factor until the far field uniform grid with spacing h is obtained. Grids of this type have been used in [23] for standard finite element methods, and in [24] for those based on least square approximations.

It can be shown that for such grids the approximation ϕ_n using linear elements satisfies

$$\left\{\int |\nabla(\phi - \phi_h)|^2\right\}^{1/2} \leq Ch\left\{\int r^{2\beta}|D^2\phi|\right\}^{1/2}. \tag{4.3}$$

Numerical evidence shows that if approximate difference quotients are used, then approximations σ_h to the stress intensity factor σ can be obtained satisfying

$$\sigma - \sigma_h = O(h^{3/2}) \tag{4.4}$$

in the case of a slit region (Figure 1.1) [25]. It is a point of practical significance that the form of the singularity must be known in order to establish the relevant difference approximation for σ_h. Path-independent J integrals [26] can be used as an alternative; however, they produce approximations that converge at lower rates, typically $O(h)$ when using linear elements.

In a singular element approach, we seek approximations in the form

$$\phi_h = \sigma_h \phi_s + \sum_j \phi_j N_j, \tag{4.5}$$

where ϕ_s is the (known) singular function, σ_h is the intensity of the singularity (to be computed), and N_j are the piecewise linear nodal functions. Since variational methods typically yield best approximations (in suitable norms), the term $\sigma_h \phi_s$ in effect subtracts out the singularity; i.e., for a suitable number σ (the exact intensity)

$$\|\phi - \sigma\phi_s\|_{2,0} < \infty, \tag{4.6}$$

and thus $\phi - \sigma\phi_s$ can be approximated by piecewise linear functions on a regular grid. Stated differently,

$$\inf_{\xi_i, \eta} \|\phi - \eta\phi_s - \sum \xi_i N_i\|_{1,0} \leq Ch\|\phi - \sigma\phi_s\|_{2,0}. \tag{4.7}$$

This means the ϕ_h computed by a finite element approximation will satisfy

$$\|\phi - \phi_h\|_{1,0} \leq Ch. \tag{4.8}$$

In addition, the intensity σ_h in (4.5) has been observed to converge at the rate found in (4.4) [25].

Substitution of (4.5) into a variational principle typically leads to a matrix problem of the form

$$\begin{bmatrix} K_{11} & K_{1s} \\ K_{s1} & K_{ss} \end{bmatrix} \begin{bmatrix} \boldsymbol{\phi} \\ \sigma_h \end{bmatrix} = \begin{bmatrix} \mathbf{f} \\ f_s \end{bmatrix}, \tag{4.9}$$

where K_{11} is a regular finite element matrix, K_{1s} contains inner products of N_j with ϕ_s, and the number K_{ss} is the inner product of ϕ_s with itself. Note that K_{s1} is $1 \times N$ where N is the number of N_j whose support intersects the support of ϕ_s. For the preceding error estimates to be uniformly valid as $h \to 0$, it is necessary that the support of ϕ_s be fixed independently of h. Thus N grows like $O(h^{-2})$ as $h \to 0$.

One of the early fears about the use of singular functions was that we might be giving up stability in order to gain greater accuracy. Although it is true that the coefficient matrix in (4.9) has a far larger condition number than the standard stiffness matrix K_{11}, this has not created problems in practice [25]. For example, in direct elimination the problems are isolated in a one-dimensional subspace. Indeed, the matrix in (4.9) admits the factorization

$$\begin{bmatrix} K_{11} & K_{1s} \\ K_{s1} & K_{ss} \end{bmatrix} = \begin{bmatrix} L_{11} & \\ L_{s1} & L_{ss} \end{bmatrix} \begin{bmatrix} V_{11} & V_{1s} \\ & V_{ss} \end{bmatrix}, \tag{4.10}$$

where

$$K_{11} = L_{11} V_{11}, \tag{4.11}$$

$$K_{1s} = L_{11} V_{1s}, \qquad K_{s1} = L_{s1} V_{11}, \tag{4.12}$$

$$K_{ss} - L_{s1} V_{1s} = L_{ss} V_{ss}. \tag{4.13}$$

Thus the bulk of the computation is in the factorization of the standard stiffness matrix K_{11} in (4.11). The bordered parts L_{s1}, V_{s1} are obtained through the (stable) backsolves (4.12), and the only part of the calculation in which the instability occurs in the 1×1 problem (4.13) (here $L_{ss} = 1$ and V_{ss}, $K_{ss} - L_{s1} V_{1s}$ are numbers).

Also, note that if the stress intensity factor σ_h is the only quantity of interest, and this is frequently the case, then half of the backsolves can be omitted. Indeed, to get σ_h from (4.9) we backsolve

$$L_{11}\mathbf{y} = \mathbf{f}, \qquad L_{ss}\omega = f_s - L_{s1}\mathbf{y} \tag{4.14}$$

for $\mathbf{y}$ and ω. Then

$$V_{ss}\sigma = \omega. \tag{4.15}$$

The remaining backsolves, i.e.,

$$V_{11}\boldsymbol{\phi} = \mathbf{y} - V_{1s}\sigma, \tag{4.16}$$

can be omitted.

The rates of convergence for ϕ_h and σ_h are the same as that observed for the case of grid refinement [12]. Thus asymptotically as $h \to 0$ there is not a great deal of difference between the two approaches, although, as noted earlier, the superiority of grid refinement emerges as adaptivity is incorporated into the approximation.

In practical terms, the singular function approach has the advantage of fewer backsolves (in the case where only the intensity is desired), but suffers the disadvantage of requiring that not only inner products of nodal functions N_j be evaluated, but also inner products of the more complicated singular functions ϕ_s. Both of these effects contribute to the lower-order terms in the overall work effect—considered as a function of $1/h$—and thus both effects are negligible in the limit as $h \to 0$.

On the other hand, for modest accuracy requirements (h not small) the saving in backsolves tends to be a significant effect. It is exactly in these cases where singular element methods have proved to be useful. To cite a specific numerical example, consider the problem despicted in Figure 4.1. This problem has been used as a test problem for a large number of different numerical

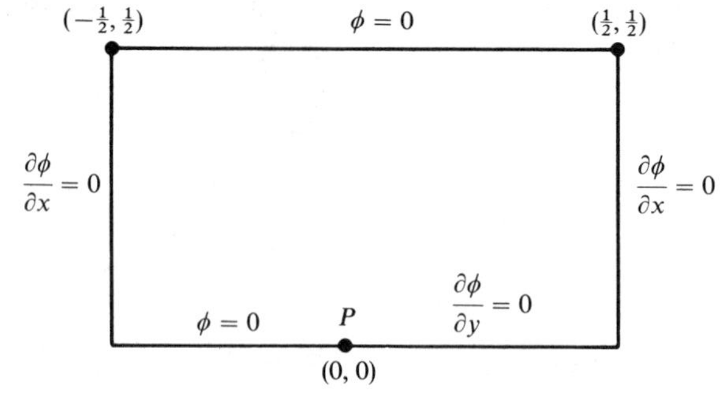

FIGURE 4.1. Test problem.

methods [25–30], and it is known that the stress intensity factor is

$$\sigma = 0.1917$$

(to the number of decimals shown). Using a single singular function on a uniform grid with $h = \frac{1}{8}$ (65 unknowns) gives

$$\sigma_h = 0.1862$$

whereas a β grid refinement with 16 unknowns (and over three times the CPU cycles and storage requirements) gives

$$\sigma_h = 0.1621$$

with finer grids; however, the differences in overall work for a given accuracy rapidly disappear [27].

Another special case where singular elements have proved useful is what could be called an "$h - p$ version" of the singular element method. Here we use more singular functions than would be actually needed, not only to subtract out the singularity but also to approximate. For example, in the problem cited, the use of eight singular functions and a uniform grid with $h = \frac{1}{4}$ gives

$$\sigma_h = 0.1916$$

(see [27] for this and other examples).

Acknowledgments. This work was supported in part by NSF under contract DMS-8601287. Also, under the National Aeronautics and Space Administration under NASA Contract No. NAS1-18107 while the author was in residence at the Institute for Computer Applications in Science and Engineering (ICASE), NASA Langley Research Center, Hampton, VA 23665-5225, U.S.A.

References

1. J.L. Swedlow (ed.), *The Surface Crack*, ASME, New York, November, 1972.
2. G.R. Irwin, *Fracture Mechanics*, Pergamon, Elmsford, NY, 1960.
3. N.I. Muskhelishvili, *Some Problems of the Mathematical Theory of Elasticity*, North-Holland, Groningen, 1953.
4. A.E.H. Love, *The Mathematical Theory of Elasticity*, Cambridge University Press, Cambridge, UK, 1927.
5. A.A. Griffith, "The phenomenon of rapture and flow in solids," *Philos. Trans. Roy. Soc. Ser. (A)*, **221** (1920), 163–179.
6. G.R. Irwin, "Analysis of stresses and strains near the end of a crack," *J. Appl. Mech.*, **24** (1957).
7. Milne-Thomson, *Theoretical Hydrodynamics*, Macmillan, New York, 1968.
8. G. Birkhoff, *Numerical Solution of Elliptic Equations*, SIAM, Philadelphia, 1971.
9. I. Babuska and B. Kellogg, *The Mathematical Foundation of the Finite Element Method*, (A.K. Aziz, ed.), Academic Press, New York, 1972.

10. Z. Schuss, *Theory and Applications of Stochastic Differential Equations*, Wiley, New York, 1980.
11. J.L. Zeman, *Approximate Analysis of Schochastic Processes in Mechanics*, Springer-Verlag, New York, 1971.
12. G. Strang and G.J. Fix, *An Analysis of the Finite Element Method*, Academic Press, New York, 1973.
13. J.F. Thompson, ZUA Warsi and C.W. Mastin, *Numerical Grid Generation*, North-Holland, Amsterdam, 1985.
14. K. Miller and N. Miller, "Moving finite elements," *SIAM J. Numer. Anal.*, **18** (1981), 1019.
15. I. Babuska and W. Rheinboldt, "Error estimates for adaptive finite element computations," *SIAM J. Numer. Anal.*, **15** (1978).
16. I. Babuska and G. Gus, "The *h-p* version of the finite element method," Tech. Note BN-1043, Institute for Physical Science and Technology, University of Maryland, College Park, 1985.
17. V.A. Kondrat'ev, "Boundary value problems for elliptic equations in domains with conic or angular points," *Trans. Moscow Math. Soc.*, **69** (1967), 227–313.
18. V.A. Kondrat'ev and O.A. Oleinik, "Boundary value problems for partial differential equations in nonsmooth domains," *Russian Math. Surveys*, **38**, No. 2, (1983), 1–86.
19. L. Williams, "Stress singularities resulting from various boundary conditions in angular corners," *J. Appl. Mech.*, **19** (1952), 526–528.
20. P. Grisvard, *Elliptic Problems in Nonsmooth Domains*, Pitman, Boston, 1985.
21. P. Tolksdorf, "On the behavior near the boundary of solutions of quasilinear equations," Preprint No. 459, Universität Bonn, 1981.
22. L. Lehman, "Developments at an analytic corner of solutions of elliptic partial differential equations," *J. Math. Mech.*, **8** (1959), 727–760.
23. I. Babuška, R.B. Kellogg, and Pitkäranta, "Dirct and inverse error estimates for finite elements with mesh refinements," *Numer. Math.*, **33** (1979), 447–471.
24. C.L. Cox and G.J. Fix, "On the accuracy of least square methods in the presence of corner singularities," *Comput. Math. Appl.*, **10** (1984), 463–475.
25. G.J. Fix, S. Gulati, and G.I. Wakoff, "On the use of singular functions with finite element approximations," *J. Comput. Phys.*, **13** (1973), 209–228.
26. Y. Lee, "Shear bands in elastic–perfectly plastic materials," Ph.D. thesis, Carnegie–Mellon University, 1981.
27. G.I. Wakoff, "Piecewise polynomial spaces and the Ritz–Galerkin method," Ph.D. thesis, Harvard University, 1970.
28. R. Wait and A.R. Mitchell, "Corner singularities in elliptic problems," *J. Comput. Phys.*, **8** (1971), 45–52.
29. E. Byskov, "Calculation of stress intensity factors using finite element methods," *Internat. J. Fracture*, **6** (1976), 159–168.
30. W.S. Blackburn, "Calculation of stress intensity factors at crack tips using special finite elements," in *Mathematics of Finite Elements and Applications* (J.R. Whiteman, ed.), Academic Press, London, 1973.
31. H. Blum and M. Dobrowolski, "On finite element methods for elliptic equations on domains with corners," *Computing*, **25** (1983), 53–63.

CHAPTER 4

The Finite Element Method for Hyperbolic Systems

O. Pironneau

4.1. Introduction

The finite element method (FEM) [32] was invented by structural analysts; to replace a continuous structure by many equivalent beams is a process that does not leave an engineer's intuition at a loss.

In contrast, the application of the method to acoustics or fluids was not performed by physicists, but by mathematicians; indeed, it is not a physical process to replace a continuous fluid by a set of fixed-size cells and the theorem of virtual work by variational formulations.

The mathematical formulation of the FEM as a Galerkin method, and the first error estimates date from around 1965 (see [9]). To our knowledge, the first major work of this type for hyperbolic systems is Lesaint [22]. There, error estimates are obtained with continuous or discontinuous elements on the symmetric Friedrichs systems, and the analogy between upwinding finite differences and discontinuous flux corrected FEM is shown. In the context of fluid mechanics, and for the advection diffusion equation in particular, several methods, some of which are still popular, appeared later: the Petrov–Galerkin methods [8, 14], Eulerian–Lagrangian methods [26, 12, 28], stream line diffusion [16, 18], and lately higher-order upwind FEM schemes [7, 10]. Not all of these have been shown to converge, but many are shown to be a well-known good finite difference scheme when the triangulation is uniform. Thus we can say that this branch of finite element theory owes much to finite difference theory.

Since intuition is sometimes not a safe guide in this field (centered finite differences work if the boundary conditions are well treated! [33]), we have chosen to present the methods in view of the error estimates available. At the end of the presentation some numerical tests, taken from the literature, will be shown.

We shall discuss three problems:

the general Friedrichs systems;
the advection diffusion equation;
the Euler equations of fluid dynamics.

Thus, we recall briefly the equations.

4.1.1. The Friedrichs Systems

Let u be a function from $Z \subset R^n$ into R^p and consider the symmetric Friedrichs systems

$$Au \equiv \sum A_i(z)u_{,i} + A_0(z)u = f(z), \qquad \forall z \in Z, \tag{1.1}$$

$$(\sum A_i n_i - M)u = 0, \qquad \forall z \in \partial Z, \tag{1.2}$$

where $A_i(z)$ are symmetric $p \times p$ matrices and n is the outer normal of ∂Z. We usually make the following hypotheses:

M, A_i Lipschitz continuous with bounded coefficients;
$\mathrm{Ker}(\sum A_i n_i - M) + \mathrm{Ker}(\sum A_i n_i + M) = R^p$;
There exists $a > 0$ such that $A_0 + A_0^T - \sum A_{i,i} - aI$ is positive definite.

With these assumptions the problem has a unique solution u. The adjoint operator A^* is defined by

$$A^*v = -\sum (A_i v)_{,i} + A_0 v.$$

Note that (1.2) is a condensed way of writing boundary conditions even on one part of the domain only; it suffices to take $M = A_i n_i$ on the other part.

4.1.2. Advection Diffusion Equation

Let φ be a function of time $t \in]0, T[$ and space $x \in \Omega \subset R^n$ and consider

$$\varphi_{,t} + \nabla \cdot (u\varphi) + a\varphi - \nu \Delta\varphi = f \quad \text{in }]0, T[\times \Omega, \tag{1.3}$$

where u is a given velocity field ($u(x, t) \in R^n$), a, f are given scalar functions, and ν is positive and usually very small. When $\nu > 0$ (1.3) is a parabolic partial differential equation and it is well posed with standard boundary conditions such as

$$\varphi(0, x) = \varphi^0(x), \qquad \forall x \in \Omega, \tag{1.4}$$

$$\varphi|_{\partial\Omega} = \varphi_\Gamma. \tag{1.5}$$

However, when $\nu = 0$, (1.5) must be replaced by

$$\varphi|_{\partial-\Omega} = \varphi_\Gamma \quad \text{where} \quad \partial - \Omega = \{x \in \partial\Omega: u(x)n(x) < 0\}, \tag{1.6}$$

where $n(x)$ is the outer normal of $\partial\Omega$ at x. For smooth data, both these problems have one and only one solution. Notice that (1.3) is a Friedrichs system when $\nu = 0$ and $a > 0$.

Finally, when $\nabla \cdot u = 0$, the following holds

$$\nabla \cdot (u\varphi) = u\, \nabla \varphi, \tag{1.7}$$

when $a = 0$ and $\nabla \cdot u = 0$, (1.3) has two interesting properties that the numerical methods should preserve

$$\varphi^0, f \geq 0 \Rightarrow= \varphi(x, t) \geq 0 \qquad \text{(maximum principle)}, \tag{1.8}$$

$$\nu = 0, \quad f = 0 \Rightarrow \int_\Omega \varphi(x, t)\, dx = \int_\Omega \varphi^0(x)\, dx \qquad \text{(mass conservation)}. \tag{1.9}$$

Two important nonlinear systems have the same difficulties as this one:

the incompressible Navier–Stokes equation [17, 29] (see (5.3), (5.4));
the petrol displacement problem of the oil industry [7] (see (4.11)–(4.14)).

4.1.3. The Euler Equations of Fluid Dynamics

This is a system

$$W_{,t} + \sum F_i(W)_{,i} = 0 \quad \text{in }]0, T[\times \Omega, \qquad \Omega \subset R^3, \tag{1.10}$$

where

$$W = [\rho, \rho u_1, \rho u_2, \rho u_3, \rho E], \tag{1.11}$$

$$F_1(W) = [\rho u_1 + p, \rho u_2, \rho u_1 u_2, \rho u_1 u_3, (\rho E + p)u_1], \tag{1.12}$$

$$F_2(W) = [\rho u_2, \rho u_1 u_2, \rho u_2 + p, \rho u_2 u_3, (\rho E + p)u_2],$$

$$F_3(W) = [\rho u_3, \rho u_1 u_3, \rho u_2 u_3, \rho u_3 + p, (\rho E + p)u_3],$$

$$E = \frac{|u|^2}{2} + \frac{p}{(\gamma - 1)\rho}. \tag{1.13}$$

The standard theory of systems of conservation laws can be applied to this system, and (1.10) implies

$$\partial_t \int_\theta W\, dx = -\int_{\partial\theta} \sum F_i(W)\nu_i\, d\gamma \tag{1.14}$$

for all subregions θ of Ω with normal ν.

The linearized underlying hyperbolic system is of the form

$$W'_t + \sum A_i(W) W'_{,i} = 0. \tag{1.15}$$

The matrix

$$P_\mu(W) = \sum \mu_i A_i \tag{1.16}$$

has real eigenvalues and there exists a change of basis $T_\mu(W)$ such that

$$T_\mu^{-1}(W)P_\mu(W)T_\mu(W) = \begin{pmatrix} u\cdot\mu - c & 0 & 0 & 0 & 0 \\ 0 & u\cdot\mu & 0 & 0 & 0 \\ 0 & 0 & u\cdot\mu & 0 & 0 \\ 0 & 0 & u\cdot\mu & 0 & 0 \\ 0 & 0 & 0 & 0 & u\cdot\mu + c \end{pmatrix}, \quad (1.17)$$

where $c^2 = \gamma p/\rho$.

This relation and (1.14) make it possible to define proper sets of boundary conditions for (1.10), the basic idea being that a boundary condition must be given for each negative eigenvalue of (1.17) when μ is the outer normal of Ω. Naturally, ρ, ρu, and ρE must also be specified at time 0.

There are only local existence theorems for this problem (the solution is known to exist on a small finite time interval), and whenever the solution is discontinuous entropy conditions must be added. Note finally that (1.15) linearized is not a symmetric Friedrichs system unless it is written in entropy variable [16].

4.2. Artificial Viscosity

Let us apply the standard Galerkin theory to discretize the advection–convection equation (1.3) with the simplest finite element, namely the continuous piecewise element. For simplicity, we assume $a = 0$.

For this purpose, Ω is divided into nonoverlapping elements (triangles in two dimensions, tetrahedra in three dimensions) T_j with the following restrictions:

$$T_i \cap T_j = \varnothing \quad \text{or 1 side (and/or face) or 1 vertex (and/or 1 edge)}, \quad (2.1)$$

$$\Omega_h = \bigcup T_i \quad \text{with the vertices of } \partial\Omega_h \text{ on } \partial\Omega \text{ and the corners of } \partial\Omega \text{ on } \partial\Omega_h. \quad (2.2)$$

Denoting by h the size of the largest edge, we assume that we have a family of such triangulations T_h as $h \to 0$ and that no angles tend to zero or π in this process; then

$$H_h = \{\varphi_h \text{ continuous}, \varphi_h | \mathrm{T}_i \text{ linear}\} \quad (2.3)$$

is an approximation of $H^1(\Omega)$ in the sense that

$$\forall \varphi \in H^1(\Omega), \qquad \exists \{\varphi_h\}, \varphi_h \in H_h \quad \text{such that} \quad \varphi_h \to \varphi. \quad (2.4)$$

Furthermore, if $\varphi \in H^2(\Omega)$ we have small global interpolation errors

$$\|\varphi_h - \varphi\|_0 < Ch^2 \|\varphi\|_2; \qquad \|\varphi_h - \varphi\|_1 \le Ch\|\varphi\|_2. \quad (2.5)$$

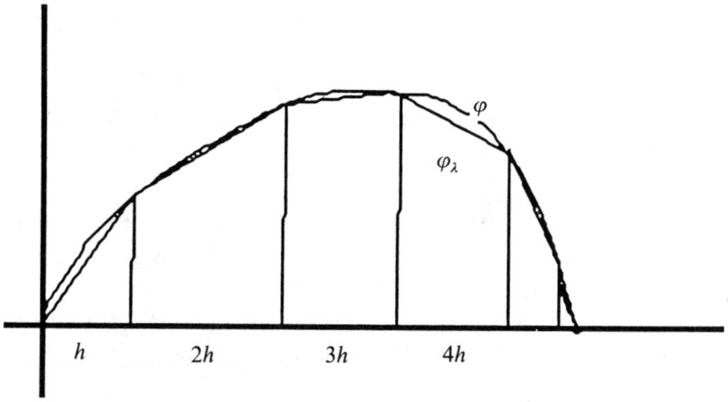

FIGURE 2.1. Lagrange continuous finite elements of order 1 use linear continuous interpolations to approximate φ. Such interpolations are obviously completely determined by the values of φ at points ih, $i = 0, \ldots, N$. Integrals like (2.10) can be computed exactly when u is polynomial because φ, w are piecewise linear. Then (2.9), (2.11) are obtained.

Problem (1.3) is approximated by

$$\int_\Omega \varphi_{h,t} w_h \, dx + \int_\Omega \nabla \cdot (u\varphi_h) w_h \, dx + \nu \int_\Omega \nabla \varphi_h \nabla w_h \, dx = \int_\Omega f w_h \, dx, \quad \forall w_h \in H_{0h}; \tag{2.6}$$

$$\varphi_h(t, \cdot) - \varphi_{\Gamma h} \in H_{0h} = \{w_h \in H_h : w_h = 0 \text{ on } \partial\Omega_h\}. \tag{2.7}$$

If we denote by $\{w^i\}$ the canonical basis functions of H_{0h} (see Figure 2.1):

$$w^i(q^j) = \delta_{ij}, \qquad \forall q^j \text{ vertex of } T_h; \quad w^i \in H_{0h}, \tag{2.8}$$

then (2.6) is a system of ordinary differential equations

$$B\frac{d\Psi}{dt} + A\Psi = F, \qquad \Psi(0) = \Psi^0, \tag{2.9}$$

with the following notations

$$B_{ij} = \int_\Omega w^i w^j \, dx, \qquad A_{ij} = \int_\Omega [\nu \nabla w^i \cdot \nabla w^j + \nabla \cdot (u w^i) w^j], \tag{2.10}$$

$$\Psi_i = \varphi_h(q^i), \qquad F_j = \int_\Omega f w^j \, dx. \tag{2.11}$$

These integrals can be computed exactly, but the quadrature formula with the Gauss points at the vertices yields a diagonal approximation for B (mass lumping).

Theorem 1. *The error between the solutions of* (1.3) *and* (2.6) *verifies*

$$\|\varphi - \varphi_h\|_{0,\Omega\times]0,T[} \leq hC\|\varphi\|_{2,\Omega\times]0,T[}. \tag{2.12}$$

A sketch of the proof is given in order to understand why it works for elliptic systems, and what is needed to extend the method to hyperbolic systems.

SKETCHED PROOF. For clarity, assume $\nabla\cdot u = 0$. Let ε be the interpolation error on the solution φ of (1.3); i.e., $\varepsilon(x) = \varphi(x) - \psi_h(x)$ where $\psi_h \in H_{0h}$, $\psi_h(q^i) = \varphi(q^i)$, $\forall q^i$ vertex.

Let ρ_h be the interpolation of the error; i.e., $\rho_h \in H_{0h}$ and $\rho_h(x) = \varphi_h(x) - \psi_h(x)$ where φ_h is the solution of (2.6). Then write (1.3) in variational form and subtract it from (2.6); it gives

$$\int_\Omega \{(\varphi_{h,t} - \varphi_{,t})w_h + \nabla\cdot[u(\varphi_h - \varphi)]w_h + \nu\nabla(\varphi_h - \varphi)\nabla w_h\}\, dx = 0. \tag{2.13}$$

Now take $w_h = \rho_h$ and use the fact that $\varepsilon = \rho_h - (\varphi_h - \varphi)$ to find

$$\begin{aligned}\tfrac{1}{2}\|\rho_h\|_{0,t}^2 + \nu\|\nabla\rho_h\|_0^2 &= \int_\Omega \{[\varepsilon_t + \nabla\cdot(u\varepsilon)]\rho_h + \nu\nabla\varepsilon\nabla\rho_h\}\, dx \\ &\leq \|\varepsilon_t + \nabla\cdot(u_\varepsilon)\|_0\|\rho_h\|_0 + \nu\|\nabla\varepsilon\|_0\|\nabla\rho_h\|_0.\end{aligned} \tag{2.14}$$

Now, there are two cases:

(a) $\|\nabla\varepsilon\|_0 \leq \|\nabla\rho_h\|_0$ then (2.14) gives $\|\rho_h\|_{0,t} \leq \|\varepsilon_t + \nabla\cdot(u\varepsilon)\|_0$;

(b) $\|\nabla\varepsilon\|_0 \geq \|\nabla\rho_h\|_0$ then (2.14) and the Poincaré inequality ($\|\rho_h\|_0$ ¡ $\lambda\|\nabla\rho_h\|_0$) give

$$\|\rho_h\|_{0,t} \leq 2(\lambda\|\varepsilon_t + \nabla\cdot(u\varepsilon)\|_0^2 + \nu\|\nabla\varepsilon\|_0^2)^{1/2}\|\nabla\varepsilon\|_0.$$

In both cases $\|\rho_h\|_0$ is bounded by $C\|\varepsilon\|_1$, which is $O(h\|\varphi\|_2)$. The total error $\|\varepsilon - \rho_h\|_0$ is now easy to evaluate to prove (2.12). Notice that an H^1 bound on the error can be found from (2.4): simply drop the first term in (2.14) and use the Poincaré inequality:

$$\nu\left(\int_{]0,T[}\|\nabla\rho_h\|^2\right)^{1/2} \leq \lambda\left(\int_{]0,T[}\|\varepsilon_{,t} + \nabla\cdot(u\varepsilon)\|_0^2\right)^{1/2} + \nu\left(\int_{]0,T[}\|\nabla\varepsilon\|_0^2\right)^{1/2}. \tag{2.15}$$

□

Comments

Formula (2.12) implies two comments:

1. It is not optimal when compared with (2.5); in contrast, (2.15) is optimal but ν appears explicitly.
2. When $\nu \to 0$, $\|\varphi\|_2$ blows up because of the boundary layers; because of the loss of the boundary condition at all points $x \in \partial\Omega$ where $u(x)\cdot n(x) > 0$ (n = outer normal) there is a term like $e^{-y/\nu}$ in φ where y is the distance to $\partial\Omega$, so $\|\varphi\|_2 = O(\nu^{-1})$.

Indeed numerical experiments show that $\|\varphi - \varphi_h\|_0$ does blow up when $\nu \to 0$. *Thus viscosity less than h cannot be handled by this method.* There is also a serious numerical difficulty with (2.9) because A stops being diagonally dominant.

The method of artificial viscosity simply adds a viscous term of order h to the hyperbolic equations to be solved, so as to make them parabolic or elliptic in order to apply the previous analysis. It performs very well when it is used in conjunction with local mesh refinement because then it is possible to take a local small h where φ is singular (it may prove costly, however, if the time scheme is explicit because of the stability condition between the time step and h) [35, 37, 40].

Baker and Jameson [3] have recently obtained a three-dimensional solution of the Euler equations around a complete aircraft by this method. Their artificial viscosity is a complicated expression that for the case of (1.3) is of the form $-\nabla\cdot(\nu^1(\nabla\varphi)) - \Delta\nu^2(\varphi)$ where ν^1 and ν^2 are nonlinear functions. The scheme also uses special quadrature formula to make it faster and more parallel. The results are impressive, considering the potential dissipation of such methods.

Angrand and Dervieux [10] also used the Lagrange linear elements with artificial viscosity to integrate (1.10). The scheme has two steps:

(a) *Predictor step.* Define a piecewise constant predictor w^p on the element Σ of T_h by (1.15) with $\theta = \Sigma$: Σ are polygonal elements constructed from T_h by taking the perpendicular to the sides from the middle points of each element T of T_h (Fig. 2.2) or the element defined by the diagonals of the triangles running through the midsides:

$$w^p|_S = |\Sigma|^{-1}\left[\sum_{T\cap\Sigma\neq 0}\int_T w^n\,dx - \alpha k\int_{\partial T} F(w^n)\cdot\nu\,d\gamma\right]. \tag{2.16}$$

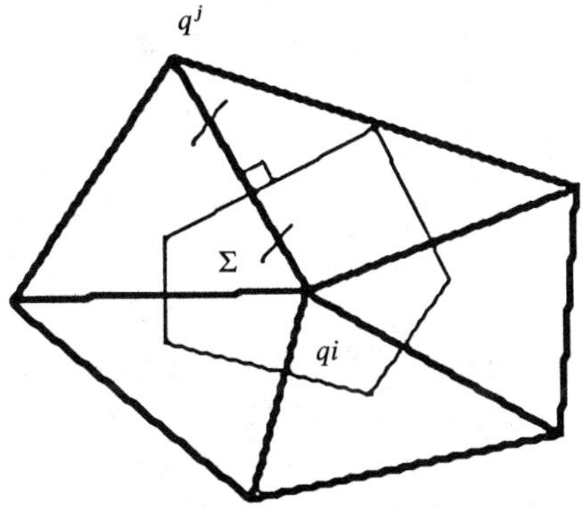

FIGURE 2.2. A new triangulation made of polygons; Σ is built by drawing the perpendicular to the sides from their middle.

(b) *Corrector step.*

$$\int_{\Omega} (w^{n+1} - w^n)k^{-1} v_h \, dx = \int_{\Omega} (\beta_1 F(w_h) \cdot \nabla v_h + \beta_2 F(w^p) \cdot \nabla v_h) \, dx$$
$$+ \int_{\Omega} hN(\nabla w_h) \sum \nabla w_{hi} \cdot \nabla v_{hi} \cdot dx$$
$$- \int_{\partial\Omega} v_h F(w_\Gamma) \cdot n \, d\gamma, \qquad \forall v_h \in H_h; \quad w_h \in H_h, \tag{2.17}$$

with $\alpha = 1 + \sqrt{5}/2$, $\beta_1 = 1 - 1/2\alpha$, $\beta_2 = 1/2\alpha$, and N a suitable diagonal matrix. The last term comes from the boundary conditions; it will be explained in the next Section.

4.3. Lagrange Finite Elements and Hyperbolic Systems

So far we have done nothing but use the standard theory as explained in [9] and [32]. In order to present the methods particular to hyperbolic systems, we step back to the beginning and summarize the basis of Lagrange finite element approximations (by Lagrange interpolation we mean all interpolations from the values of the function at points, or *nodes*, in contrast with Hermite interpolations, which use derivatives). The material of this section is organized as follows.

First we summarize the properties needed to apply the standard FEM as we have seen it in the proof of Theorem 1. Then we shall see that some of these properties are in fact not essential, and that it is possible to generalize the result to some hyperbolic problems (the Friedrichs systems). Then we will introduce the concept of upwinding to recover the well-known finite difference methods for these hyperbolic problems.

4.3.1. Introduction

A partial differential equation is a relation between a function $z \to \varphi(z)$ and its derivatives on a domain Z:

$$F(\varphi, \nabla\varphi, \nabla(\nabla\varphi), \ldots, z) = 0, \qquad \forall z \in Z. \tag{3.1}$$

Boundary conditions

$$B(\varphi, \nabla\varphi, \ldots, z) = 0 \qquad \text{on } \partial Z \text{ or part of } \partial Z \tag{3.2}$$

are usually necessary to have a unique solution of (3.1).

A problem is *well posed* in the classical sense if (3.1), (3.2) has a unique solution in $C^m(Z)$, where m is the order of the highest derivative appearing in (3.1) or (3.2).

For example, the advection equation

$$\varphi_{,t} + u\nabla\varphi + a\varphi = f, \qquad \forall t \in]0, T[, \quad \forall x \in \Omega \subset R^n, \tag{3.3}$$

has a unique solution in $C^1(]0, T[) \times \Omega$ when u_i, a, $u_{i,j}$, f are Lipschitz continuous, and φ is given at time zero:

$$\varphi(0, x) = \varphi^0(x), \qquad \forall x \in \Omega. \tag{3.4}$$

and when $\nabla \cdot u = 0$, $u \cdot n = 0$ on $\partial\Omega$ (for instance, otherwise a boundary condition on $\partial - \Omega$ is also needed). There is a simple constructive proof of the preceding statement because (3.3) has an analytical solution

$$\varphi(x, t) = \varphi^0(X(x, t; 0)) \exp\left[\int_{]0,t[} a(X(x, t; \tau))\, d\tau\right]$$
$$+ \int_{]0,t[} f(X(x, t; \tau) \exp\left[-\int_{]\tau,t[} a(X(x, t, \sigma))\, d\sigma\right] d\tau, \tag{3.5}$$

where $X(x, t; \cdot)$ is the solution of

$$\frac{dX}{d\tau} = u(X, \tau), \qquad \forall \tau \in]0, t[; \quad X(t) = x. \tag{3.6}$$

In the finite difference methods approximations of (3.1), (3.2) are taken on a *finite* number of *points* of Z. In contrast, the FEM are based on interpolations of φ, F, B on *spaces* of *finite* dimensions.

To do that we must first replace the pointwise equalities (3.1), (3.2) by functional equalities

$$\int_Z F(\varphi, \nabla\varphi, \ldots, z)w(z)\, dz = 0, \qquad \forall w \in W(z); \quad \varphi \in H(Z). \tag{3.7}$$

This is called a *weak formulation* of (3.1). Because of integration by parts, a problem has many weak formulations. For example, (3.3) may be replaced by

$$\int_\Omega \varphi_{,t} w\, dx + \int_\Omega u\nabla\varphi w\, dx + \int_\Omega a\varphi w\, dx = 0, \qquad \forall w \in L^2(\Omega), \tag{3.8}$$

and also by

$$\int_\Omega \varphi_{,t} w\, dx - \int_\Omega \varphi u \nabla w\, dx + \int_{\partial\Omega} \varphi u \cdot n w\, d\gamma + \int_\Omega a\varphi w\, dx = 0,$$
$$\forall w \in H^1(\Omega). \tag{3.9}$$

Each time a weak formulation is studied, we must specify the spaces of the tests functions w and of the solution φ and make sure that it has a unique solution.

In the proof of Theorem 1 two essential properties have been used:

w was replaced by φ; this is possible only if the space of test functions in the variational formulation is a subspace of the space in which the approximate solution is searched;

when $w = \varphi$ the variational formulation was a norm for φ (see (2.14)).

Thus the standard theory of the FEM works best when we have succeeded in finding a *symmetric* weak formulation of the problem, i.e., something like

$$\int_\Omega G(\varphi, \nabla\varphi, \ldots, w, \nabla w, \ldots)\, dz = 0, \qquad \forall w \in H(Z); \quad \varphi \in H(Z), \quad (3.10)$$

where G is coercive, i.e., greater than some norm of φ when $w = \varphi$, and φ and w belong to the *same* space. Indeed, in such cases, the Lax–Milgram theorem and its nonlinear extentions show the existence and uniqueness of φ under reasonnable assumptions on G. The Galerkin approximation of the problem is

$$\int_\Omega G(\varphi_h, \nabla\varphi_h, \ldots, w_h, \nabla w_h, \ldots)\, dz = 0, \qquad \forall w_h \in H_h; \quad \varphi_h \in H_h. \quad (3.11)$$

where H_h is an approximation of $H(Z)$ in the sense that

$$\forall w \in H(Z), \qquad \exists \{w_h\}_h,\ w_h \in H_h \quad \text{such that} \quad \lim_{h\to 0} \|w - w_h\| = 0. \quad (3.12)$$

Then for computer implementation (3.12) is usually linearized, and a solution is sought by iteratively solving the underlying linear system. From a practical point of view one last property is required: the linear systems must have a unique solution.

So three properties are required:

1. Coercivity of G.
2. Symmetry of spaces: $H = W$.
3. Well-posed underlying linear systems.

Although these properties are not strictly necessary with them, it is reasonable to embark on a computer solution of the problem and also on a proof of convergence of φ_h to φ.

4.3.1.1. *Will FEM Work on Hyperbolic Systems?*

Except in special situations like the wave equation, hyperbolic partial differential equations do not have symmetric weak formulation. If it was so, then the underlying linear operators would be self-adjoint; that is clearly not the case for the advection equation. But Property 1 is needed only to the extent that it implies uniqueness and existence of φ_h; for hyperbolic systems this can be proved by other means. Weak formulations with Property 2 ($H = W$)

can usually be found, but it is not always straightforward and this is mostly due to the boundary conditions.

Going back to our example, the advection equation (3.3), Lesaint [21] studied the space–time variational formulation

$$\int_{\Omega\times]0,T[} [\varphi_{,t}w - w_{,t}\varphi + u\cdot\nabla w\varphi - u\nabla\varphi w + 2a\varphi w]\,dx\,dt$$

$$+ 2\int_{\Omega} \varphi(T,x)w(T,x)\,dx = 0,$$

$$\forall w \in H^1(]0,T[\times\Omega); \quad \varphi \in H^1(]0,T[\times\Omega). \tag{3.13}$$

It has Properties 2 and 3 in the sense that if A denotes the linear operator underlying (3.13), $w = \varphi$ implies

$$\langle\varphi, A\varphi\rangle = 2\int_{\Omega\times]0,T[} a\varphi^2\,dx\,dt + 2\int_{\Omega} \varphi^2(T,x)\,dx > 0, \tag{3.14}$$

when $a > 0$. And for this reason it is still possible to apply the general framework (Theorem 2).

However, to show the limits of the Lagrange–Galerkin method as a canvas for building finite elements for hyperbolic systems we consider time-independent, one-dimensional solutions of (3.13) on $\Omega =]0,1[$.

$$\int_{]0,1[} (u\varphi_{,x}w - uw_{,x}\varphi + 2a\varphi\omega)\,dx = 0, \qquad \forall\omega \in H^1(\Omega); \quad w(0) = 0, \tag{3.15}$$

$$\varphi \in H^1(\Omega); \quad \varphi(0) = 1.$$

If $H^1(]0,1[)$ is replaced in (3.15) by the space of piecewise linear continuous functions H_h on a partition $\{x_0 = 0, x_1 = h, x_2 = 2h, \ldots, x_N = Nh = 1\}$ of $]0,1[$ then, when u and a are constant, (3.15) becomes:

$$u\frac{(\varphi_{i+1} - \varphi_{i-1})}{h} + a\frac{(\varphi_{i+1} + 4\varphi_i + \varphi_{i-1})}{3} = 0, \qquad i = 1,\ldots,N-1, \tag{3.16}$$

$$u\frac{(\varphi_N - \varphi_{N-1})}{h} + a\frac{(2\varphi_N + \varphi_{N-1})}{3} = 0. \tag{3.17}$$

Equation (3.16) can be rewritten as

$$u\frac{(\varphi_{i+1} - \varphi_{i-1})}{2h} + a\varphi_i + \left(\frac{ah^2}{3}\right)\frac{(\varphi_{i+1} - 2\varphi_i + \varphi_{i-1})}{h^2} = 0. \tag{3.18}$$

So we expect such FEM to behave like the centered finite difference schemes with added artificial viscosity of order ah^2. In particular, it will not work if $a = 0$. In many situations, however, artificial viscosity of this type is good enough, so we recall the general convergence result obtained by Lesaint [21].

4.3.2. General Results for Symmetric Systems of Friedrichs

Theorem 2. *Consider* (1.1)–(1.2) *and its approximation*

$$\int_Z (Au_h \cdot v_h + u_h \cdot A^* v_h)\, dz + \int_{\partial Z} Mu_h \cdot v_h \, d\gamma = 2 \int_Z f v_h \, dz, \quad \forall v_h \in X_h; \quad (3.19)$$

$$u_h \in X_h = \{v_h \in V_h \colon (\textstyle\sum A_i n_i - M) v_h = 0 \text{ at the nodes of } \partial Z\}, \quad (3.20)$$

where A^* is the adjoint of A and V_h, the space of continuous piecewise polynomial functions of degree m on a triangulation of Z. Then

$$\|u - u_h\|_{0,Z} \le C(a) h^m \|u\|_{m+1,Z}. \quad (3.21)$$

Remark 1

1. This result is not optimal because we could expect h^{m+1}.
2. It does not hold in the presence of shocks because $\|u\|_{m+1}$ is infinite.
3. It does not hold when $a = 0$ because $C \to +\infty$ as $a \to 0$.

Remark 2. The same result holds when X_h is replaced by V_h only because (3.19) also contains the boundary conditions.

Remark 3. When there are no shocks (3.19) is equivalent (but not equal) to

$$\int_Z Au_h v_h \, dx - \int_{\partial Z} (A_i n_i - M) u_h v_h \, d\gamma = \int_Z f v_h \, dz, \qquad \forall v_h \in X_h.$$

It could also be written as

$$\int_Z A^* u_h v_h \, dx + \int_{\partial Z} (A_i n_i - M) u_h v_h \, d\gamma = \int_Z f v_h \, dz, \qquad \forall v_h \in X_h.$$

In both formulas the middle integrals are in fact zero when u_h is replaced by u.

Lesaint also showed that the Gauss quadrature formula with positive coefficients can be used in (3.19). These must be exact for polynomials of degree $2m - 1$ for the volume integrals and degree $2m$ for the boundary integrals; otherwise they destroy the convergence rate (3.21).

If an explicit scheme is required then a quadrature formula of degree one must be used for the volume integrals of (3.19); this brings down the error estimates (3.21) to $\sqrt{h}$.

On uniform finite difference like grids, when the elements are quadrilaterals and a suitable quadrature formula is used, then we can show [24] that (3.19) becomes a collocation method at the Gauss points q^i

$$(Au_h - f)(q^i) = 0.$$

EXAMPLE (Application of Theorem 2). Consider the stationary advection equation

$$u\nabla\varphi + a\varphi = f \quad \text{in } \Omega,$$

$$\varphi \text{ given on } \partial - \Omega = \{x \in \partial\Omega: u(x)\cdot n(x) < 0\}.$$

Then (3.19) with $M = |u\cdot n|$ is

$$\int_\Omega u\nabla\varphi_h w_h\,dx - \int_\Omega \varphi_h\nabla\cdot(uw_h)\,dx + \int_{\partial\Omega}\varphi_h u\cdot n w_h\,d\gamma + \int_\Omega 2a\varphi_h w_h\,dx$$
$$= \int_\Omega 2fw_h\,dx,$$

or equivalently,

$$\int_\Omega (u\nabla\varphi_h w_h + a\varphi_h w_h)\,dx = \int_\Omega fw_h\,dx, \qquad \forall w_h \in X_h;$$

$$\varphi_h \in X_h = \{w_h \text{ continuous, } P^m \text{ and zero on } \partial - \Omega\}.$$

Theorem 2 shows convergence of order m when $a > 0$. Note that when $m = 1$ Theorem 1 applies.

4.3.3. The Galerkin Method and Total Derivatives

The previous method does not work for the advection equation when $a = 0$.
One way though is to set

$$\psi(x, t) = \varphi(x, t)e^{-\lambda t},$$

then (3.3) becomes

$$\psi_{,t} + u\nabla\psi + (a + \lambda)\psi) = fe^{\lambda t},$$

and the previous method can be applied. However, the long term behavior of φ cannot be found this way. For Friedrichs systems like the advection equation, i.e., systems that have only one characteristic speed, there is a very efficient method based on the concept of total derivative.

Consider, for example, (3.3) again. It can be rewritten as

$$\frac{d}{d\tau}\left[\varphi(X(x, t; \tau))\exp\left[\int_{[0,\tau]} a(X(x, t; \sigma))\,d\sigma\right]\right]_{\tau=t}\exp\left[-\int_{[0,t]} a(X(x, t; \tau))\,d\tau\right]$$
$$= f. \tag{3.22}$$

Now, $d/d\tau$ is replaced by a finite difference approximation, for example, with backward differences between t and $t - k$:

$$\frac{1}{k}\left[\varphi(X(x, t; \tau)) - \varphi(X(x, t; t-k))\exp\left[-\int_{[t-k,t]} a(X(x, t; \sigma))\,d\sigma\right]\right] = f(x, t). \tag{3.23}$$

From its definition (3.6) we known that $X(x, t; t) = x$ so if we denote by $X^{n-1}(x)$ an approximation of $X(x, t; t-k)$, the solution of (3.6) at $t-k$, we get the following scheme:

$$\varphi^n(x) = \varphi^{n-1}(X^{n-1}(x))(1 - a^{n-1}(x)k) + f^n(x)k, \tag{3.24}$$

for which a symmetric variational formulation is easy to find:

$$\int_\Omega \varphi^n w\, dx = \int_\Omega \varphi^{n-1}(X^{n-1}(x))(1 - a^{n-1}(x)k)w(x)\, dx + k \int_\Omega f(x)w(x)\, dx,$$
$$\forall w \in L^2(\Omega); \quad \varphi^n \in L^2(\Omega). \tag{3.25}$$

Let V_h be a finite element space made of continuous piecewise polynomials of degree m; then (3.25) is approximated by

$$\int_\Omega \varphi_h w_h\, dx = \int_\Omega [\varphi_h^{n-1}(X^{n-1}(x))(1 - a^{n-1}(x)k) + kf^n(x)]w_h(x)\, dx,$$
$$\forall w_h \in V_h; \quad \varphi_h \in V_h. \tag{3.26}$$

Notice that there is a simple interpretation of this algorithm:

1. A transport step. The computation of $\varphi_h^{n-1} \circ X^{n-1}$.
2. A L^2-projection step. $\varphi_h^n = \prod [\varphi_h^{n-1} \circ X^{n-1}(1 - a^{n-1}k)] + kf^n$.

Error estimates have been obtained by Douglas and Rusell [12], Pironneau [26], and Priestley [28].

Theorem 3 ([12]). *For smooth data the error between the solutions of* (3.26) *and* (3.3) *satisfies*

$$\|\varphi^n - \varphi_h^n\|_0 \le C\left[h^m + k + \min\left\{\frac{h^{m+1}}{k}, h^m\right\}\right]. \tag{3.27}$$

The effects of quadrature formulas on the error estimate are still an open problem; experience shows that positive Gauss formulas of order $2m$ are good enough, however, in one dimension there are counterexamples that show that quadrature errors can be fatal in pathological cases (see Figures 3.1 and 3.2).

Extensions of this method to the vector case (linear or not) are easy for systems with only one characteristic speed (see [26–28] for the incompressible Navier–Stokes equations and the petrol displacement problem). They can even handle shocks [4], when there is more than one characteristic speed, then the transport step must be replaced by a multidimensional Riemann problem; these methods will be studied later.

For the incompressible Navier–Stokes equation there is an error estimate in [26] that is valid even when $\nu \to 0$ and is similar to (3.27) with $m = 1$.

Löhner *et al.* [35] have proposed a method that could be interpreted in this framework. Although the authors base their method on a Taylor expansion in time of the solution, it amounts also to choosing a Runge–Kutta scheme

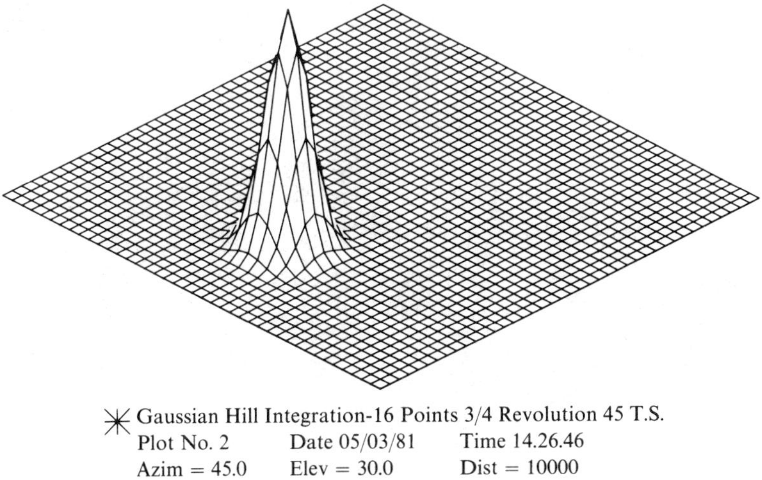

(a)

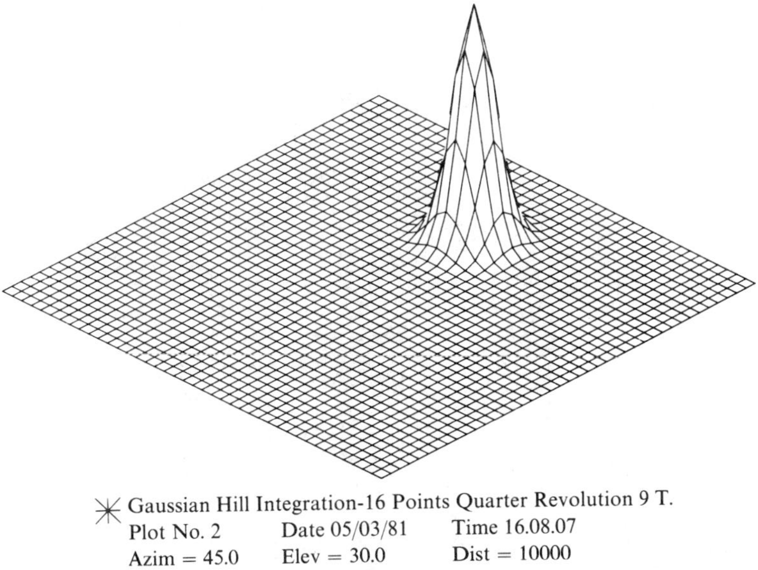

(b)

FIGURE 3.1. Advection equation in a rotating flow. The method of (3.26) has been used to obtain these plots. The discretization is done with Q^2 quadrilateral elements and no mass lumping; it takes 50 time steps to make a turn. The plots show φ after half a turn (top) and after a full turn: no visible difference (computed by Hasbani).

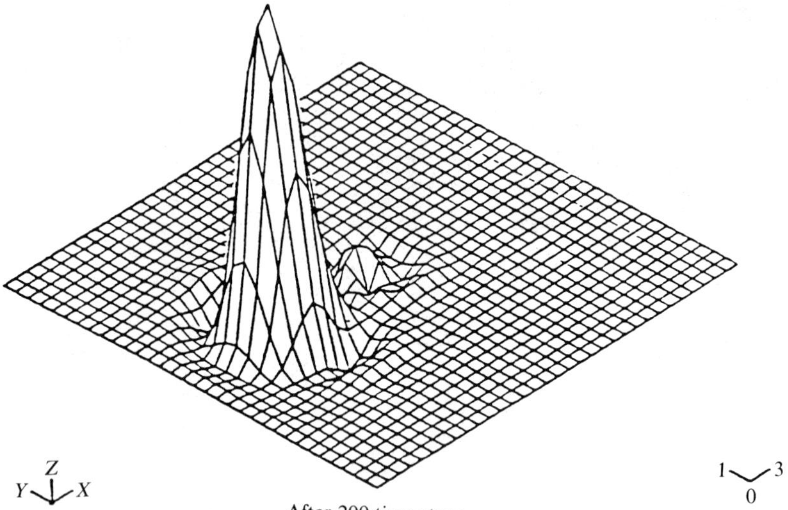

After 200 time steps

(a)

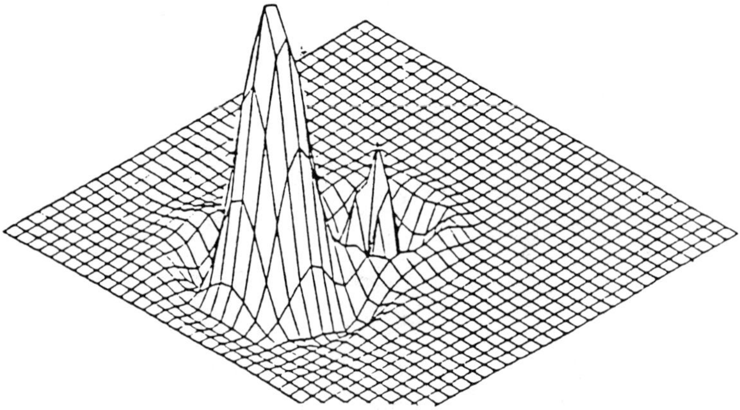

Max(U) = 1.052
Min(U) = −0.248
After 500 time steps

(b)

FIGURE 3.2. Quadrature effects. This computation is the same as in Figure 3.1 but with more time steps. The effect of the quadrature formula on the right-hand side of (3.26) is visible in places where the velocity is small. These pictures are taken from Priestley [28].

for X:

$$X^{n-1}(x) = x - X'k + X''\frac{k^2}{2}$$

$$= x - u^{n-1}(x)k + (\partial_t u^{n-1}(x) + u^{n-1}(x)\,\nabla u^{n-1}(x))\frac{k^2}{2};$$

and make a Taylor expansion of the solution

$$\varphi^n(x) = \varphi^{n-1}\left(x - u^{n-1}(x)k + (\partial_t u^{n-1}(x) + u^{n-1}(x)\,\nabla u^{n-1}(x))\frac{k^2}{2}\right)$$

$$= \varphi^{n-1}(x) - \left[u^{n-1}(x)k + (\partial_t u^{n-1}(x) + u^{n-1}(x)\,\nabla u^{n-1}(x))\frac{k^2}{2}\right]\nabla\varphi^{n-1}(x)$$

$$+ \nabla\nabla\varphi^{n-1}(x)\colon u^{n-1}(x) \otimes u^{n-1}(x)\frac{k^2}{2}.$$

However, it can also be interpreted as an artificial viscosity method and to some extent as a streamline diffusion method (discussed later). Spectacular results were obtained by Löhner with this method on the Euler equations for a time-dependent problem with shocks.

4.4. Discontinuous Elements

4.4.1. Lesaint's Method

In order to introduce upwinding, Lesaint [21] used discontinuous elements and modified the variational formulation. With discontinuous elements, integration by parts gave new integrals at the interface of the elements. It is these terms that are modified. To present the method we consider the scalar case (u is real valued).

Going back to the systems of Friedrichs, we approximate (1.1), (1.2) by (the sum extends over all elements of T_h):

$$\frac{1}{2}\sum\left\{\int_T \left(Au_h v_h + u_h A^* v_h\right) dx + \int_{\partial T} A_i \nu_{T_i} u_h v_h \, d\gamma\right.$$

$$\left. - \int_{\partial - T - \partial - T \cap \partial\Omega} (A_i \nu_{T_i} - M')[u_h] v_h \, d\gamma\right\} - \int_{\partial\Omega} \tfrac{1}{2}(A_i \nu_{T_i} - M) u_h v_h \, d\gamma$$

$$= \int_\Omega f v_h \, dx, \qquad \forall v_h \in V_h; \quad u_h \in V_h, \tag{4.1}$$

$$\partial - T = \{x \in \partial T\colon A_i(x)\cdot \nu_i(x) < 0\},$$

where ν_T is the outer normal of T at ∂T and $[u_h] = u_h' - u_h$ on $\partial T \cap \partial T'$, and $u_h' = u_h|_{T'}$ where T' is the element adjacent to T. In (4.1) M' is an extension of M inside Ω and V_h is an approximation of $H^1(\Omega)^p$ by possibly discontinuous piecewise polynomial functions. Notice that (1.1), (1.2) leads naturally to (4.1) except for the middle integral on the left containing $[u_h]$. Notice also that it reduces to (3.19) when u_h is continuous.

When $M = \pm A_i \nu_{T_i}$ with the sign so that M is positive, (4.1) can also be written as

$$\sum \left\{ \int_T A u_h v_h \, dx - \int_{\partial - T} (A_i \nu_{T_i} - M)[u_h] v_h \, d\gamma \right\} = \int_\Omega f v_h \, dx, \qquad \forall v_h \in V_h;$$

with the convention that $A_i \nu_{T_i} - M$ be zero outside Ω. Or also as

$$\sum \left\{ \int_T A^* u_h v_h + \int_{\partial T} A_i \nu_{T_i} u_h^- v_h \, d\gamma \right\} = \int_\Omega f v_h \, dx, \qquad \forall v_h \in V_h;$$

where u^- denotes the upstream value of u with respect to $A_i \nu_{T_i}$.

On this last formula the upwinding is clearly seen: *Lesaint's method consists in taking the upstream value of the function u_h on the integrals involving interelements on the dual variational formulation.*

Theorem 4 ([21]). *If $M' + M'^T$ is positive semidefinite at all points and M' is continuous across all ∂T then the error between the solutions of* (1.1), (1.2), *and* (4.1) *verifies*

$$\|u - u_h\|_0 \le C h^m \|u\|_{m+1,\Omega}, \tag{4.2}$$

where

$$V_h = \{v_h \text{ such that } v_h|_T \text{ is a polynomial of degree } m, \forall T \in T_h\}. \tag{4.3}$$

Remark. This result can be extended to the case $m = 0$; on uniform meshes the error is then $O(h)$. The boundary conditions $(A_i \nu_{T_i} - M)u_h = 0$ can be included in V_h if desired. On irregular meshes, Johnson and Pitkaranta [20] have also shown that it can be improved to $h^{m+1/2}$ even when $m = 0$.

EXAMPLE. Consider the stationary advection equation

$$u \nabla \varphi + a \varphi = f \quad \text{in } \Omega, \tag{4.4}$$

$$\varphi = 0 \quad \text{on } \partial - \Omega, \tag{4.5}$$

and let us apply the theory with $M' = |u \cdot n|$. Replacing (4.5) by

$$(A \cdot n - M)\varphi = (u \cdot n - |u \cdot n|)\varphi = 2 \min\{u \cdot n, 0\} \varphi = 0 \quad \text{on } \partial\Omega, \tag{4.6}$$

we find that (1.2) is satisfied. Then define $M' = |u \cdot \nu_T|$; (4.1) becomes

$$\begin{aligned} \frac{1}{2} \sum \Bigg\{ & \int_T [(u \nabla \varphi_h + 2a\varphi_h) w_h - \varphi_h \nabla \cdot (u w_h)] \, dx \\ & + \int_{\partial - T - \partial - T \cap \partial\Omega} 2|u \cdot \nu_T| [\varphi_h] w_h \, d\gamma + \int_{\partial T - \partial T \cap \partial\Omega} u \cdot \nu_T \varphi_h w_h \, d\gamma \\ & + \frac{1}{2} \int_{\partial\Omega} |u \cdot n| \varphi_h w_h \, d\gamma \Bigg\} = \int_\Omega f w_h \, dx, \qquad \forall w_h \in V_h, \end{aligned} \tag{4.7}$$

where $\partial - T$ is the part of ∂T on which $u \cdot \nu_T$ is negative. It can be rewritten as

$$\sum \left\{ \int_T (u \nabla \varphi_h + a\varphi_h) w_h \, dx - \int_{\partial - T - \partial - T \cap \partial\Omega} [\varphi_h] w_h \, d\gamma \right\}$$

$$= \int_\Omega f w_h \, dx + \int_{\partial - \Omega} u \cdot n \varphi_h w_h \, d\gamma. \tag{4.8}$$

Theorem 4 shows convergence at rate m the order of the polynomial on T for φ_h and w_h. Again it requires $a > 0$. For $m = 0$ and a uniform grid in one dimension the method is equivalent to the upwind scheme, when $u > 0$

$$u \frac{(\varphi_i - \varphi_{i-1})}{h} + a\varphi_i = f_i. \tag{4.9}$$

4.4.2. Generalization to Nonlinear Problems

Probably the first generalization to nonlinear systems is due to Fortin and Thomasset [13] (see also [29]).

It was done for the incompressible Navier–Stokes equations.

Looking at (4.8), we see that the basic ingredient that makes Lesaint's method work when $M = |A_i \nu_{T_i}|$ is that we add to the variational formulation integrals over the interelements $\partial - T$ of the jump of the convective terms so as to introduce upwinding.

Thus, for the Navier–Stokes equation, we work with

$$\int_\Omega u_{h,t} v_h \, dx + \sum \left\{ \int_T (u_{hi} u_h)_{,i} v_h - \int_{\partial - T} u_h \cdot \nu_T [u_h] \cdot v_h \, d\gamma \right\} = \int_\Omega f v_h \, dx,$$

$$\forall v_h \in V_h = \{v_h \text{ continuous at the midsides: } v_{h|T} \in P^m, \nabla \cdot v_h = 0\} \tag{4.10}$$

Chavent and Jaffre [7] have proposed a generalization for the oil displacement problem. This problem is basically a nonlinear advection diffusion equation coupled with an elliptic equation, but the system also has discontinuous initial data for S:

$$\varphi S_{,t} + \nabla \cdot \psi(p, S, \nabla S) = 0 \quad \text{in } \Omega \times]0, T[, \tag{4.11}$$

$$-\nabla \cdot [a(S) \nabla p] = b(S) \quad \text{in } \Omega, \forall t, \tag{4.12}$$

$$S(0, x) = S^0(x) \text{ has a shock}, \tag{4.13}$$

$$S \text{ or } \frac{\partial S}{\partial n} \quad \text{and} \quad p \text{ or } \frac{\partial p}{\partial n} \qquad \text{given at all } x \in \partial\Omega. \tag{4.14}$$

The discrete variational formulation for (4.10) proposed in [7] is

$$\int_\Omega \varphi S_{h,t} w_h \, dx - \int_\Omega \psi(p_h, S_h, \nabla S_h) \nabla w_h \, dx + \sum \int_{\partial T} \psi_h^-(p_h, S_h, \nabla S_h) \cdot \nu_T w_h \, d\gamma$$

$$= 0, \tag{4.15}$$

where $\psi_h^-(p_h, S_h, \nabla S_h)$ is the value of $\psi(p_h, \sigma, \nabla\sigma)$ at the minimizer σ of

$s \to \text{sign}(s^+ - s^-)\psi_h(p_h, s, \nabla s) \cdot \nu$ on the interval defined by S^+ and S^-, the two values of S_h on the two adjacent elements T and T' of ∂T. This complex formula gives the solution of the Riemann problem on ∂T. It also gives the downstream value if ψ is monotone, which is exactly what we need for Lesaint's method.

The authors of [7] have also suggested a slope limiter and showed that the resulting scheme is total variation diminishing in one dimension.

4.4.3. Mixed Continuous/Discontinuous Element

The drawback of the method described previously is that there are very many degrees of freedom per element when $m > 0$. In three dimensions it is too much. For instance, with tetrahedra there are four degrees per tetrahedron and there are at least five tetrahedra in a brick; that means 20 degrees of freedom per brick, whereas the continuous piecewise linear element has only one on the average! Thus, several authors have suggested keeping the basic idea of Lesaint but in a continuous finite element framework. Baba and Tabata [2] is among the first work in that direction. It is extensively used in [36].

If φ_h is continuous piecewise linear on T_h, then we can construct φ'_h piecewise constant, yet quite close to φ_h as follows in two dimensions:

Define a cell Σ_i around each vertex q^i by drawing the diagonals or the perpendiculars to each side from their middle (Figure 2.2).
Define φ'_h piecewise constant on the cells Σ_i by $\varphi'_h|_{\Sigma_i} = \varphi_h(q^i)$.

We shall write the mapping $\varphi_h \to \varphi'_h$ as

$$\varphi'_h = \Pi_\Sigma \varphi_h.$$

Then we use Lesaint's method on the convective terms, written in terms of φ'_h, but retain the usual formulation for the other terms.

When mass lumping is applied as in [10] in order to get an explicit scheme, then this method is also a zero-order Lesaint method on the triangulation made by the Σ.

Following Dervieux [10], we discretize (1.10) by

$$\int_\Omega \Pi_\Sigma w_{h,t} \Pi_\Sigma v_h \, dx + \int_{\partial\Sigma} F_h(w_h) \cdot \nu_\Sigma v_h \, d\gamma - \int_\Omega F_h(w_h) \cdot \nabla v_h \, dx = 0, \qquad \forall v_h \in V_h, \tag{4.16}$$

$$W_h \in V_h = \{v_h \text{ continuous, piecewise linear on } T_h\}, \tag{4.17}$$

where F_h is an upwind approximation of F. Then taking $w_h(q^i) = \delta_{ij}$ yields

$$w_h(q^i)_{,t} = |\Sigma_i|^{-1} \int_{\partial\Sigma} F_h(w_h) \cdot \nu_\Sigma \, d\gamma. \tag{4.18}$$

The formula for F_h is an adaptation of the finite difference ideas [30]. Each side of Σ_i is perpendicular to an edge of T_h. If the two ends are the vertices

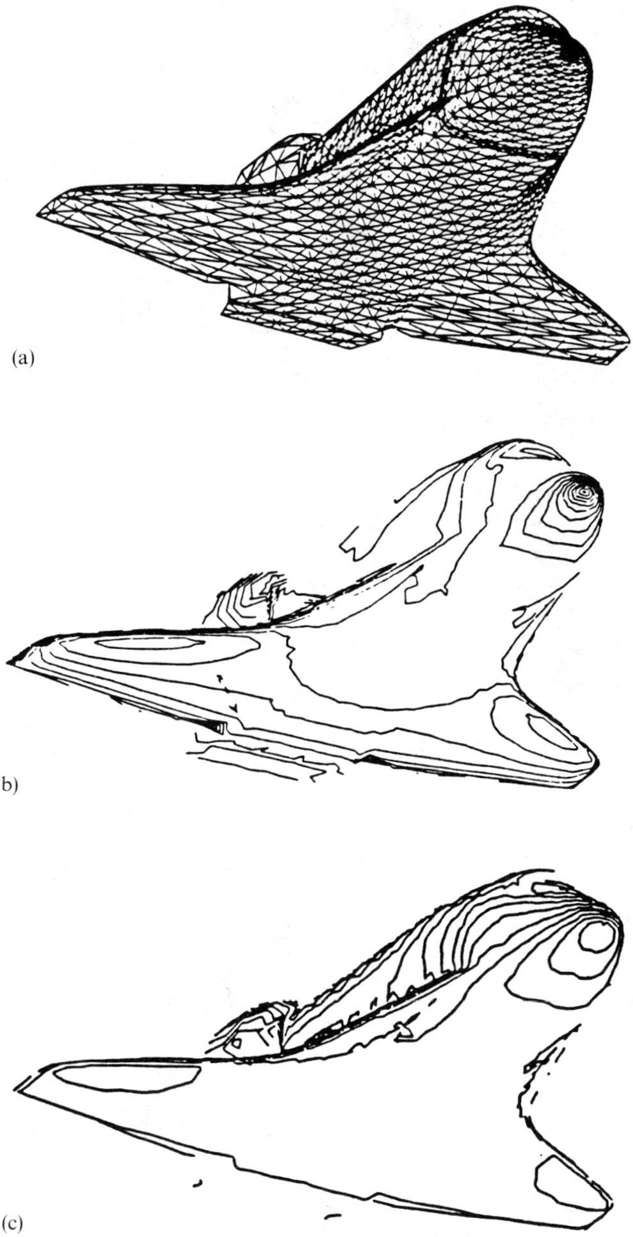

(a)

(b)

(c)

FIGURE 4.1. Hypersonic flow simulation by FEM using the first-order accurate scheme of Section 1. The first picture shows the trace of the mesh on the body of the shuttle. The second picture shows the Mach lines at Mach number 1.3 and 15 degrees of incidence, and the third one for Mach 25 and 40 degrees. These plots are taken from Angrand *et al.* [1]. © Copyright American Institute of Aeronautics and Astronautics; reprinted with permission.

q^i and q^j and if $W^i = w_h(q^i)$, $W^j = w_h(q^j)$, then

$$F_h(w_h)\cdot v_\Sigma = F\left(\frac{(W^i + W^j)}{2}\right)\cdot v_\Sigma + \tfrac{1}{2}\left|F'\left(\frac{(W^i + W^j)}{2}\right)\cdot v_\Sigma\right|(W^i - W^j). \quad (4.19)$$

Other formulas have been tested, and higher-order methods can be devised on these lines. According to Lesaint's result we cannot prove more than $O(h)$ for the L^2-error on quasi-uniform meshes or $\sqrt{h}$ on irregular meshes according to Johnson *et al.* [19, 20].

4.5. Petrov–Galerkin Formulations and Streamline Diffusion

Another train of thought from Christie *et al.* [8] to Johnson *et al.* [19, 16, 20] and Hughes [15] studies how upwinding can be introduced into the discretization of the convection terms of the variational formulation by taking a different space for the test functions W_h and for the solution function φ_h (see Figure 5.1). (When the test space and the solution space are not the same (2.6) is called a *Petrov–Galerkin formulation.*)

If this procedure is applied only to the convection terms, the net result of such a modification can also be viewed as a special quadrature formula for the convective term. In this fashion, Hughes [15] obtained one of the simplest upwind finite element schemes for (2.6) by simply shifting in the direction of the flow of the quadrature points for the second integral. If H_{0h} is made of

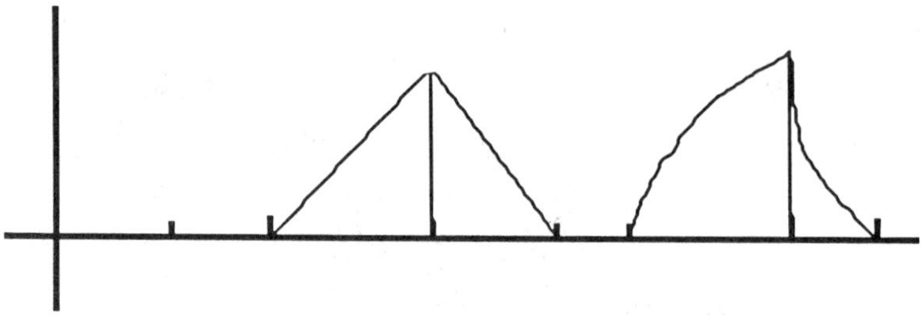

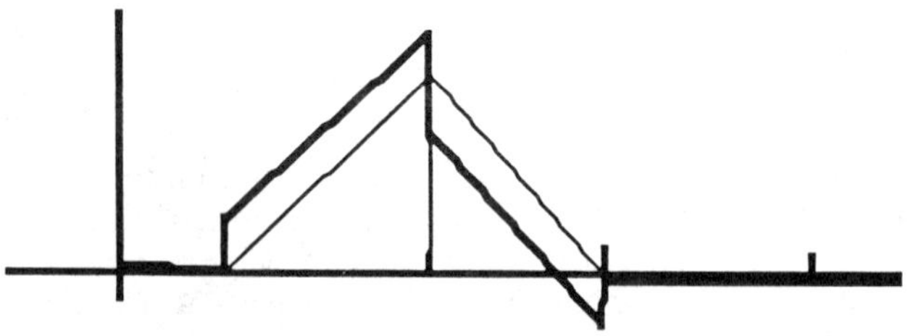

FIGURE 5.1. The basis function of H_{0h} in (2.6) is modified and more weight is put in the upstream direction (b). w_h (/thin) and $w_h + u\nabla w_h$ (/thick).

piecewise linear functions on triangles or tetrahedra then information on the element upstream can be used to compute $\nabla(\cdot u\varphi_h)$ and [2, 5]. Some error estimates can be found in [39].

Coming back to the Petrov–Galerkin formulations, one possible method of constructing a modified H_{0h} is to change w_h into $w_h + hu\nabla w_h$.

Then the stationary convection equation

$$u\nabla\varphi + a\varphi = f; \qquad \varphi = 0 \quad \text{on } \partial - \Omega$$

becomes

$$\int_\Omega (u\nabla\varphi_h + a\varphi_h)(w_h + hu\nabla w_h)\,dx = \int_\Omega f(w_h + hu\nabla w_h)\,dx. \tag{5.1}$$

Now in view of the fact that $u\nabla\varphi_h$ is $d\varphi_h/du$, it is clear that this Petrov–Galerkin formulation is also an artificial diffusion method because

$$h\int_\Omega u\nabla\varphi_h u\nabla w_h\,dx = h\int_\Omega \left(\frac{d\varphi_h}{du}\right)\left(\frac{dw_h}{du}\right)dx = -h\int_\Omega \left(\frac{d^2\varphi_h}{du^2}\right)w_h\,dx. \tag{5.2}$$

However, diffusion has been added only in the direction of the flow so it is called a *streamline diffusion method.*

The method has been studied by Johnson [18] for the linear advection problem, the incompressible Navier–Stokes equation, and Burger's equation [16]. It has been tested thoroughly by Hughes and Mallet [16] and applied also to the compressible Navier–Stokes equations.

For the incompressible Euler equations in two dimensions:

$$u_{,t} + u\nabla u + \nabla p = f, \qquad \nabla\cdot u = 0 \quad \text{in } \Omega \times]0, T[, \tag{5.3}$$

$$u(0, x) = u^0(x) \quad \text{in } \Omega; \quad u\cdot n|_{\partial\Omega} = 0. \tag{5.4}$$

The method proposed by Johnson is a space–time finite element method: Find $u_h \in V_h$ and $p_h \in Q_h$ such that

$$\int_\Omega (u_{h,t} + u_h\nabla u_h + \nabla p_h)\cdot(v_h + h(v_{h,t} + u_h\nabla v_h + \nabla q_h))\,dx$$

$$-\int_{\partial - T} [u_h]v_h|n_{,t} + n\cdot u_h|\,d\gamma = \int_\Omega f(v_h + h(v_{h,t} + u\nabla v_h + \nabla q_h)\,dx, \tag{5.5}$$

$$\begin{aligned}&\forall v_h \in V_h = \{v_h = \nabla\times\Psi_h,\ \Psi_h \text{ continuous}, P^{m+1} \text{ in } x \text{ and } P^m \text{ in } t\},\\ &q_h \in Q_h = \{q_h \text{ continuous}, q_h P^m \text{ in } x \text{ and } t\}.\end{aligned} \tag{5.6}$$

Here u_h is discontinuous on ∂T because $u_h = \nabla\times\varphi_h$ so there is a Lesaint-like term in addition to the streamline diffusion terms.

Theorem 5 ([18]). *For smooth data the solution of* (5.3)–(5.4) *and* (5.5)–(5.6) *verifies* ($m > 0$):

$$\max\{\|u(t,\cdot) - u_h(t,\cdot)\|_0 + h\|\nabla p - \nabla p_h\|_{L^1(\Omega)}\} \le Ch^{m+1/2}. \tag{5.7}$$

Although this method may not be ideal from a practical point of view because it is rather high in degree it illustrates well that the streamline upwinding methods can be analyzed theoretically even in the nonlinear cases.

4.6. Conclusion

We have discussed four families of methods:

artificial viscosity;
Euler–Lagrange characteristics and finite element methods;
discontinuous finite element upwinding;
streamline upwinding.

For each method we have presented some of the error estimates available for linear and nonlinear problems and have discussed some of the numerical results published.

Notations

Ω: open subset of R^n.
$\partial\Omega$: boundary of Ω.
$L^2(\Omega)$: space of square integrable functions.
$H^1(\Omega) = \{\varphi \in L^2(\Omega): \nabla\varphi \in L^2(\Omega)^n\}$.
$C^m(\Omega)$: space of m continuously differentiable functions, here $\psi = \nabla\varphi$ is defined in the sense of distribution i.e., roughly.
$\int_\Omega \psi w \, dx = -\int_\Omega \varphi \, \nabla \cdot w \, dx$, $\forall w$ differentiable, zero on the boundary.

$$\|\varphi\|_{0,\Omega} = \left(\int_\Omega \varphi^2 \, dx\right)^{1/2}, \qquad \|\varphi\|_{m,\Omega} = \left(\Sigma_{0,m} \int_\Omega |\nabla^i \varphi|^2 \, dx\right)^{1/2}.$$

T_h: triangulation of Ω.
h: length of the largest edge of the elements of T_h.
u_1, u_2: velocity components.
ρ: density.
E: energy.
p: pressure.
S: entropy.
P^m: space of polynomials of degree $\leq m$.

References

1. F. Angrand, V. Billey, A. Dervieux, J. Periaux, C. Pouletty, and B. Stoufflet, "2-D and 3-D Duler flow calculations with a second-order accurate Galerkin finite element method," *AIAA 18th Fluid Dynamics and Plasmadynamics and Lasers Conference* (*Cincinnat, Ohio, July* 16–18, 1985).
2. K. Baba and T. Tabata, "On a conservative upwind finite element scheme for convective diffusion equations," *RAIRO Anal. Numér.* **15**, No. 1 (1981), 3–25.

3. T.J. Baker and A. Jameson, "A novel finite element method for the calculation of inviscid flow over a complete aircraft." (R. Glowinski, ed.), *Proceedings of the Conference on Finite Elements in Flow Problems* (*Nice*, 1986).
4. Y. Brenier and G. Cohen, "Transport of contour lines with mixed finite elements for 2 phase flows," in *Numerical Methods for Transient Coupled Problems* (K.W. Lewis, ed.), Pineridge Press, 1984, pp. 741–755.
5. M.O. Bristeau, R. Glowinski, J. Periaux, P. Perrier, and O. Pironneau, "On the numerical solution of nonlinear problems in fluid dynamics by least squares and finite element methods," in *Proceedings of FENOMECH'78*, *Comp. Meth. Appl. Mech. Eng.*, 17/18 (1979), part 3." Formulations for convection dominated flows with particular emphasis on the incompressible Navier–Stokes equations," *Comput. Methods Appl. Mech. Engrg.*, **32** (1982), 199–259.
6. A.N. Brooks and T. Hughes, "Streamline upwind/Petrov–Galerkin formulations for convection dominated flows with particular emphasis on the incompressible Navier–Stokes equations." *Comput. Methods in Appl. Mech. Engrg.*, **32** (1982), 199–259.
7. G. Chavent and J. Jaffre, "Mathematical models and finite elements for reservoir simulation," *Stud. Math. Appl.* (1986), to appear.
8. I. Christie, D.F. Griffiths, A.R. Mitchell, and O.C. Zienckiewicz, "Finite element methods for second-order differential equations with significant first derivatives," *Internat. J. Numer. Methods Engrg.*, **10** (1976), 1389–1396.
9. P. Ciarlet: *The Finite Element Method*, North-Holland, Amsterdam, 1979.
10. A. Dervieux, "Steady Euler simulations using unstructured meshes," in *Computational Fluid Dynamics*, Von Karman Institute for Fluid Dynamics, Lecture Series 1985-04, March 26–29, 1985.
11. A. Dervieux, J.A. Desideri, F. Fezoui, B. Palmerio, and J.R. Rosenblum, "Euler calculations by upwind finite element methods," GAMM Workshop, 1986, to appear. See also *Numerical Methods for the Euler Equations of Fluid Dynamics* (Dervieux *et al.*, eds.) SIAM, Philadephia, 1985.
12. J. Douglas and T. Russell, "Numerical methods for convection dominated diffusion problems based on combining the methods of characteristics and the finite element methods,." *SIAM J. Numer. Anal.*, **19** (1982), 871–885.
13. M. Fortin and F. Thomasset, "Mixed finite element methods for incompressible flow problems," *J. Comput. Phys.*, **31**, No. 1 (1979), 113–145.
14. J.C. Heinrich, P.S. Huyakorn, O.C. Zienckiewicz, and A.R. Mitchell, "An upwind finite element scheme for the two-dimensional convective equation," *Internat. J. Numer. Methods Engrg.*, **11** (1977), 131–143.
15. T.J.R. Hughes, "A simple finite element scheme for developping upwind finite elements," *Internat. J. Numer. Methods Engrg.*, **12** (1978), 1359–1365.
16. T.J.R. Hughes, M. Mallet, and L.P. Franca, "Entropy-stable methods for compressible fluids: Application to high Mach number flows with shocks," *Finite Element Methods for Nonlinear Problems* (Bergan, Bathe, Wunderlich, eds.), Springer-Verlag, Berlin, 1986.
17. P. Hood and C. Taylor, "A numerical solution of the Navier–Stokes equations using the finite element technique," *Comput. & Fluids*, **1** (1973), 73–100.
18. C. Johnson, "Streamline diffusion methods for problems in fluid mechanics," in *Finite Elements in Fluids*, Vol. 6, (R.H. Gallagher, G.F. Carey, J.T. Oden, and O.C. Zienkiewicz, eds.), Wiley, New York, 1985.
19. C. Johnson and A. Szepessy, "A shock-capturing streamline diffusion finite element

method for a nonlinear hyperbolic conservation law," Dept. of Mathematics, Chalmers University of Technology and the University of Göteborg, S-412 96, Goteborg, Sweden, 1986.

20. C. Johnson and J. Pitkaranta, "An analysis of the discontinuous Galerkin method for a scalar hyperbolic equation," *Math. Comput.* **46**, No. 173, (1986), 1–26.
21. P. Lesaint, "On a finite element for solving the neutron transport equation." *Math Aspects of FEM in PDE*" (C. de Boor, ed.), Academic Press, New York, 1974, pp. 89–123.
22. P. Lesaint, "Sur la résolution des systèmes hyperboliques du premier ordre par des méthodes d'éléments finis," Thèse d'Etat, Université Pierre et Marie Curie, 1975.
23. P. Lesaint and P.A. Raviart, "Finite element collocation methods for first-order systems," *Math. Comp.*, **33**, No. 147 (1979), 891–918.
24. A. Mizukami and T.J.R. Hughes, "A Petrov–Galerkin finite element method for convection-dominated flows—an accurate upwinding technique for satisfying the maximum principle," *Comput. Methods Appl. Mech. Engrg.*, **50** (1985), 181–193.
25. M. Mallet, "A finite element method for computational fluid dynamics," Ph.D. Thesis, Standard, University 1985.
26. O. Pironneau, "On the transport diffusion algorithm and its applications to the Navier–Stokes equations," *Numer. Math.*, **38** (1982), 309–332.
27. T. Russell, "An incomplete iterated characteristic finite element method for the miscible displacement problem." Ph.D. Thesis, University of Chicago, 1980.
28. A. Priestley, "Lagrange and characteristic Galerkin methods for evolution problems," D.Phil. Thesis, Oxford University 1986.
29. F. Thomasset, *Implementation of Finite Element Methods for the Navier–Stokes Equations*, Springer Lecture Notes in Computational Physics, 1981.
30. B. Van LEER, "Computational methods for ideal compressible flows, VKI Lecture Series, 04, 1983.
31. S.F. Wornom and M.M. Hafez, "Calculation of quasi-one-dimensional flows with shocks," *Comput. & Fluids*, **14**, No. 2 (1986), 131–140.
32. O.C. Zienkiewicz, *The Finite Element Method in Engineering Science*, McGraw-Hill, New York, 1977.
33. O.C. Zienkiewicz and J.C. Heinrich, "Quadratic finite element scheme for two-dimensional convective transport problem," *Internat. J. Numer. Methods Engrg.*, **11** (1977), 1831–1844.
34. O.C. Zienkiewicz, K. Morgan, J. Peraire, M. Vahdati, and R. Lohner, "Finite elements for compressible gas flow and similar systems," *Proceedings of the 7th International Conference on Computational Methods*, (*Versailles*, 1985). (R. Glowinski, ed.), North-Holland, Amsterdam 1986.
35. R. Löhner, K. Morgan, and O. Zienkiewicz, "The solution of nonlinear systems of hyperbolic equations by the finite element method," *Internat. J. Numer. Methods Fluids* **4** (1984), 1043–1063.
36. F. Fezoui, B. Stoufflet, J. Periaux, and A. Dervieux, "Implicit high-order upwind finite element schemes for the Euler equations." *GAMNI*, *Conference* (*Atlanta*, 1986).
37. J.T. Oden, S.J. Robertson, T. Strouboulis, P. Devloo, L.W. Spradey, and H.V. Mc Connaughey, "Adaptive and moving mesh finite element methods for flow interaction problems," *6th International Symposium on Finite Element in Flow Problems* (*Antibe*, 1986) (M.O. Briteau *et al.*, eds.).

38. K.W. Morton and B.W. Scotney, "Petrov–Galerkin methods and diffusion–convection problems in 2D," The Math. of FEM and Appl V, Proc. MAFELAP 1984, (Whiteman, ed.), Academic Press, New York, 1985, pp. 343–366.
39. K.W. Morton, *Finite element methods for non-self-adjoint problems*, Lecture Notes in Mathematics, Vol. 965 (P.R. Turner, ed.), Springer-Verlag, New York, 1982, pp. 113–148.
40. B. Palmerio, V. Billey, A. Dervieux, and J. Periaux "Self-adaptive mesh refinements and FEM for solving the Euler equation," in *Numerical Methods for Fluid Dynamics*, Vol. II (K. Morton ed.), Clarendon Press, Oxford, 1986, pp. 369–388.
41. Scotney, "Petrov–Galerkin methods and diffusion–convection problems in 2D," The Mathematics of FEM and Appl. V Proc., MAFELAP 1984 (Whiteman, ed.), Academic Press, New York, 1985, pp. 343–366.

CHAPTER 5

Iterative Methods for Elliptic Finite Element Equations on General Meshes

R.A. Nicolaides and Shenaz Choudhury

5.1. Introduction

It is fair to say that the development of iterative solution techniques for all kinds of discretized partial differential equations remains a vigorous branch of numerical analysis. Perhaps the greater part of the effort has gone into multigrid algorithms, the next most common topic being preconditioning methods. Traditionally applied to elliptic problems, multigrid methods have also recently been successfully applied to solving the hyperbolic equations of gas dynamics (see [23] for a survey).

Iterative methods in general have yet to penetrate mainstream (elliptic) finite element methodology, where direct solvers are the rule. This situation is partly historical but is also due to the difficulty of using multigrid methods in situations of great geometrical complexity such as those routinely encountered in structural mechanics. It is not easy to automate the construction of several increasingly coarse embedded meshes that conform to the geometry of an arbitrary domain. Even when this is possible, problems with smoothing or other multigrid components may remain. In spite of these difficulties progress has been made toward a general implementation in [25]. Where three-dimensional problems must be solved, there does seem to be considerable interest in iterative methods, even for routine structural mechanics. The storage and time requirements of direct methods for such problems are sufficiently large that serious consideration of competing methods is a virtual necessity.

Aside from classical multigrid methods—which we will not consider further here—what choices remain? Basically there are two: general first- or second-order recursions with some form of preconditioner and *algebraic multigrid.* Both groups have their origins in the classical iterative methods. In the first class a special role is assigned to the conjugate gradient method, and in the second the Gauss–Seidel method is employed. The purpose of this paper is to survey some recent developments in this field. Section 5.2 introduces notation and conventions used in the analysis of iterative methods, indicates criteria for selecting good methods, and defines the conjugate gradient method on which many later results depend. Section 5.4 introduces the idea of preconditioning of conjugate gradients and defines an important class of preconditioners based on approximate Gauss elimination. Section 5.3 is about algebraic multigrid and its applicability to finite element problems. Section 5.5 introduces three newer methods. These are a *deflation* method for improving the convergence of conjugate gradients, an *element by element* iterative method, and some recent ideas based on classical substructuring, which may have applications to parallel computing. Section 5.6 gives a few computational results illustrating the properties of the various methods and also contains some further general remarks.

5.2. Preliminaries

The equation to be solved is

$$Ku = f, \tag{2.1}$$

where K is an $N \times N$ (symmetric) positive definite matrix. (Reference [13] contains a survey of iterative methods for nonsymmetric problems.) In keeping with the aims of the paper, (2.1) is assumed to come from the finite element discretization of an elliptic equation or system. In practice, most such problems concern the equation

$$\operatorname{div}(A \operatorname{grad} v) = g \tag{2.2}$$

in a bounded domain of $\mathbb{R}^2$ or $\mathbb{R}^3$ with appropriate boundary conditions. A denotes a positive definite second- or fourth-order tensor, frequently piecewise constant and/or highly anisotropic. The displacement equations of linear elasticity have the form (2.2), where v now denotes the displacement vector and A denotes the elasticity tensor. We will usually stay with the scalar case of (2.2), using it as a model for the vector case (but see Bercovier's example in Section 5.6.2). Concerning the discretization of (2.2) we shall once and for all assume that the finite element space chosen uses only function values as nodal parameters on the mesh. This is an essential restriction for most of the more efficient methods we consider later: efforts to circumvent it tend to involve extra constraints. The important point is that efficient methods are not usually designed to deal with the general case and can fail if applied indiscriminately.

Iterative methods (im) produce a sequence of approximations $u^{(k)}$ to the solution u of (2.1). We define the *error* $\varepsilon^{(k)} = u - u^{(k)}$ and the *residual* $r^{(k)} = f - Ku^{(k)}$. An important relation is the *residual equation*

$$K\varepsilon^{(k)} = r^{(k)}. \tag{2.3}$$

A basic idea of im is that if $u^{(k)}$ is a given approximation to u, then we can write

$$u = u^{(k)} + \varepsilon^{(k)}. \tag{2.4}$$

Although $\varepsilon^{(k)}$ is unknown, we can use (2.3) to compute an approximation $e^{(k)}$

and then define

$$u^{(k+1)} = u^{(k)} + e^{(k)}. \tag{2.5}$$

Surprisingly, crude choices of $e^{(k)}$ lead to convergent iterations. For example, even approximating (2.3) by

$$\left(\frac{1}{\alpha}\right) I e^{(k)} = r^{(k)}, \tag{2.6}$$

where α is a carefully chosen scalar and I is the identity matrix gives a (slowly) convergent algorithm: $u^{(k)} \to u$ as $k \to \infty$. This method, the simplest of all iterative methods, was apparently first suggested in 1910 [34] and is known as the (first-order) *Richardson method.* Its convergence can be improved by allowing α to vary with k. Using (2.6) in (2.5) we get the *iteration formula*

$$u^{(k+1)} = u^{(k)} + \alpha r^{(k)}, \tag{2.7}$$

and from (2.3), (2.4), and (2.7) the *error equation*

$$\varepsilon^{(k+1)} = (I - \alpha K)\varepsilon^{(k)}. \tag{2.8}$$

The *Jacobi method* defines $e^{(k)}$ by

$$De^{(k)} = r^{(k)}, \tag{2.9}$$

where D denotes the diagonal of K. For this, the error equation is

$$\varepsilon^{(k+1)} = (I - D^{-1}K)\varepsilon^{(k)}. \tag{2.10}$$

In the method of *steepest descents* introduced by Cauchy in 1847, (2.7) is replaced by

$$u^{(k+1)} = u^{(k)} + \alpha_k r(k) \tag{2.11}$$

for which

$$\varepsilon^{(k+1)} = (I - \alpha_k K)\varepsilon^{(k)}, \tag{2.12}$$

where $\alpha_k = (r^{(k)}, r^{(k)})/(r^{(k)}, Kr^{(k)})$, the parentheses denoting ordinary inner products of vectors. α_k chosen in this way has the property of minimizing $(u, Ku) - 2(u, f)$, the energy functional associated with (2.1), down the gradient at the point $u^{(k)}$.

The classical *Gauss–Seidel method* defines $e^{(k)}$ as

$$e^{(k)} = \left(\frac{1}{a_{kk}}\right) i_k i_k^T r^{(k)} \tag{2.13}$$

where the subscripts on the right are evaluated mod N, and i_k denotes the vector with 1 in position k, and 0 in all other positions. Letting $k = 1, 2, \ldots, N$ in turn in (2.13) and (2.5) gives one iteration of the Gauss–Seidel method. To generalize this to Successive Over-Relaxation (SOR), the second term in (2.13) is multiplied by ω, the relaxation factor.

Each of the methods defined is a *first-order* im. This refers to the fact that as in, e.g., (2.12) successive errors (and iterates) satisfy a first-order difference equation. For these and other methods it is often possible to determine a number $0 < \rho < 1$ such that

$$|\varepsilon^{(k)}| \le C\rho^k |\varepsilon^{(0)}|, \tag{2.14}$$

where C is a constant, and $|\cdot|$ denotes some norm on $\mathbb{R}^N$. Provided (2.14) is a sharp inequality, ρ can serve as a performance measure for a method; to be useful it must be complemented by a measure of the work needed to compute each iterate.

In the literature on im it is customary at least to evaluate ρ for the test problem consisting of the standard five-point approximation to the Laplacian on a uniform mesh of spacing $h = 1/(n + 1)$ in a square of side 1 in the plane with Dirichlet boundary conditions. Each im has its own analysis, usually involving substantial mathematics. Here, we shall merely list some results valid for $h \to 0$:

$$\text{First-order Richardson: } \rho = 1 - Ch^2,$$

$$\text{Same: variable } \alpha_k\text{: } \rho = 1 - Ch,$$

$$\text{Jacobi: } \rho = 1 - Ch^2,$$

$$\text{Steepest descents: } \rho = 1 - Ch^2,$$

$$\text{Gauss–Seidel: } \rho = 1 - \mathrm{Ch}^2,$$

$$\text{SOR: } \rho = 1 - Ch.$$

Proof of these results may be found in the standard references [14, 38, 40].

The constant C that appears is not necessarily the same for each method but in every case is independent of h.

In general, if $\rho = 1 - Ch^\beta$ then $O(h^{-\beta})$ iterations are required to compute each new decimal digit of the solution. Thus, we see that, e.g., SOR requires only about h times the number of iterations of Gauss–Seidel for a given accuracy for the simple test problem. But SOR is much more difficult to use because the parameter ω must be determined somehow, and the number of iterations can increase substantially if it deviates even slightly from its exact theoretical value.

The preceding observation suggests that we should list some criteria for an im to be acceptable. Basic ones are the following:

1. There should be a high rate of convergence (i.e., small ρ).
2. Performance should be essentially independent of A in (2.2).
3. Users should not have to supply parameter or interact with the program.
4. A method should achieve an order of magnitude saving in storage and computation time over Gaussian elimination.

Point 1 is required in order that the iteration can be terminated with a reasonable guarantee that the current approximation is a good one. It is very difficult to decide whether a slowly convergent iteration has in fact "converged." Concerning Point 2, the methods described previously may fail for highly anisotropic or highly inhomogeneous materials. For example, if the square is bisected parallel to the y axis and the material tensor A is diag(1, 1) in the left side and diag(κ, κ) where $0 < \kappa < 1$ on the right, then with certain boundary conditions, κ appears in the preceding expressions for ρ as $\rho = 1 - C\kappa h^{\beta}$. This gives an unacceptably poor rate of convergence if $\kappa \ll 1$. Incidentally, related difficulties can arise even with isotropic and homogeneous problems if there are high mesh aspect ratios or sudden changes in the mesh spacing. Then κ measures the mesh ratios, etc. Point 3 reflects the fact that users are just that and presumably do not want to become experts in im. Point 4 is the main rationale for considering im in the first place: symmetric banded elimination for the model two-dimensional problem has an operation count of $N^2/2$ and requires $N^{3/2}$ storage locations. Other demands could be made; for example, it could be required that im impose absolutely no more constraints on the user than Gauss elimination, but this is not usually feasible.

We now generalize the previous ideas to *second-order* im. The basic formula for these is a generalization of (2.7),

$$u^{(k+1)} = \omega_k u^{(k)} + \omega_k' u^{(k-1)} + \alpha_k \omega_k r^{(k)}, \tag{2.15}$$

where ω_k and α_k are scalar parameters, and $\omega_k + \omega_k' = 1$. The problem here is to choose the parameters to have a good convergence rate. As with the first-order case, they may be chosen as constants (the second-order Richardson method, due to Frankel [15]), as dependent on k (the semi-iterative method [38]) or by a variational approach (conjugate gradients). We shall not consider the first two cases any further here, because the correct choice of the parameters requires a knowledge of the smallest and largest eigenvalues of K—information that is rarely available. For the details see [21, 38, 40]. Instead, consider the third case, the conjugate gradient method, where $\omega_0 = 1$ and

$$\alpha_k = \frac{(r^{(k)}, r^{(k)})}{(r^{(k)}, Kr^{(k)})}, \tag{2.16}$$

and

$$\frac{1}{\omega_k} = 1 - \frac{1}{\omega_{k-1}} \frac{\alpha_k}{\alpha_{k-1}} \frac{(r^{(k)}, r^{(k)})}{(r^{(k-1)}, r^{(k-1)})}. \tag{2.17}$$

This choice for the two parameters of the basic formula (2.15) is obtained by imposing the two orthogonality conditions

$$(r^{(k+1)}, r^{(j)}) = 0, \qquad j = k, \quad k - 1, \tag{2.18}$$

on the new residual $r^{(k+1)}$. This is not the standard way of writing the conjugate gradient algorithm, but it is mathematically equivalent to it (see Appendix) and is, perhaps, conceptually simpler. The basic result that follows rather surprisingly from (2.15)–(2.17) is that $r^{(k+1)}$ is orthogonal to *all* the previous residuals, not just to the two previous ones, by construction. This can be easily proved by induction, making essential use of the symmetry of K. Thus, after at most N steps the residual will be zero, and the current iterate will be u itself. In practice, N is usually too large for the method to be used in this way. What occurs frequently is that the residual decreases sufficiently rapidly that an acceptable approximation is produced after considerably fewer steps. In fact, for second-order homogeneous and isotropic equations, the required number of iterations is usually closer to $\sqrt{N}$ in two dimensions ($\sqrt[3]{N}$ in three dimensions). As with the first-order im the speed of convergence may slow significantly if the material properties deviate much from isotropy or homogeneity.

Applied to the standard test problem it can be proved [21] that for conjugate gradients, the factor $\rho = 1 - Ch$. This implies the $O(\sqrt{N})$ iterations per digit of accuracy mentioned earlier. Conjugate gradients is the easiest way to get this $1 - Ch$ factor without supplying special problem-dependent iteration parameters. Each of the methods so far considered either converges more slowly than conjugate gradients or needs such parameters. This is perhaps the main reason for the popularity of the method. On the other hand, this is not very fast convergence. It would be more satisfactory to have $\rho = 1 - C\sqrt{h}$, say, giving $O(N^{1/4})$ iterations per digit, or even faster rates, provided that could be obtained at small cost.

It should be mentioned that making the constant C small is another way to improve convergence. This seems to be much more difficult than adjusting the exponent of h, to judge from the very small number of methods that achieve it. In the conjugate gradient setting, such a situation occurs sometimes when there are relatively few distinct eigenvalues. These distinct eigenvalues may range from, say, $O(1)$ to $O(h^{-2})$, giving $\rho = 1 - Ch$, but if there are m of them it can be proved that at most m iterations will produce the exact solution of the linear system, apart from round-off. In general, there are a large number of distinct eigenvalues so that the estimate in terms of h is more realistic, but transformations that reduce the number of distinct eigenvalues and make this estimate irrelevant could be sought.

Each iteration of conjugate gradients consists of forming matrix–vector products and dot products of vectors that are $O(N)$ operations. Since the number of digits significant in approximating the solution to the model differential equation is $O(\log_{10} N)$, the total work count is $O(N^{3/2} \log_{10} N)$. The storage is $O(N)$ because only the nonzeros of K are needed. This is to be compared with $N^2/2$ work and $N^{3/2}$ storage for the solution of the model problem by direct methods.

5.3. Algebraic Multigrid (amg)

The algebraic multigrid (amg) technique attempts to extend the basic ideas of regular multigrid methods to a more general class of problems. No continuous problem underlies the given algebraic system of equations that is to be solved and, in particular, no grids are involved. The outline given here is based on [35], which contains more details and references.

5.3.1. Multigrid Principles

Standard multigrid algorithms are based on the following principles: consider the residual equation (2.3)

$$K\varepsilon = r. \tag{3.1}$$

If K is a discrete partial differential operator, and r a corresponding source or load vector representing a suitably smooth function, then ε can probably be well approximated by defining a coarser problem

$$\hat{K}\hat{\varepsilon} = \hat{r}, \tag{3.2}$$

where ^ denotes a coarsening of the objects involved. For K the simplest example is the discrete five-point Laplacian operator on a uniform mesh of side $h = 1/2^m$ in a unit side square. If $K := K(h)$ denotes this approximation, then $\hat{K} := K(2h)$. For $\hat{r}$ numerous approximations can be used, of which the simplest is $\hat{r} := r$ at points of the coarse $(2h)$ mesh. If (3.2) can be solved, then $\hat{\varepsilon}$ should approximate ε on the coarse mesh points. Then it can be interpolated (extended) to the fine mesh by a rectangular matrix E taking coarse vectors to fine ones, $\varepsilon \simeq E\hat{\varepsilon}$, so that (2.4) gives

$$u^{(\text{new})} = u^{(\text{old})} + E\hat{\varepsilon}. \tag{3.3}$$

This algorithm requires specifying (i) the coarse grid operator $\hat{K}$, (ii) the extension operator E, and (iii) a smooth residual r.

For standard finite elements there is a self-evident definition of E, coming from using the coarse grid trial (shape) functions to interpolate to the fine grid. To find $\hat{K}$ we substitute the coarse trial functions into the energy functional and minimize in the usual way. It then turns out that

$$\hat{K} = E^T K E, \tag{3.4}$$

so that $\hat{K}$ can be expressed in terms of K and E and does not have to be chosen independently. In a very similar way, the correct choice for $\hat{r}$ is

$$\hat{r} = E^T r. \tag{3.5}$$

Thus, the components of (3.2) are fully defined once E is specified.

The construction of a smooth r is achieved by the application of a few iterations of a classical iterative method to an arbitrary starting approximation. On (i, j) meshes, relaxation methods, such as Gauss–Seidel in its point or line versions, are frequently successful.

In practice, (3.2) is itself reduced by smoothing and coarsening, and so on recursively until an easily solved coarse equation, usually containing just a handful of unknowns, is reached. This recursion is rather irrelevant to the main mathematical properties of multigrid methods. Its significance is practical: without it, (3.2) cannot be efficiently solved.

The finite element algorithm just described was first defined and its (*W cycle*) convergence analyzed in [30, 31] following earlier finite difference work. Many improvements and additions to this analysis have since been made; [20] is a recent reference on the theoretical aspects of multigrid methods for elliptic problems. Here, we wish to write out the error formula for the two grid method in order to motivate some of the amg concepts later. Substituting (3.2) into (3.3) and subtracting both sides from the exact solution u give

$$\varepsilon^{(\text{new})} = (I - E\hat{K}^{-1}E^T K)\varepsilon^{(\text{old})}. \tag{3.6}$$

The key observation is that if

$$\varepsilon^{(\text{old})} = E\eta \tag{3.7}$$

for *some* η then since $\hat{K} = E^T K E$,

$$\varepsilon^{(\text{new})} = 0; \tag{3.8}$$

i.e., we will have the exact solution on completing the coarse grid operation. This conclusion is independent of any particular choice of E. Of course, there is no reason in general to suppose that (3.7) will hold, but we can try to relate the smoothing to the choice of coarse grid so that it is nearly true. Then we can expect to get a rapidly convergent method; amg attempts explicitly to achieve this correspondence.

5.3.2. amg Components

In standard multigrid, the coarse meshes are assumed given and a smoothing algorithm that enables smoothed residuals and errors to be adequately represented on these meshes must be found. In the amg context where no meshes exist, the reverse idea is adopted: the smoothing algorithm is chosen first, and on the basis of what it achieves—strongly dependent on the properties of K—the coarse operators are constructed. This is done by using the method of Section 5.3.1, i.e., by choosing a coarse to fine extension operator E and defining the other operations in terms of it as in (3.4)–(3.5). This must be done recursively for several coarsenings.

In practice, this idea is not easy to implement, and attention has so far been restricted to symmetric positive definite matrices K satisfying the additional conditions

$$k_{ij} \le 0, \quad i \ne j, \qquad \sum_{j=1}^{N} k_{ij} \ge 0, \quad i, j = 1, 2, \ldots, N. \tag{3.9}$$

Numerical results show that the method continues to converge when these conditions are slightly violated.

Only the Gauss–Seidel method is used as an amg smoother in [35]. Relaxation at a single point is given by

$$u_i = -\left(\frac{1}{k_{ii}}\right)\left(\sum_{j\ne i}^{(1)} k_{ij}u_j + \sum_{j\ne i}^{(2)} k_{ij}u_j - f_i\right), \tag{3.10}$$

where $\sum^{(1)}$ denotes summation over points already updated in this pass and

$\sum^{(2)}$ over those not yet modified. Smoothing occurs for *strongly connected* subsets of the unknowns, where the *strength* is measured by the relative magnitude of the off-diagonals $|k_{ij}|$.

Next, note that for the errors ε_i, (3.10) gives

$$\varepsilon_i = -\left(\frac{1}{k_{ii}}\right)\left(\sum_{j\neq i}^{(1)} k_{ij}\varepsilon_j + \sum_{j\neq i}^{(2)} k_{ij}\varepsilon_j\right). \tag{3.11}$$

Equation (3.11) can be further simplified by ignoring the notational distinction between $\sum^{(1)}$ and $\sum^{(2)}$ and by setting to zero the terms with only weak (i.e., $|k_{ij}|$ relatively small) connections. Let S_i denote the set of strong connections of point i. Then

$$\varepsilon_i \cong -\left(\frac{1}{k_{ii}}\right) \sum_{\substack{j\in S_i\\ j\neq i}} k_{ij}\varepsilon_j. \tag{3.12}$$

Let C denote the set of points designated as coarse and F the remainder. Interpolation to an F point is by linear combinations of values at C points. Let $S_i(E) \subseteq S_i \cap C$ denote the set to be used for interpolation to point $i \in F$. Equation (3.12) now becomes

$$\varepsilon_i \cong -\left(\frac{1}{k_{ii}}\right)\left(\sum_{j\in S_i(E)} k_{ij}\varepsilon_j + \sum_{\substack{j\notin S_i(E)\\ j\neq i}} k_{ij}\varepsilon_j\right). \tag{3.13}$$

Equation (3.13) is a good approximation to the actual smoothing formula we are using. Recalling (3.7) we want to use (3.13) as the interpolation formula too. But the points in the second sum are not interpolation points for $i \in F$. Clearly, we must choose C so that they can be expressed in terms of $S_i(E) \cup \{i\}$ values. Several ideas have been proposed. Good results [35] have been obtained by choosing C so that points in the second sum in (3.13) are strongly connected to the set $S_i(E)$. For these points the following approximations are now made:

$$\varepsilon_j \cong \left(k_{ji}\varepsilon_i + \sum_{l\in S_i(E)} k_{jl}\varepsilon_l\right)\Bigg/\left(k_{ji} + \sum_{l\in S_i(E)} k_{jl}\right). \tag{3.14}$$

Substituting (3.14) into (3.13) gives $\varepsilon_i \cong \sum_{j\in S_i(E)} v_{ij}\varepsilon_j$ for $i \in F$, where

$$v_{ij} = -\frac{k_{ij} + c_{ij}}{k_{ii} + c_{ii}}, \tag{3.15}$$

and

$$c_{ij} = \sum_{\substack{l\notin S_i(E)\\ l\neq i}} \frac{k_{il}k_{lj}}{k_{li} + \sum_{m\in S_i(E)} k_{lm}}. \tag{3.16}$$

For $i \in C$ the interpolated value is just the coarse value existing at i. This defines E, and the rest of the problem setup follows automatically, as in Section 5.3.1.

It remains to give explicit rules for the construction of C and F. It is very important that C contains a relatively small number of points, because $|C|$ is the size of the coarse problem. A low rate of coarsening would produce a large number of successively coarser levels, causing problems with storage as well as efficiency. Finding suitable sets C is thus crucial to the operation of the algorithm. The algorithm given in [35] for finding C is, interestingly, considerably more complicated than the amg algorithm itself, involving much graph-theoretical manipulation and empirical testing. We will not describe it here. Full details including flowcharts and rules for picking parameters are in [35].

5.3.3. Applications

Some figures from [35] give an idea of the behavior of the amg algorithm described previously. Several examples, including severely anisotropic and nonhomogeneous scalar second-order elliptic problems, are reported. The meshes were structured, although of course this fact was not explicitly used. The performance is uniformly good on all the test problems, once the setup phase is complete. The latter appears to cost about one to two times as much as the solution itself, so that presumably a number of solutions would have to be done to neutralize this cost. Convergence factors are quite remarkable, being well below 0.1 for the Poisson equation on an $h = 1/64$ mesh, giving an expectation of at least one new correct digit each iteration, and less than 0.12 for all the test problems. For the work actually to solve a model problem with a given precision, see Section 5.6.4.

5.4. Preconditioning and Preconditioners

In this section, we will survey a class of techniques for improving the convergence of im. Although these techniques can be used quite generally, we will consider only their application to *conjugate gradients* (cg). It has long been known that cg works best for matrices that have small condition numbers $\lambda_{\max}/\lambda_{\min}$, the number of iterations required per digit of accuracy being proportional to the square root of the condition number. Convergence is also improved if there are many nearly identical eigenvalues. *Preconditioning* refers to the general strategy of transforming K to reduce its condition number or cluster or otherwise redistribute its eigenvalues.

5.4.1. Using Preconditioners

Suppose we have constructed a representation of K in the form

$$K = LL^T + R, \tag{4.1}$$

where L is lower triangular and R can be interpreted as the error of the approximate factorization of $K \cong LL^T \equiv M$. Then we can solve

$$M^{-1}Ku = M^{-1}f \tag{4.2}$$

instead of (2.1) with the expectation of a smaller condition number or of better clustered eigenvalues for $M^{-1}K$ than for K. A minor point is that $M^{-1}K$ is not symmetric, so that cg may not converge for (4.2) as it stands. But using the definition of M it can be transformed into

$$L^{-1}KL^{-T}v = L^{-1}f, \qquad v = L^T u, \tag{4.3}$$

where the coefficient matrix is symmetric positive definite. This appears slightly inconvenient in that it does not refer to the original variable u, but it is easy to express the iteration in terms of approximations $u^{(k)}$ convergent directly to u. Indeed, (2.15) gives for (4.3)

$$v^{(k+1)} = \omega_k v^{(k)} + \omega_k' v^{(k-1)} + \alpha_k \omega_k s^{(k)}, \tag{4.4}$$

where $s^{(k)}$ denotes the residual of (4.3),

$$\alpha_k = \frac{(s^{(k)}, s^{(k)})}{(s^{(k)}, L^{-1}KL^{-T}s^{(k)})}, \tag{4.5}$$

and ω_k is defined by the recursion formula corresponding to (2.17). Defining $u^{(k)} = L^{-T}v^{(k)}$, (4.4) gives

$$u^{(k+1)} = \omega_k u^{(k)} + \omega_k' u^{(k-1)} + \alpha_k \omega_k L^{-T} s^{(k)}, \tag{4.6}$$

and (4.3) gives

$$s^{(k)} = L^{-1}(f - KL^{-T}L^T u^{(k)}) = L^{-1} r^{(k)}, \tag{4.7}$$

so that (4.6) can be expressed as

$$u^{(k+1)} = \omega_k u^{(k)} + \omega_k' u^{(k-1)} + \alpha_k \omega_k M^{-1} r^{(k)}. \tag{4.8}$$

Similarly, by (4.7)

$$\alpha_k = \frac{(r^{(k)}, M^{-1}r^{(k)})}{(M^{-1}r^{(k)}, KM^{-1}r^{(k)})}, \tag{4.9}$$

and ω_k continues to be given by the formula (2.17),

$$\frac{1}{\omega_k} = 1 - \frac{1}{\omega_{k-1}} \frac{\alpha_k}{\alpha_{k-1}} \frac{(r^{(k)}, M^{-1}r^{(k)})}{(r^{(k-1)}, M^{-1}r^{(k-1)})}. \tag{4.10}$$

To summarize, (4.8)–(4.10) are the formulas for preconditioned cg with preconditioner M. It is clear that the only extra work consists of forming $M^{-1}r^{(k)}$ at each iteration.

This formulation shows that any convenient M can be used as preconditioner and that knowing its factored form (LL^T) is theoretically superfluous. Some very simple preconditioners can exploit this, for example, diagonal preconditioning in which the preconditioner is D, the diagonal of K. More complicated choices of M, required for faster convergence, should also be cheap to form, store, and invert. This motivates the definition of M as the product of incomplete factors in the next subsection. These factors preserve the sparsity pattern of K, so that only $O(N)$ operations are needed for the forward and back substitutions to compute $M^{-1}r^{(k)}$. The factors themselves are computed, once and for all, in $O(N)$ operations, and require only $O(N)$ storage.

5.4.2. Incomplete Factorizations

The most important class of preconditioners is based on an idea known as *incomplete factorization.* This idea was first suggested by Varga [39], although a Russian paper [5] the same year—according to some remarks in [37]—contains the algorithm of [11], which itself contains a detailed implementation of Varga's idea. These authors consider the application of preconditioning to first-order im for five- or nine-point difference formulas in rectangles. The application to cg was suggested much later by Meijerink and Van der Vorst [28], who also give other extensions.

We will quickly review the standard matrix formulation of Gaussian elimination without pivoting. Let $K^{(i)}$ be the ith stage matrix with pivot row i, and columns $1, 2, \ldots, i-1$ already eliminated. Then with $K^{(1)} = K$,

$$K^{(i+1)} = L^{(i)}K^{(i)}, \qquad i = 1, 2, \ldots, N-1, \tag{4.11}$$

where $L^{(i)}$ is unit lower triangular, having multipliers

$$\frac{-k_{ji}^{(i)}}{k_{ii}^{(i)}}, \qquad j = i+1, \ldots, N, \tag{4.12}$$

in column i, and zeros in the other off-diagonal positions. Then

$$K^{(N)} = L^{(N-1)} \ldots L^{(1)}K \equiv U, \tag{4.13}$$

where U is upper triangular. It follows by a simple direct calculation that

$$K = LU, \tag{4.14}$$

where L is unit lower triangular with the negatives of (4.12) in the same places below the diagonal.

Instead of (4.11), consider the more general recursion with $\kappa^{(1)} = K$,

$$\kappa^{(i+1)} = l^{(i)}\kappa^{(i)} - R^{(i)}, \qquad i = 1, 2, \ldots, N-1, \tag{4.15}$$

where $l^{(i)}$ eliminates column i from $\kappa^{(i)}$, and $R^{(i)}$, called the *local error matrix*, is any $N \times N$ matrix not introducing fill and such that the $(i+1)$th step is well defined. Denoting by $\Pi_{j,k}$ the product $l^{(j)}l^{(j-1)} \ldots l^{(k)}$ and by $\Pi^{k,j}$ its inverse, it follows from (4.15) that

$$\Pi_{N-1,1}K = \kappa^{(N)} + \Pi_{N-1,2}R^{(1)} + \Pi_{N-1,3}R^{(2)} + \cdots + R^{(N-1)}, \tag{4.16}$$

where $\kappa^{(N)}$ is upper triangular. Hence, defining $\mathscr{L} = \Pi^{1,N-1}$ and $\mathscr{U} = \kappa^{(N)}$, (4.16) gives

$$K = \mathscr{L}\mathscr{U} + \Pi^{1,1}R^{(1)} + \Pi^{1,2}R^{(2)} + \cdots + \Pi^{1,N-1}R^{(N-1)} \tag{4.17}$$

$$= \mathscr{L}\mathscr{U} + R, \tag{4.18}$$

where R denotes the combination of the local error matrices. Equation (4.18) shows that $\mathscr{L}\mathscr{U}$ is the Gaussian decomposition of $K - R$. If the $R^{(i)}$ are well chosen, then $\mathscr{L}\mathscr{U}$ can approximate LU of (4.14) and be used as a preconditioner.

A significant specialization of (4.17) occurs when $R^{(i)}$ is zero in and above its ith row, so that $R^{(i)}$ modifies the exact elimination result only below the pivot row. In this case, (4.17) becomes

$$K = \mathscr{L}\mathscr{U} + R^{(1)} + R^{(2)} + \cdots + R^{(N-1)} \tag{4.19}$$

because

$$\Pi^{1,i}R^{(i)} = (l^{(1)})^{-1}(l^{(2)})^{-1}\ldots(l^{(i)})^{-1}R^{(i)}, \tag{4.20}$$

and the successive inverse elimination matrices use only pivot rows 1 through i of $R^{(i)}$. Thus R is just the sum of the local error matrices.

There are two more or less standard ways to choose R, both controlling the fill in $\mathscr{L}$ and $\mathscr{U}$. In the first approach, $R^{(i)}$ is simply defined to contain the fill from the ith step of the elimination. This is the method of [28]. It follows that wherever K has nonzero entries, R has zero entries and thus, since

$$\mathscr{L}\mathscr{U} = K - R, \tag{4.21}$$

K and $\mathscr{L}\mathscr{U}$ agree on the nonzeros of K. In the second approach [11, 12] a modification to the R just defined, consisting of adding to the diagonal of $R^{(i)}$ at each stage the negative of its row sums, so that the resulting matrix has zero row sums, is made. Thus, we keep agreement of the off-diagonal nonzero terms of K with those of the approximate factorization, while giving up agreement of the diagonal terms in favor of having the row sums of K and $\mathscr{L}\mathscr{U}$ equal. The second approach is usually considerably better than the first in terms of convergence of *preconditioned* cg (pcg).

In both of these methods, the zero patterns of $\mathscr{L}$ and $\mathscr{U}$ match those of the lower and upper triangular parts of K. This suggests seeking $\mathscr{L}$ and $\mathscr{U}$ directly in this form, by multiplying them together and equating to K. The extra nonzero terms (analogous to fill) arising in this product are put into R. In a similar way, a Cholesky-type decomposition can be obtained. This approach is usual in the finite difference literature and is especially helpful when there is an (i, j) mesh available. However, it is possible to show that the different approaches lead to the same approximate factors.

The preceding approximate factorizations are called *incomplete* factorizations and designated as *incomplete LU* (ILU) or *incomplete Cholesky* (IC). If the zero row sum feature is incorporated, the methods are usually called the *modified incomplete* LU (MILU) or *Cholesky* (MIC) *factorizations.*

5.4.3. Theoretical Results

There are two groups of results available for ILU methods. The first concerns the existence and stability of the approximate factorizations, and the second gives estimates for the condition numbers of the preconditioned matrices, from which the rate of convergence of pcg can be found. In this section we will state representative theorems of this kind.

It follows from (4.21) that $\mathscr{L}\mathscr{U}$ is the decomposition of $K - R$. In the ILU setting, if K is positive definite then R will be symmetric with 0 diagonal and so indefinite. If K also has nonpositive off-diagonal entries, then in the MILU case R will have a positive diagonal, negative off-diagonal entries, and zero row sums. Hence, it is positive semidefinite. In both cases, therefore, it follows that the matrix actually factored is less positive definite than K. It is conceivable—and can actually happen for merely positive definite matrices—that the construction of $\mathscr{L}$ and $\mathscr{U}$ can break down. On the other hand, if $K - R$ is positive definite not only does the factorization exist, it is also stable.

In [28] existence and relative stability are proved for M matrices, i.e., matrices K such that $k_{ij} \leq 0$ for $i \neq j$ and $K^{-1} \geq 0$. For such matrices, it is known that Gauss factorization is well defined.

Theorem 1 ([28]). *Let K be an M matrix. Then the ILU factorization algorithm is well defined for K. Moreover, the factorization is relatively stable in the sense that the ILU pivots are at least as large in magnitude as those of the exact LU process, and $|\mathscr{L}| \leq |L|$ elementwise where L is the exact lower triangular factor of K.*

For MILU we have another theorem.

Theorem 2. *Let K satisfy the conditions*

$$k_{ij} \leq 0, \quad i \neq j, \qquad k_{ii} > 0, \quad i, j = 1, 2, \ldots, N, \tag{4.22}$$

and

$$k_{ii} > -\sum_{j \neq i} k_{ij}. \tag{4.23}$$

Then for MILU, $K - R$ is positive definite.

Condition (4.23) can be replaced with the weaker $\geq$ condition with at least one strict inequality, provided each row of K contains a nonzero element after the diagonal [1]. A class of matrices satisfying this, and also the conditions of Theorem 1, is given by linear finite element discretizations of the scalar problem (2.2) with Dirichlet boundary conditions, provided all of the triangles are acute. Other than this, there does not seem to be any large class of symmetric finite element problems satisfying the conditions of either theorem. But in practice both methods have been successful in cases where these conditions are not too strongly violated. Also, convection–diffusion equations discretized by upwind schemes often satisfy the conditions of Theorem 1 [29].

The condition number of the preconditioned matrix $M^{-1}K$ can be computed for model problems. Chandra [7] proved the following result:

Theorem 3 ([7]). *For the Laplacian operator $-\Delta$ with Dirichlet boundary conditions in a unit side square in $\mathbb{R}^2$, discretized by linear elements on a standard triangulation of side h and preconditioned by ILU, there exist constants C_1 and*

C_2 *such that*

$$C_1 h^{-2} \leq \operatorname{cond}(M^{-1}K) \leq C_2 h^{-2}. \tag{4.24}$$

From (4.24), it follows that ILU pcg does not have an improved rate of convergence over regular cg in terms of powers of h. Nevertheless, it is observed in applications that ILU pcg does give a more efficient algorithm than cg alone in the sense that a smaller number of iterations and a smaller amount of work are needed to solve with given accuracy. But as h decreases, the required number of iterations to compute a new digit increases as h^{-1}. If we have a method where this figure increases like some smaller power of h, then as h reduces, at some point the second method will become cheaper. Where this point actually occurs is practically unpredictable and has to be found experimentally. For the model problem of Theorem 3 it is proved (for finite differences) in [18] that a slight variation of MILU has $\operatorname{cond}(M^{-1}K) = O(h^{-1})$ giving $O(h^{-1/2})$ iterations per digit, potentially a large saving over ILU. The variation consists of adding a quantity of $O(h^2)$ to the diagonal during the factorization. We will call this MILU+ for brevity. In practice, the same performance is seen for MILU.

Theorem 4 [18]. *For the problem in Theorem* 3, *for MILU+, constants* D_1, D_2 *exist such that*

$$D_1 h^{-1} \leq \operatorname{cond}(M^{-1}K) \leq D_2 h^{-1}. \tag{4.25}$$

In the case of an arbitrary mesh and for other generalizations the $O(h^{-1})$ condition number can be obtained, although once again only by introducing one or more problem-dependent parameters; [1] contains an extensive account of these developments.

5.4.4. Further Developments

This subsection gives several refinements and extensions of the basic algorithms. First, we will consider procedures that increase the density of nonzero elements in $\mathscr{L}$ and $\mathscr{U}$; Reference [29] contains a suggestion for doing this with finite differences. The ILU procedure, suitable for (i, j) difference schemes, is to define a prior set P of entries in $\mathscr{L}$ where nonzeros are permitted to occur at least including the nonzero positions of K. In all other positions, $\mathscr{L}_{ij} = 0$ by definition. $\mathscr{U}$ has the transposed nonzero positions. Then we generate the permitted entries in $\mathscr{L}$ and $\mathscr{U}$ by

$$\begin{aligned}
\mathscr{U}_{ii} &= k_{ii} - \sum_{k=1}^{i-1} \mathscr{L}_{ik}\mathscr{U}_{ki}, \\
\mathscr{L}_{ij} &= \left(\frac{1}{\mathscr{U}_{jj}}\right)\left(k_{ij} - \sum_{k=1}^{j-1} \mathscr{L}_{ik}\mathscr{U}_{kj}\right), \\
\mathscr{U}_{ij} &= k_{ij} - \sum_{k=1}^{i-1} \mathscr{L}_{ik}\mathscr{U}_{kj}.
\end{aligned} \tag{4.26}$$

These are the usual recurrences, but with the entries not in P omitted. For mesh calculations with, e.g., five-point formulas, a convenient choice for P consists of K's nonzero bands together with some nearby ones [29].

Clearly, this technique is not very useful for general meshes. A more suitable method given in [19] is to define $\mathscr{L}^{(s)}\mathscr{U}^{(s)}$ recursively by defining $P^{(s)}$ to be the nonzero set associated with the product $\mathscr{L}^{(s-1)}\mathscr{U}^{(s-1)}$. $P^{(0)}$ is defined to be P, so $s = 0$ corresponds to the original approximate factorization. In general, as s increases $P^{(s)}$ contains more and more nonzero positions, and eventually all of them, so that an exact factoring would be required; Gustafsson [19] reports that small values of s give the best overall results.

Another problem that arises in solving general positive definite systems is that of negative pivots in ILU and MILU. Conjugate gradients is proved to converge only for symmetric positive definite matrices. If $\mathscr{U}$ does not have a positive diagonal a problem may arise. Two papers [24, 27] deal with the avoidance of these negative or small pivots. In [24] the proposed remedy is to add a sufficiently large number to the diagonal culprit in $\mathscr{U}$. This just corresponds to a further addition into the $R^{(i)}$ for this step. The argument usually advanced is that if not too many of these corrections have to be made, their effect on the rate of convergence of pcg will be small. Perhaps because of its *ad hoc* nature no rigorous estimate of the effect of this modification seems to be known. In [27] the prior addition of a matrix αI to K is advocated. However, [29] points out that this is a global change being made to correct what can be conceived as a local problem. Moreover, a problem-dependent parameter has once again crept in. In spite of these objections, some good results have been reported in both of the preceding references.

5.5. Other Methods

In this section, we will very briefly metion three methods that have recently been proposed. Much less is known about them than the methods considered previously, but all have some new feature that may be of interest. They are *deflated* cg, the *element by element method*, and *domain decomposition*. In each case we can do little more than describe the algorithm and mention whatever else seems relevant.

5.5.1. Deflated cg

The deflated cg method, described more fully in [32], gives another way to improve the convergence of cg or pcg. It is closer in spirit to multigrid methods than to incomplete factorizations but is suitable for general meshes.

Recalling (2.15) we can generalize it to

$$u^{(k+1)} = \omega_k u^{(k)} + \omega_k' u^{(k-1)} + \alpha_k \omega_k (r^{(k)} - Ec^{(k)}), \tag{5.1}$$

where E is $N \times m$ ($m \ll N$) and has a meaning similar to the E in Section 5.3. The idea behind (5.1) is that $Ec^{(k)}$ "deflates" certain constituents of the residual, particularly those for which the regular algorithm is ineffective. $c^{(k)}$ is chosen to minimize

$$(K(r^{(k)} - Ec^{(k)}), r^{(k)} - Ec^{(k)})$$

leading to

$$E^T K E c^{(k)} = E^T K r^{(k)}, \tag{5.2}$$

which must be solved for $c^{(k)}$ at each iteration. Since $m \ll N$ this is feasible. From (5.1) and (5.2) we find

$$r^{(k+1)} = \omega_k r^{(k)} + \omega_k' r^{(k-1)} - \alpha_k \omega_k K(I - E\mathbf{K}^{-1}E^T K) r^{(k)} \tag{5.3}$$

where $\mathbf{K} \equiv E^T K E$ (cf. (3.4)). Let $P = E\mathbf{K}^{-1}E^T K$. In (5.3), $K(I - P)$ is positive semidefinite since it is symmetric and equals

$$K^{1/2}(I - K^{1/2} P K^{-1/2}) K^{1/2},$$

where the bracketed matrix is an orthogonal projection matrix, and $K^{1/2}$ is positive definite. If $E^T r^{(0)} = 0$ then it follows by induction from (5.3) that

$$E^T r^{(k)} = 0, \qquad k = 1, 2, \ldots, N. \tag{5.4}$$

This shows that the residual is always orthogonal to the column space of E and so we can try to set up a cg iteration in the subspace null(E^T), which presumably will converge faster than in the whole space for a good choice of E. To set up the iteration, all we have to do is pick the two coefficients in (5.3)—which are arbitrary up until now—to force the usual orthogonalities

$$(r^{(k+1)}, r^{(j)}) = 0, \qquad j = k, \quad k - 1,$$

from which it follows as before that

$$(r^{(k+1)}, r^{(j)}) = 0, \qquad j = 0, 1, \ldots, k.$$

To make $E^T r^{(0)} = 0$, pick v arbitrarily, let $s = f - Kv$ and solve

$$\mathbf{K} d = E^T s$$

for d. Then define

$$u^{(0)} = v + Ed.$$

$r^{(0)}$ now has the required property.

Deflation is used in a very similar way to improve a preconditioned cg algorithm [32].

5.5.2. Convergence

Success with the deflation technique is dependent on a good choice of E. A basic strategy for problems with smooth coefficients and not too much anisotropy is to divide the domain into m disjoint subdomains of approximately equal areas and to associate one column of E with each subdomain. The jth column will be zero in every position except those corresponding to the unknowns in the jth subdomain, where it is equal to 1. Equation (5.4) then implies that the residuals always have zero mean in each subdomain. In the general case, the choice of E depends on the properties of K and will have to be made using the ideas required to pick the corresponding operator for algebraic multigrid.

If the maximum area of the subdomains is small, then **K** will be of large order, although convergence will be rapid. If the minimum area is too large, then **K** will be a small matrix but will not much help the convergence in relation to cg. For the preceding *piecewise constant* choice of E, it is proved in [32] that for second-order equations of the type (2.2) with smooth coefficients and meshes, the error multiplier $\rho = 1 - Ch/d$, where d^2 is the area of the largest subdomain. If there are severe anisotropies or discontinuities, their effect will show up in C and slow the convergence. E has to be chosen somewhat differently for these cases. For the Poisson equation with Dirichlet conditions and a uniform mesh [32] proves that choosing $d = O(h^{3/5})$ gives a cost of $O(N^{6/5})$ flops per digit. This cost is a slight theoretical improvement over say MILU+, for which the corresponding figure is $O(N^{5/4})$.

The deflation algorithm has the advantage of more general applicability than ILU-type methods because it is based on partial differential equation theory rather than on Gauss elimination. It would apply immediately to linear elasticity, for example. On the other hand, good choices for E are problem-dependent for degenerate cases. General software for generating E in such cases could be developed, however.

See Section 5.6.3 for some numerical results.

5.5.3. Element-by-Element Method

This section contains a brief description of an algorithm recently proposed by Hughes *et al.* [22]. Unlike the methods of the previous sections, this one is specifically for finite element equations because it uses the fact that the stiffness matrix is a sum of element stiffness matrices. In [22] the steady state solution of the ordinary differential equation system

$$W\frac{du}{dt} = Ku - f, \qquad u(0) = 0, \tag{5.5}$$

is approximated by the iteration

$$(W - \delta t K)u^{(k+1)} = Wu^{(k)} - \delta t f, \qquad u^{(0)} = 0. \tag{5.6}$$

In this section only it is assumed that K is *negative* definite. W is a positive definite matrix chosen to enhance the convergence to steady state, e.g., $W = -\text{diag}(K)$, and δt is the time step. It follows from (5.6) after some manipulation that

$$u^{(k+1)} = u^{(k)} - \delta t W^{-1/2} V W^{-1/2} r^{(k)},$$

where, as usual, $r^{(k)} = f - Ku^{(k)}$ and

$$V = (I - \delta t W^{-1/2} K W^{-1/2})^{-1}.$$

Noting for finite element systems that

$$K = \sum K_i,$$

where K_i are element matrices it follows that

$$V = \left(I - \delta t \sum_i W^{-1/2} K_i W^{-1/2}\right)^{-1}.$$

Hughes *et al.* [22] suggest the approximation

$$\left(I - \delta t \sum_i W^{-1/2} K_i W^{-1/2}\right)^{-1} \approx \prod_{i=1}^{\mu} (I - \delta t W^{-1/2} K_i W^{-1/2})^{-1}$$
$$\equiv V_1. \tag{5.7}$$

Each of the inverses on the right is easy to compute, in essence reducing to the inversion of a matrix of order equal to that of K_i. It is also suggested that the approximations (5.7) and

$$V_2 \equiv \prod_{i=\mu}^{1} (I - \delta t W^{-1/2} K_i W^{-1/2})^{-1}$$

be used alternately. The product $V_3 \equiv V_1 V_2$ is symmetric, a potentially desirable property in certain cases. For this, the algorithm becomes

$$u^{(k+1)} = u^{(k)} - \delta t W^{-1/2} V_3 W^{-1/2} r^{(k)}. \tag{5.8}$$

Of course, the convergence will be improved if partial assemblies are carried out, although the cost of the inversions will increase.

Some further refinements are to use the well-known BFGS update to improve the "search direction" $W^{-1/2} V_3 W^{-1/2}$ in (5.8) and to use an accurate line search in the improved direction. Numerical results for a plane strain problem are reported in [22].

The line search/BFGS combination actually brings the iteration closer to a pcg technique. In fact, V_3 can be used as a cg preconditioner. This is investigated in [33].

5.5.4. Domain Decomposition/Substructuring

This section describes some work that is the subject of much current attention [4, 6]. We can give only a general impression of the underlying ideas.

Substructuring refers to the technique of ordering the variables of a finite element system in such a way that the coefficient matrix takes the form

$$\begin{bmatrix} K_{11} & & & & K_{1s} \\ & K_{22} & & & K_{2s} \\ & & \ddots & & \vdots \\ & & & K_{s-1s-1} & K_{s-1s} \\ K_{1s}^T & K_{2s}^T & \cdots & K_{s-1s}^T & K_{ss} \end{bmatrix}. \tag{5.9}$$

The diagonal blocks are usually square matrices, although this is not necessary [17]. Usually, the variables belonging to each diagonal block are associated with some disjoint physical subdomains, and those belonging to the last block column are variables associated with the interfaces between (and disjoint from) the subdomains. Such an ordering is frequently convenient on physical grounds, where the subdomains may represent different parts of a physical structure. These orderings also seem attractive for parallel processing. The recent interest is mostly motivated by this last factor. Specifically, the (block) last row of (5.9) can be *simultaneously* eliminated, giving a block upper triangular system with coefficient matrix

$$\begin{bmatrix} K_{11} & & & & K_{1s} \\ & K_{22} & & & K_{2s} \\ & & \ddots & & \vdots \\ & & & K_{s-1s-1} & K_{s-1s} \\ & & & & C \end{bmatrix}, \tag{5.10}$$

where

$$C = K_{ss} - K_{1s}^T K_{11}^{-1} K_{1s} - \cdots - K_{s-1s}^T K_{s-1s-1}^{-1} K_{s-1s}. \tag{5.11}$$

Back substitution with this matrix can be achieved with another simultaneous operation once the last (vector) unknown representing the interface variables is found. It is the solution of the latter equation, say

$$Cu_B = g, \tag{5.12}$$

that is the bottleneck for the parallel implementation of this method. We are confronted with a familiar situation: if s is large, each K_{ii} can be of small order and easy to invert. But C will be of large order and expensive to form and invert. If s is small, K_{ii} can be large and more difficult to invert, whereas C will be small, fairly easily formed, and easily inverted.

For parallel implementation it is mostly the first case that is important. However, the gains from parallelism can be lost because of the problems of solving (5.12) for large C. To circumvent this, several suggestions have recently been made for solving (5.12) iteratively. It can be proved without difficulty that for the type of problem we are considering, C is positive definite so that cg is naturally suggested. The residual can be simultaneously computed using (5.11) and the $L_i U_i$ factors of K_{ii}. The main problem is then to speed the convergence of the cg iteration. The important new point is that the elements of C are not explicitly given, so that preconditioners of the earlier sections cannot be easily used.

Some recent work has dealt with the case of the Poisson equation on a uniform mesh in a rectangular domain divided horizontally by one or more mesh lines into subdomains. For this case, Fourier techniques can be used to find the eigenvalues and eigenvectors of C explicitly. On this basis, Dryja [10] proposes the choice of preconditioner M as

$$M = \sqrt{L},$$

where $L = (1/h^2)\{-1 \quad 2 \quad -1\}$ with suitable boundary conditions and proves that $M^{-1}C$ has eigenvalues independent of h. Golub and Mayers [16] generalize this to

$$M = \sqrt{L^2 + 4L}.$$

It is unclear whether these have any use for arbitrary substructures of a given domain, but it seems unlikely. On the other hand, both generalize formally to the case of several horizontal strips.

Another preconditioner that has been suggested is, for two subdomains,

$$M = K_{33} - 2K_{13}^T K_{33}^{-1} K_{13}.$$

Reference [3] contains a full account of the motivation for this and its use.

5.6. Numerical Examples

This section reports some numerical results for cg, ILU and MILU pcg, deflated cg, and amg. It is not intended that any method be selected as "best" from these results. Each has its strengths and weaknesses, which these results do not fully reveal. The idea is just to give some feeling for what efficiency can be expected in a few special cases. Also, only methods for which a fair amount of published data exists, or for which the authors have personal experience, are reported.

5.6.1. cg and pcg

The first set of numerical examples are from [7] and deal with the Poisson equation for two- and three-dimensional problems with Dirichlet boundary conditions. The domains are the square $(0, 1) \times (0, 1)$ (two dimensions) and the unit cube (three dimensions), with $h = 1/64$ (two dimensions) and $h = 1/16$ (three dimensions). The initial distribution is "random" (distribution unknown), and the termination criterion is $\|\varepsilon^{(k)}\|_2 \leq 10^{-6}\|\varepsilon^{(0)}\|_2$. Computations were all done in single precision on a PDP10. Results are shown in Tables 6.1 and 6.2. Flop counts are approximate, and setup times are not included. For comparison, solution of the two-dimensional problem by a banded Cholesky algorithm would require about 9.3×10^6 flops. Table 6.1 shows that for the

TABLE 6.1. Two Dimensions.

	cg	ILU	MILU
# itns	180	47	27
# flops/10^{-6}	3.8	1.5	0.86

TABLE 6.2. Three Dimensions.

	cg	ILU	MILU
# itns	47	18	21
# flops/10^{-6}	0.95	0.49	0.57

TABLE 6.3

	cg	ILU	MILU
# itns	109	33	23
# flops/10^{-6}	2.7	1.3*	0.92
α	-2.0	-1.97	-1.08

* This particular run took 1.37 cpu seconds.

given accuracy, MILU achieves the goal of an order of magnitude speed improvement over the direct method. In relation to cg, Table 6.1 shows that ILU and MILU, respectively, require about 0.39 and 0.23 of the computer time of the unpreconditioned algorithm. For smaller h, the MILU preconditioned cg algorithm would probably show larger gains in relation to the ILU case.

Table 6.2 shows a rather worse performance for MILU than ILU. Presumably, this can be attributed to the coarseness of the mesh. More evidence is required on this point.

Reference [7] contains comparisons with other preconditioners that we have not discussed, because they either are sensitive to problem-dependent parameters or depend on a special mesh structure being available.

A more recent set of calculations was performed by Concus *et al.*, [8]. We will give some results from this paper. The square $(0, 1) \times (0, 1)$ with $h = 1/51$ is used for Table 6.3, which reports results for the two-dimensional Poisson equation with Dirichlet boundary conditions. The initial approximation was "random" with entries in $[-1, 1]$ (presumably uniformly distributed). Iterations were terminated when $\|r^{(k)}\|_\infty \leq 10^{-6}\|r^{(0)}\|_\infty$. Double-precision arithmetic on an IBM 3081 was used. In the last row of Table 6.3, α gives the numerically computed exponent of h in the condition number of $M^{-1}K$. The improvement in the condition number of MILU over ILU is essentially the theoretical one, namely a factor of h. On the other hand, Tables 6.1 and 6.3

show that the condition number is not the whole story—the performance of ILU is too good to be explained this way. It is the clustering of the eigenvalues of $M^{-1}K$ that is responsible for the good behavior of ILU. References [7] and [8] contain direct computations of the spectrum of $M^{-1}K$ that support this.

Table 6.3 shows that ILU and MILU need, respectively, about 0.48 and 0.34 of the standard cg computation time. These are a little worse than the corresponding figures for Table 6.1. Presumably this is accounted for by the different termination rules and initial approximations.

Reference [8] contains two more difficult computations, one with a piecewise constant material tensor and another with pure Neumann boundary conditions. In the former case, the square with lower-left corner at (1/4, 1/4) and upper-right corner at (3/4, 3/4) is given the material constant 1000, and the remainder of the original square is given its previous value 1. The rest of the details are as previously for the Poisson equation. The latter example also has piecewise constant coefficients although not with wide variations, and a term σu, which ensures unique solvability. σ is relatively small and piecewise constant. For this case only, $h = 1/43$. This example is due to [38] and is used in [19] as well. Tables 6.4 and 6.5 show the results.

The strong effect of the eigenvalue distribution is evident from Table 6.4, in which the ratio of the estimated condition numbers is greater than 10^3, and yet the ratio of the number of iterations is less than 1.5. On the other hand, MILU is the more efficient algorithm in both cases. Table 6.5 shows that the third example is the most difficult. Probably the Neumann condition is the major reason for this. A banded Cholesky algorithm would require about 1.8×10^6 flops for solving the third problem, and Table 6.5 shows that MILU is nothing like the desired order of magnitude better than this. h is rather large for this example, and greater relative savings would be seen if it were decreased.

Concus *et al.* [8] deal specifically with preconditioning of block tridiagonal matrices such as those arising from five point difference formulas with mesh lines parallel to the coordinate axes. Over 30 preconditioners are compared for such problems, some of them quite sophisticated. It is very interesting to note that for Problem 3 only one preconditioner was more than twice as good

TABLE 6.4

	ILU	MILU
# itns	47	32
# flops/10^{-6}	1.9	1.3
Est. cond. no.	46770	40

TABLE 6.5

	ILU	MILU
# itns	74	53
# flops/10^{-6}	2.2	1.6

as MILU, for Problem 2 none was twice as good, and for the Poisson problem none was more than 3.3 times as good. Moreover, none of them required less storage, most of them needing quite a bit more, as well as more complex programming. This is in addition to the fact that it is not clear how to apply the majority of the algorithms to general mesh problems correctly.

5.6.2. Bercovier's Example

In this section, we shall show some results from [2] of a calculation for which negative pivots are encountered in the ILU factorization but for which formal application of pcg gives good results. According to [2] the results obtained this way are "far better" than those using either of the remedies in [24] and [27]. The example is that of an orthotropic cantilevered beam, 10 units long by 1 unit deep, in plane strain. The load f is applied at the free end. Discretization is by bilinear elements on a uniform 10×3 mesh. Letting 1 and 2 denote the principal directions of the orthotropic material, and $\varepsilon_1, \varepsilon_2, \varepsilon_{12}$ and $\sigma_1, \sigma_2, \sigma_{12}$ the corresponding strains and stresses, the elasticity matrix is defined by

$$D = \begin{bmatrix} 10^3 & 30 & 0 \\ 30 & 1 & 0 \\ 0 & 0 & 10 \end{bmatrix}.$$

Three cases are considered, in which the angle β between the x axis and direction 1 is 0°, 45°, and 90°. The negative pivots occur in the latter two cases. The results are given in Tables 6.6, 6.7, and 6.8. Unfortunately, [2] does not

TABLE 6.6 ($\beta = 0°$)

	cg	ILU
# itns	133	7
$\|r\|_2^2$	0.00080	0.00014

TABLE 6.7 ($\beta = 45°$)

	cg	ILU
# itns	>300	14
$\|r\|_2^2$	—	0.00003

TABLE 6.8 ($\beta = 90°$)

	cg	ILU
# itns	>300	19
$\|r\|_2^2$	—	0.0025

give much more detail than that reproduced here, so it is difficult to draw specific conclusions. Clearly, though, further investigation is warranted.

5.6.3. Deflation

Relatively few computations have so far been carried out with deflation; here we record its performance against cg on the model Poisson problem with Dirichlet bc. The discrete equations were solved by using n square subdomains each containing n nodes of the triangulation for $n = 9, 16, 25, 36, 49, 64, 81$, where n is the number of interior nodes along cross sections and $N = n^2$. For Table 6.9, the initial error was a smooth function, and iterations were terminated when the rms error was reduced below 10^{-6} of its initial value. The last two lines show times in relation to the cg time for $n = 64$. The first two lines are well fitted by the formulas $2.7n$ and $6\sqrt{n}$, respectively. The exponent in the second of these is accounted for by the theory in [32]. Thus, the number of iterations is rising much more slowly for the deflated case, as is the overall time. The times given include factoring times for the deflation matrix and other overheads.

Comparison with MILU is not easy, but there seems to be a small time advantage ($\approx 20\%$) with deflation applied with the preceding choice of E for the model problem. The E chosen is not the optimum one for this problem—slightly smaller subdomains are needed for that—but the difference is small. It must be mentioned that deflation requires quite a bit less work per iteration than MILU, because most of the operations carried out at the full mesh level are additions. Also, there is no problem with deflation's "breaking down" on more general problems. But deflation requires choosing E properly, and presently good choices are only known for a restricted range of problems.

5.6.4. amg

An estimate of the work required to solve the model Poisson problem can be inferred from [35]. There, for $h = 1/64$ a convergence factor $\rho = 0.054$ is reported, so that to reduce an initial error by 10^{-6} requires

$$\frac{6}{|\log_{10} 0.054|} \approx 5 \text{ iterations ("cycles").}$$

According to [35] 85 flops/mesh point/iteration are used by the amg algo-

TABLE 6.9

$n =$	9	16	25	36	49	64	81
cg itns	25	43	67	96	130	171	216
dcg itns	17	24	29	36	41	45	52
cg time	0.004	0.02	0.06	0.18	0.45	1	2.0
dcg time	0.003	0.01	0.03	0.08	0.16	0.30	0.56

rithm, giving a work count of about

$$1.7 \times 10^6 \text{ flops}$$

to reduce the error below $10^{-6}|\varepsilon^{(0)}|$.

Setup costs consists of

1. computing the interpolation weights;
2. forming the coarse grid operators;
3. constructing the coarse mesh sets C.

The third of these requires more work than the first two combined; Ruge and Stüben [35] give a figure of about 5–9 amg iterations for 1–3. Total storage is about three times that required for storage of the problem itself. Thus, for the model problem at least, the setup costs are less than twice the solution cost.

5.6.5. Comments

As already stated, no attempt should be made to order the methods on the basis of these results. Nevertheless, it is worth pointing out where each method can be expected to do well and not so well. First, it seems probable from the amg and MILU results for the model problem that neither has any clear advantage over the other, since the factor of about 2 in favor of MILU could easily change with the method of accounting. Moreover, if h becomes smaller, there should be an advantage with amg. The setup time for MILU is an order of magnitude less than for amg, and, of course, the programming is far less complex. Similar comments apply to deflation vis-à-vis amg, for both speed and storage for the model problem.

In more difficult examples, it seems that amg deteriorates less rapidly than MILU, although the latter does what can be considered a reasonable job in most cases. Again, there may be an advantage in convergence speed for amg when h becomes relatively small, although the setup costs remain to be neutralized. For problems with more general geometry, and for three-dimensional problems, MILU seems not to have been adequately tested. For three-dimensional problems, amg also has not been tested so far.

Appendix

In this appendix, we summarize the properties of the classical conjugate gradient algorithm, and obtain the second-order form used in the paper. For solving

$$Ku = f,$$

where K is symmetric positive definite and $N \times N$, the algorithm is the following:

$$
\begin{aligned}
&u^{(0)} \text{ is arbitrary}, \qquad r^{(0)} = f - Ku^{(0)}, \qquad p^{(0)} = r^{(0)}, \\
&u^{(k+1)} = u^{(k)} + \gamma_k p^{(k)}, \qquad k = 0, 1, \ldots, \\
&p^{(k+1)} = r^{(k+1)} + \beta_k p^{(k)}, \qquad k = 0, 1, \ldots.
\end{aligned} \tag{A1}
$$

The coefficients are given by

$$\gamma_k = \frac{(r^{(k)}, p^{(k)})}{\pi_k}, \qquad \beta_k = \frac{-(r^{(k+1)}, Kp^{(k)})}{\pi_k},$$

where

$$\pi_k = (p^{(k)}, Kp^{(k)}).$$

The *direction vectors* $p^{(k)}$ and the residuals $r^{(k)}$ satisfy the following orthogonality relations:

$$
\begin{aligned}
&(1) \quad (p^{(k)}, Kp^{(j)}) = 0, \qquad j = 0, 1, \ldots, k-1, \\
&(2) \quad (r^{(k)}, r^{(j)}) = 0, \qquad j = 0, 1, \ldots, k-1.
\end{aligned} \tag{A2}
$$

From (A2) it follows that $r^{(N)} = 0$. Some earlier residual can be zero in certain cases, but we do not need to worry about this here. From [26] we have the following bound for the "energy" $E(\varepsilon^{(k)}) = (\varepsilon^{(k)}, K\varepsilon^{(k)})$ of the kth error:

Theorem

$$E(\varepsilon^{(k)}) \le 4\rho^{2k} E(\varepsilon^{(0)}),$$

where $\rho = (1 - \sqrt{\kappa})/(1 + \sqrt{\kappa})$ *and* $\kappa = \lambda_{\min}/\lambda_{\max}$.

For the model Poisson problem, the eigenvalue ratio $\kappa = O(h^2)$, giving $\rho = 1 - Ch$ as mentioned in Section 5.2.

To obtain the second-order form of the algorithm, apparently used first in [36] (see also [9]), eliminate $p^{(k)}$ from (A1) using

$$p^{(k)} = \beta_{k-1} p^{(k-1)} + r^{(k)},$$

and

$$\gamma_{k-1} p^{(k-1)} = u^{(k)} - u^{(k-1)}.$$

The result is

$$u^{(k+1)} = \omega_k u^{(k)} + \omega'_k u^{(k-1)} + \alpha_k \omega_k r^{(k)}, \tag{A3}$$

where $\omega_k + \omega'_k = 1$, $\omega'_k = -\gamma_k \beta_{k-1}/\gamma_{k-1}$, and $\alpha_k \omega_k = \gamma_k$. We define $\beta_{-1} = 0$ for consistency with (A1). From (A3) it follows that

$$r^{(k+1)} = \omega_k r^{(k)} + \omega'_k r^{(k-1)} - \alpha_k \omega_k K r^{(k)}. \tag{A4}$$

Since the coefficients in (A4) involve the vectors $p^{(k)}$, we shall determine them afresh, directly from (A4). To do this, note that the property (A2) implies in particular that

$$(r^{(k+1)}, r^{(k)}) = (r^{(k+1)}, r^{(k-1)}) = 0.$$

In conjunction with (A4) the first implies

$$\alpha_k = \frac{(r^{(k)}, r^{(k)})}{(r^{(k)}, Kr^{(k)})}. \tag{A5}$$

Similarly, the second implies

$$0 = \omega_k'(r^{(k-1)}, r^{(k-1)}) - \alpha_k \omega_k (r^{(k-1)}, Kr^{(k)}),$$

$$\therefore 0 = \omega_k'(r^{(k-1)}, r^{(k-1)}) - \alpha_k \omega_k (Kr^{(k-1)}, r^{(k)}),$$

by the symmetry of K. This last term can be rewritten, using (A4) with $k := k - 1$ and taking its inner product with $r^{(k)}$. Then

$$0 = \omega_k'(r^{(k-1)}, r^{(k-1)}) + \left(\frac{\alpha_k \omega_k}{\alpha_{k-1}\omega_{k-1}}\right)(r^{(k)}, r^{(k)})$$

from which it follows that

$$\frac{1}{\omega_k} = 1 - \frac{1}{\omega_{k-1}} \frac{\alpha_k}{\alpha_{k-1}} \frac{(r^{(k)}, r^{(k)})}{(r^{(k-1)}, r^{(k-1)})}. \tag{A6}$$

(A3) with (A5)–(A6) and the initial condition $\omega_0 = 1$ now define the second-order form of the algorithm.

It can be proved by induction that the property (A2) is preserved for (A3), (A5)–(A6) with $\omega_0 = 1$. There remains the question of whether the $u^{(k)}$ of the standard form and $u^{(k)}$ of the second-order form are indeed equal. In fact, all of the iterates of the two methods are identical (for exact arithmetic). This can also be proved by induction, although the calculations are longer than for the proof of (A2).

Acknowledgments. This work was supported by the Air Force Office of Scientific Research, under Grant AFOSR-84-0137, and by the National Aeronautics and Space Administration, under Contract NAS1-18107 to ICASE, Hampton, VA 23665, U.S.A.

References

1. O. Axelsson and V.A. Barker, *Finite Element Solution of Boundary Value Problems*, Academic Press, Orlando, 1984.
2. M. Bercovier and A. Rosenthal, "Using the conjugate gradient method with preconditioning for solving FEM approximations of elasticity problems," *Engrg. Comput.*, **3** (1986), 77.
3. P.E. Bjorstad and O.B. Widlund, "Iterative methods for the solution of elliptic problems on regions partitioned into substructures," Technical Report 136, Courant Institute of Mathematical Sciences, New York University, 1984.

4. J.H. Bramble, J.E. Pasciak, and A.H. Schatz, "The construction of preconditioners for elliptic problems by substructuring," *Math. Comp.*, **47** (1986), 103.
5. N.I. Buleev, "A numerical method for the solution of two-dimensional and three-dimensional equations of diffusion," *Mat. Sb.*, **51** (1960), 227.
6. T.F. Chan and D.C. Resasco, "A survey of preconditioners for domain decomposition," Research Report YALEU/DCS/RR-414, Department of Computer Science, Yale University, New Haven, 1985.
7. R. Chandra, "Conjugate gradient methods for partial differential equations," Ph.D. dissertation, Department of Computer Science, Yale University, New Haven, 1978.
8. P. Concus, G.H. Golub, and G. Meurant, "Block preconditioning for the conjugate gradient method," *SIAM J. Sci. Statist. Comput.*, **6** (1985), 220.
9. P. Concus, G.H. Golub, and D.P. O'Leary, "A generalized conjugate gradient method for the numerical solution of elliptic partial differential equations," *Sparse Matrix Computations* (J.R. Bunch and D.J. Rose, eds.), Academic Press, New York, 1976, p. 309.
10. M. Dryja, "A capacitance matrix method for the Dirichlet problem on polygonal region," *Numer. Math.*, **39** (1982), 51.
11. T. Dupont, R.P. Kendall, and H.H. Rachford, Jr., "An approximate factorization procedure for solving self-adjoint elliptic difference equations," *SIAM J. Numer. Anal.*, **5** (1968), 559.
12. T. Dupont, "A factorization procedure for the solution of elliptic difference equations," *SIAM J. Numer. Anal.*, **5** (1968), 753.
13. H.C. Elman, "Iterative methods for large, sparse, nonsymmetric systems of linear equations," Ph.D. dissertation, Department of Computer Science, Yale University, New Haven, 1982.
14. G.E. Forsythe and W.R. Wasow, *Finite-Difference Methods for Partial Differential Equations*, Wiley, New York, 1960.
15. S.P. Frankel, "Convergence rates of iterative treatments of partial differential equations," *Math. Tables Aids Comput.*, **4** (1950), 65.
16. G.H. Golub and D. Mayers, "The use of pre-conditioning over irregular regions," Lecture at the *6th International Conference on Computing Methods in Applied Sciences and Engineering* (*Versailles*, 1983).
17. M.D. Gunzburger and R.A. Nicolaides, "On substructuring algorithms and solution techniques for the numerical approximation of partial differential equations," *Appl. Numer. Math.*, **2** (1986).
18. I. Gustafsson, "A class of first-order factorization methods," *BIT*, **18** (1978), 142.
19. I. Gustafsson, "Modified incomplete Cholesky methods," *Preconditioning Methods: Analysis and Applications* (D.J. Evans, ed.), Gordon and Breach, New York, 1983, p. 265.
20. W. Hackbusch, *Multi-Grid Methods and Applications*, Springer-Verlag, Berlin, 1985.
21. L.A. Hageman and D.M. Young, *Applied Iterative Methods*, Academic Press, New York, 1981.
22. T.J.R. Hughes, I. Levit, and J. Winget, "Element-by-element implicit algorithms for heat conduction," *J. Engrg. Mech.*, **109** (1983), 576.
23. A. Jameson, "Multigrid algorithms for compressible flow calculations," MAE Report 1743, Department of Mechanical and Aerospace Engineering, Princeton University, Princeton, NJ, 1985.

24. D.S. Kershaw, "The incomplete Cholesky conjugate gradient method for the iterative solution of systems of linear equations," *J. Comput. Phys.*, **26** (1978), 43.
25. R. Löhner and K. Morgan, "An unstructured multigrid method for elliptic problems," *Internat. J. Numer. Methods Engrg.* **24** (1987), 101.
26. D.G. Luenberger, *Linear and Nonlinear Programming*, 2nd ed., Addison-Wesley, Reading, MA, 1984.
27. T.A. Manteuffel, "An incomplete factorization technique for positive definite linear systems," *Math. Comp.*, **34** (1980), 473.
28. J.A. Meijerink and H.A. van der Vorst, "An iterative solution method for linear systems of which the coefficient matrix is a symmetric M-matrix," *Math. Comp.*, **31** (1977), 148.
29. J.A. Meijerink and H.A. van der Vorst, "Guidelines for the usage of incomplete decompositions in solving sets of linear equations as they occur in practical problems," *J. Comput. Phys.*, **44** (1981), 134.
30. R.A. Nicolaides, "On the l^2 convergence of an algorithm for solving finite element equations," *Math. Comp.*, **31** (1977), 892.
31. R.A. Nicolaides, "On some theoretical and practical aspects of multigrid methods," *Math. Comp.*, **33** (1979), 933.
32. R.A. Nicolaides, "Deflation of conjugate gradients with applications to boundary value problems," *SIAM J. Numer. Anal.*, **24** (1987), 355.
33. B. Nour-Omid and B.N. Parlett, "Element preconditioning using splitting techniques," *SIAM J. Sci. Statist. Comput.*, **6** (1985), 761.
34. L.F. Richardson, "The approximate arithmetical solution by finite differences of physical problems involving differential equations, with an application to the stresses in a masonry dam," *Philos. Trans. Roy. Soc. London Ser. A*, **210** (1910), 307, and *Proc. Roy. Soc. London Ser. A*, **83** (1910), 335.
35. J. Ruge and K. Stüben, "Efficient solution of finite difference and finite element equations by algebraic multigrid," in *Proceedings of the Multigrid Conference (Bristol*, 1983).
36. H. Rutishauser, "Theory of gradient methods," in *Refined Iterative Methods for Computation of the Solution and the Eigenvalues of Self-Adjoint Boundary Value Problems*, (by M. Engeli, T. Ginsburg, H. Rutishauser, and E. Stiefel), Birkhäuser, Basel, 1959, p. 24.
37. H.L. Stone, "Iterative solution of implicit approximations of multidimensional partial differential equations," *SIAM J. Numer. Anal.*, **5** (1968), 530.
38. R.S. Varga, *Matrix Iterative Analysis*, Prentice-Hall, Englewood Cliffs, N.J., 1962.
39. R.S. Varga, "Factorization and normalized iterative methods," in *Boundary Value Problems in Differential Equations* (R.E. Langer, ed.), University of Wisconsin Press, Madison, 1960, p. 121.
40. D.M. Young, *Iterative Solution of Large Linear Systems*, Academic Press, New York, 1971.

CHAPTER 6

Mathematical Aspects of Finite Element Methods for Incompressible Viscous Flows

Max D. Gunzburger

One of the most successful and well-developed mathematical theories concerning finite element methods (FEM) is that connected with incompressible flow problems. The success of this theory lies not only in the accumulated elegant mathematical results, but also in its impact on practical computations. The outstanding monograph by Girault and Raviart [15] give a rigorous account of this theory, and to this day remain the definitive sources.

In this survey we examine certain mathematical aspects of finite element methods for the approximate solution of incompressible flow problems. Our principal goal is to present some of the important mathematical results that are relevant to practical computations. In so doing we also discuss useful algorithms. Because of space limitations we focus on the steady primitive variable formulation. Moreover, even within this narrow context, we concentrate on only one of the very many different known approaches. Some other approaches are discussed in, e.g., [15, 24].

We state at the outset that we make no attempt to be comprehensive in our coverage or in our attributions. To anyone who takes offense, we sincerely apologize.

6.1. The Primitive Variable Formulation

Let Ω denote a bounded, possibly multiply connected, domain in $\mathbb{R}^d$, $d = 2$ or 3, and let Γ denote its boundary. As a prototype for incompressible flow problems we consider the Navier–Stokes equations

$$\mathbf{u} \cdot \operatorname{grad} \mathbf{u} + \operatorname{grad} p = \nu \Delta \mathbf{u} + \mathbf{f} \quad \text{in } \Omega, \tag{1.1}$$

together with the incompressibility constraint

$$\operatorname{div} \mathbf{u} = 0 \quad \text{in } \Omega, \tag{1.2}$$

and the boundary condition

$$\mathbf{u} = \mathbf{0} \quad \text{on } \Gamma, \tag{1.3}$$

where $\mathbf{u}$ is the velocity field, p the pressure, $\mathbf{f}$ the given body force, and ν the given constant kinematic viscosity. In (1.1) the constant density has been absorbed into the pressure. Whenever $\mathbf{u}$ and p represent nondimensionalized variables, then ν is the inverse of the Reynolds number Re.

After our detailed discussion of the approximation of solutions of (1.1)–(1.3) by finite element methods, we will consider other incompressible flow formulations, especially as they concern boundary conditions other than (1.3).

6.1.1. Function Spaces, Norms, and Forms

In order to introduce a Galerkin-type weak formulation through which a finite element approximation is determined, we first need to define some function spaces, associated norms, and forms involving functions belonging to those spaces. Lucid and more detailed accounts concerning these spaces may be found in, e.g., [1, 15, 21].

First, we denote by $L^2(\Omega)$ the space of functions that are square integrable over Ω and that is equipped with the inner product and norm

$$(p, q) = \int_\Omega pq \quad \text{and} \quad \|q\|_0 = (q, q)^{1/2},$$

respectively. We then define the constrained space

$$L_0^2(\Omega) = \left\{ q \in L^2(\Omega) \,\middle|\, \int_\Omega q = 0 \right\}.$$

Thus $L_0^2(\Omega)$ consists of square integrable functions with zero mean over Ω. This space is used in connection with the pressure; such a constraint is needed since it is clear from (1.1)–(1.3) that the pressure can be determined only up to an arbitrary constant. Other constraints, e.g., fixing the pressure at a given point, may be used instead without effecting any appreciable change in the results discussed later. Next we define the Sobolev spaces

$$H^k(\Omega) = \{q \in L^2(\Omega) | D^s q \in L^2(\Omega) \text{ for } s = 1, \ldots, k\},$$

where D^s denotes any and all derivatives of order s. Thus $H^k(\Omega)$ consists of square integrable functions all of whose derivatives of order up to k are also square integrable. $H^k(\Omega)$ comes equipped with the norm

$$\|q\|_k = \left\{ \|q\|_0^2 + \sum \|D^s q\|_0^2 \right\}^{1/2},$$

where the summation extends over all possible derivatives of order k or less. Clearly $H^0(\Omega) = L^2(\Omega)$. Of particular interest is the space $H^1(\Omega)$ consisting of functions with one square integrable derivative and the subspace

$$H_0^1(\Omega) = \{q \in H^1(\Omega) | q = 0 \text{ on } \Gamma\},$$

whose elements have one square integrable derivative over Ω and which vanish on the boundary Γ. These spaces have the associated norm

$$\|q\|_1 = \left\{ \|q\|_0^2 + \sum_{i=1}^{d} \left\| \frac{\partial q}{\partial x_i} \right\|_0^2 \right\}^{1/2}. \tag{1.4}$$

We note that for functions belonging to $H_0^1(\Omega)$ the seminorm

$$|q|_1 = \left\{ \sum_{i=1}^{d} \left\| \frac{\partial q}{\partial x_i} \right\|_0^2 \right\}^{1/2} \tag{1.5}$$

is actually a norm equivalent to (1.4) and thus, for such functions, (1.5) may be used instead of (1.4).

For vector-valued functions we use the spaces

$$\mathbf{H}^k(\Omega) = [H^k(\Omega)]^d = \{\mathbf{v} \,|\, v_i \in H^k(\Omega) \text{ for } i = 1, \ldots, d\},$$

and

$$\mathbf{H}_0^1(\Omega) = [H_0^1(\Omega)]^d = \{\mathbf{v} \,|\, v_i \in H_0^1(\Omega) \text{ for } i = 1, \ldots, d\}.$$

For example, $\mathbf{H}^k(\Omega)$ consists of vector-valued functions each of whose components belongs to $H^k(\Omega)$. $\mathbf{H}^k(\Omega)$ is equipped with the norm

$$\|\mathbf{v}\|_k = \left\{ \sum_{i=1}^{d} \|v_i\|_k^2 \right\}^{1/2};$$

alternately, $\mathbf{H}_0^1(\Omega)$ has the norm

$$|\mathbf{v}|_1 = \left\{ \sum_{i=1}^{d} |v_i|_1^2 \right\}^{1/2}.$$

Also, the inner product for functions belonging to $\mathbf{L}^2(\Omega) = \mathbf{H}^0(\Omega) = [L^2(\Omega)]^d$ is given by

$$(\mathbf{u}, \mathbf{v}) = \int_\Omega \mathbf{u} \cdot \mathbf{v},$$

where no ambiguity resulting from using the same notation for the inner product of both scalar and vector-valued functions is possible.

We now define the bilinear forms

$$a(\mathbf{u}, \mathbf{v}) = \nu \int_\Omega \operatorname{grad} \mathbf{u} : \operatorname{grad} \mathbf{v} \qquad \text{for all} \quad \mathbf{u}, \mathbf{v} \in \mathbf{H}_0^1(\Omega), \tag{1.6}$$

and

$$b(\mathbf{v}, q) = -\int_\Omega q \operatorname{div} \mathbf{v} \qquad \text{for all} \quad \mathbf{v} \in \mathbf{H}_0^1(\Omega) \quad \text{and} \quad q \in L_0^2(\Omega), \tag{1.7}$$

and the trilinear form

$$c(\mathbf{w}, \mathbf{u}, \mathbf{v}) = \int_\Omega \mathbf{w} \cdot \operatorname{grad} \mathbf{u} \cdot \mathbf{v} \qquad \text{for all} \quad \mathbf{u}, \mathbf{v}, \mathbf{w} \in \mathbf{H}_0^1(\Omega). \tag{1.8}$$

In (1.6) and (1.8) we have that $(\operatorname{grad} \mathbf{u})_{ij} = \partial u_j / \partial x_i$,

$$\operatorname{grad} \mathbf{u} : \operatorname{grad} \mathbf{v} = \sum_{i,j=1}^{d} \frac{\partial u_i}{\partial x_j} \frac{\partial v_i}{\partial x_j} \quad \text{and} \quad \mathbf{w} \cdot \operatorname{grad} \mathbf{u} \cdot \mathbf{v} = \sum_{i,j=1}^{d} w_j \frac{\partial u_i}{\partial x_j} v_i.$$

Using the bilinear form $b(\cdot, \cdot)$, we can define the subspace

$$Z = \{v \in \mathbf{H}_0^1(\Omega) | b(\mathbf{v}, q) = 0 \text{ for all } q \in L_0^2(\Omega)\},$$

which consists of (weakly) *divergence-free functions*, i.e., functions whose divergence is orthogonal to all $L_0^2(\Omega)$ functions. Certainly, any divergence-free function, in the strong sense, belongs to Z.

6.1.2. A Galerkin-Type Weak Formulation

The most commonly used weak formulation of (1.1)–(1.3) is the following. Given $\mathbf{f} \in \mathbf{L}^2(\Omega)$, we seek $\mathbf{u} \in \mathbf{H}_0^1(\Omega)$ and $p \in L_0^2(\Omega)$ such that

$$a(\mathbf{u}, \mathbf{v}) + c(\mathbf{u}, \mathbf{u}, \mathbf{v}) + b(\mathbf{v}, p) = (\mathbf{f}, \mathbf{v}) \quad \text{for all} \quad \mathbf{v} \in \mathbf{H}_0^1(\Omega), \tag{1.9}$$

$$b(\mathbf{u}, q) = 0 \quad \text{for all} \quad q \in L_0^2(\Omega). \tag{1.10}$$

By virtue of (1.10) we see that the solution $\mathbf{u}$ belongs to Z.

We note that $\mathbf{L}^2(\Omega)$ is not the largest function space for the data $\mathbf{f}$ such that the problem (1.9)–(1.10) makes sense; indeed, all that is required of the data is that the right-hand side of (1.9) be bounded, and this is possible for some functions that are not square integrable. However, for our purposes, $\mathbf{f} \in \mathbf{L}^2(\Omega)$ is sufficiently general.

It can be easily verified that whenever a pair $\mathbf{u}$, p satisfies (1.9)–(1.10) and is sufficiently smooth to allow for the appropriate integrations by parts, then $\mathbf{u}$, p is also a solution of (1.1)–(1.3). Of course, (1.9)–(1.10) admit solutions that are not sufficiently smooth to be solutions of (1.1)–(1.3); hence the terminology *weak formation* and *generalized solution* are applied to (1.9)–(1.10) and their solution, respectively. On the other hand, it is also clear that any solution of (1.1)–(1.3), i.e., a *strong solution*, satisfies (1.9)–(1.10).

For the weak formulation (1.9)–(1.10), the boundary condition (1.3) is an *essential* one, i.e., it must be imposed on the candidate solution functions. In Section 6.4.3 we will discuss the *natural* boundary conditions associated with the weak formulation (1.9)–(1.10).

We will not enter into details concerning the existence, uniqueness, continuous dependence on data and regularity of solutions of (1.9)–(1.10). Such results may be found in, e.g., the definitive treatise of Teman [23]. Furthermore, many of these results are similar to those discussed later for the approximate problem.

6.2. The Finite Element Problem and the Div-Stability Condition

6.2.1. The Discrete Finite Element Problem

Once the Galerkin formulation (1.9)–(1.10) is established, the approximate problem that determines the finite element solution is defined in the usual manner. First, we choose the approximating finite element spaces, or more precisely, a family of finite element spaces, $\mathbf{V}^h$ and S^h for the velocity and pressure, respectively. Here h is a parameter that is usually related to the size of the grid associated with the finite element partitioning of Ω. Then we require that (1.9)–(1.10) hold for functions belonging to these finite-dimensional spaces, i.e., we seek $\mathbf{u}^h \in \mathbf{V}^h$ and $p^h \in S^h$ such that

$$a(\mathbf{u}^h, \mathbf{v}^h) + c(\mathbf{u}^h, \mathbf{u}^h, \mathbf{v}^h) + b(\mathbf{v}^h, p^h) = (\mathbf{f}, \mathbf{v}^h) \quad \text{for all} \quad \mathbf{v}^h \in \mathbf{V}^h, \tag{2.1}$$

and

$$b(\mathbf{u}^h, q^h) = 0 \qquad \text{for all} \quad q^h \in S^h. \tag{2.2}$$

If $\mathbf{V}^h$ and S^h are subspaces of the underlying infinite-dimensional spaces of (1.9)–(1.10), i.e., if $\mathbf{V}^h \subset \mathbf{H}_0^1(\Omega)$ and $S^h \subset L_0^2(\Omega)$, then the finite element solution defined by (2.1)–(2.2) is said to be *conforming*. Otherwise, i.e., if $\mathbf{V}^h \not\subset \mathbf{H}_0^1(\Omega)$ and/or $S^h \not\subset L_0^2(\Omega)$, the method is said to be *nonconforming*. We will restrict our attention to examples of the former.

Once we choose specific bases for $\mathbf{V}^h$ and S^h, (2.1)–(2.2) are equivalent to a *nonlinear system of algebraic equations.* Indeed, if $\{q_j(\mathbf{x})\}$, $j = 1, \ldots, J$, and $\{\mathbf{v}_k(\mathbf{x})\}$, $k = 1, \ldots, K$, denote bases sets for S^h and $\mathbf{V}^h$, respectively, we may then write

$$p^h = \sum_{j=1}^{J} \alpha_j q_j(\mathbf{x}) \quad \text{and} \quad \mathbf{u}^h = \sum_{k=1}^{K} \beta_k \mathbf{v}_k(\mathbf{x})$$

for some constants α_j, $j = 1, \ldots, J$, and β_k, $k = 1, \ldots, K$. Substituting into (2.1)–(2.2) then yields

$$\sum_{k=1}^{K} a(\mathbf{v}_k, \mathbf{v}_l)\beta_k + \sum_{k,m=1}^{K} c(\mathbf{v}_m, \mathbf{v}_k, \mathbf{v}_l)\beta_k\beta_m + \sum_{j=1}^{J} b(\mathbf{v}_l, q_j)\alpha_j$$
$$= (\mathbf{f}, \mathbf{v}_l) \qquad \text{for} \quad l = 1, \ldots, K, \tag{2.3}$$

and

$$\sum_{k=1}^{K} b(\mathbf{v}_k, q_i)\beta_k = 0 \qquad \text{for} \quad i = 1, \ldots, J, \tag{2.4}$$

which constitute a nonlinear algebraic, in fact, quadratic, system of $J + K$ equations for the $J + K$ unknowns $\alpha_j, j = 1, \ldots, J$, and β_k, $k = 1, \ldots, K$. Note that the discrete continuity equation (2.2) yields the $J \times K$ rectangular *linear* system (2.4).

6.2.2. The div-Stability Condition

In the positive definite case, e.g., for the equations of linear elasticity, the mere inclusion of the finite element spaces within the underlying function spaces is essentially sufficient to assure that the approximations are well defined and are as accurate as possible for the type of finite elements functions being used. Here the inclusions $\mathbf{V}^h \subset \mathbf{H}_0^1(\Omega)$ and $S^h \subset L_0^2(\Omega)$ are not by themselves sufficient to produce stable, meaningful approximations. We find ourselves in the realm of what are known as *mixed finite element methods.*

There are a number of conditions that the elements belonging to the finite element spaces should satisfy. Most of them, e.g., the boundedness of the various bilinear and trilinear forms, are easily satisfied by conforming finite element spaces. The one condition that presents a problem has the following mathematical realization:

Given any $q^h \in S^h$,

$$\sup_{\mathbf{0} \neq \mathbf{v}^h \in \mathbf{V}^h} \left\{ \frac{b(\mathbf{v}^h, q^h)}{\|\mathbf{v}^h\|_1} \right\} \geq \gamma \|q^h\|_0, \tag{2.5}$$

where the constant $\gamma > 0$ may be chosen independent of h and of the particular choice of $q^h \in S^h$.

This condition may be equivalently expressed in the form:

Given any $q^h \in S^h$ there exists a nonzero $\mathbf{v}^h \in \mathbf{V}^h$ such that

$$b(\mathbf{v}^h, q^h) \geq \gamma \|q^h\|_0 \|\mathbf{v}^h\|_1, \tag{2.6}$$

where the constant $\gamma > 0$ may be chosen independent of h and of the particular choice of $q^h \in S^h$.

Of course, for each q^h a different $\mathbf{v}^h$ may be used in order to satisfy (2.6).

The condition (2.5), or equivalently (2.6), is variously known as the *Ladyzhenskaya–Babuška–Brezzi* (LBB) or the *inf–sup* condition, the latter designation following from the third equivalent form:

There exists a $\gamma > 0$, independent of h, such that

$$\inf_{0 \neq q^h \in S^h} \sup_{\mathbf{0} \neq \mathbf{v}^h \in \mathbf{V}^h} \left\{ \frac{b(\mathbf{v}^h, q^h)}{\|\mathbf{v}^h\|_1 \|q^h\|_0} \right\} \geq \gamma. \tag{2.7}$$

We will refer to any of the equivalent statements (2.5)–(2.7) as the condition for *div-stability*. Note that these have no relation to the nonlinearity of the Navier–Stokes equations and, in fact, the possible problems its satisfaction poses are shared by the linear equations of Stokes flow.

Associated with the finite element spaces $\mathbf{V}^h$ and S^h and the bilinear form $b(\cdot, \cdot)$ we have the subspace

$$Z^h = \{\mathbf{v} \in \mathbf{V}^h | b(\mathbf{v}^h, q^h) = 0 \text{ for all } q^h \in S^h\}.$$

of *discretely divergence-free functions*. In general, $\mathbf{Z}^h \not\subset \mathbf{Z}$, even when $\mathbf{V}^h \subset \mathbf{H}_0^1(\Omega)$ and $S^h \subset L_0^2(\Omega)$, i.e., discretely solenoidal functions are not necessarily solenoidal. This is, of course, entirely analogous to the finite difference case, e.g., a function satisfying a difference approximation to the incompressibility constraint is not in general solenoidal. A measure of the "angle" between the spaces $\mathbf{Z}^h$ and $\mathbf{Z}$ is given by

$$\Theta = \sup_{\substack{\mathbf{z}^h \in \mathbf{Z}^h \\ \|\mathbf{z}^h\|_1 = 1}} \inf_{\mathbf{z} \in \mathbf{Z}} \|\mathbf{z} - \mathbf{z}^h\|_1. \tag{2.8}$$

In general, $0 \leq \Theta \leq 1$, which is easily seen by observing that for $\mathbf{Z}^h \in \mathbf{Z}, \Theta = 0$, and that by choosing $\mathbf{z} = 0, \Theta = 1$.

Note that because of (2.2), the approximate velocity $\mathbf{u}^h \in \mathbf{Z}^h$, i.e., $\mathbf{u}^h$ is discretely solenoidal. However, since in general $\mathbf{Z}^h \not\subset \mathbf{Z}$, div $\mathbf{u}^h \neq 0$. Loosely speaking, the div-stability condition (2.5) ensures, as $h \to 0$ at least, that discretely solenoidal functions tend to solenoidal functions.

6.2.3. Error Estimates and Other Results Concerning the Approximate Solution

We now present some of the available mathematical results concerning the solution $\mathbf{u}^h$, p^h of the finite element problem (2.1)–(2.2). Here we *assume* that the chosen finite element spaces $\mathbf{V}^h$ and S^h satisfy the div-stability condition (2.5). Subsequently, we will look into the issue of verifying that condition. The summary presented is based on the detailed analysis found in [11, 14, 15, 19].

First, for any $\mathbf{f} \in L_0^2(\Omega)$, (2.1)–(2.2) has a solution $\mathbf{u}^h$, p^h, provided that the div-stability condition (2.5) holds. However, we can prove that the solution is unique only for "sufficiently small" data $\mathbf{f}$ or "sufficiently large" viscosity ν. More precisely, let

$$\kappa = \sup_{\mathbf{u}^h, \mathbf{v}^h, \mathbf{w}^h \in \mathbf{V}^h} \left\{ \frac{a(\mathbf{w}^h, \mathbf{u}^h, \mathbf{v}^h)}{|\mathbf{u}^h|_1 |\mathbf{v}^h|_1 |\mathbf{w}^h|_1} \right\}.$$

For standard choices of finite element spaces κ can be shown to be independent of h and, in fact, depends only on $\Omega \subset R^d$ and d. Then, we can show that (2.1)–(2.2) has a unique solution whenever

$$\frac{\kappa}{\nu^2} \sup_{\mathbf{v}^h \in \mathbf{V}^h} \left\{ \frac{\int_\Omega \mathbf{f} \cdot \mathbf{v}^h}{|\mathbf{v}^h|_1} \right\} \leq 1.$$

This condition is very similar to the one that is needed to show the uniqueness of the solution of (1.9)–(1.10) and in fact the latter implies the former; i.e., whenever (1.9)–(1.10) can be shown to have a unique solution, then, provided the div-stability condition is satisfied, (2.1)–(2.2) also has a unique solution.

When we can show that (1.9)–(1.10) has a unique solution, it can also be shown that the finite element solution of (2.1)–(2.2) converges to that solution. In addition, something can be said about the convergence of the finite element solution even when (1.9)–(1.10) does not possess a unique solution. For details, see [15].

Error estimates can also be derived. Provided that the div-stability condition is satisfied, we have that

$$\|\mathbf{u} - \mathbf{u}^h\|_1 \leq C_1 \inf_{\mathbf{v}^h \in \mathbf{V}^h} \|\mathbf{u} - \mathbf{v}^h\|_1 + C_2 \Theta \inf_{q^h \in S^h} \|p - q^h\|_0, \tag{2.9}$$

and

$$\|p - p^h\|_0 \leq C_3 \inf_{\mathbf{v}^h \in \mathbf{V}^h} \|\mathbf{u} - \mathbf{v}^h\|_1 + C_4 \inf_{q^h \in S^h} \|p - q^h\|_0, \tag{2.10}$$

where Θ is defined in (2.8) and C_i, $i = 1, \ldots, 4$, are constants independent of h. These estimates are optimal for the "graph norm" $\|\mathbf{u}\|_1 + \|p\|_0$ of functions belonging to $\mathbf{H}_0^1(\Omega) \times L_0^2(\Omega)$ in the sense that the rate of convergence of the finite element solution, measured in this norm, is the same as that of the best approximation to $\mathbf{u}$ and p out of $\mathbf{V}^h$ and S^h, respectively.

If the solution of (1.9)–(1.10), or, more precisely, of the linearized adjoint problem corresponding to (1.9)–(1.10), is sufficiently regular, then we can obtain an improved velocity error estimate in the $\mathbf{L}^2(\Omega)$-norm, namely

$$\|\mathbf{u} - \mathbf{u}^h\|_0 \leq C_5 h \|\mathbf{u} - \mathbf{u}^h\|_1, \tag{2.11}$$

where again C_5 is independent of h.

We see that once the div-stability condition is satisfied, the error in the finite element approximation depends only on the ability to approximate in the chosen finite element subspaces. In general, (2.9)–(2.10) indicates that the velocity and pressure errors are coupled. Furthermore, we find that it is efficient to equilibrate the rates of convergence of the two terms on the right-hand side of (2.9)–(2.10). For this reason, we would like to use, for example, polynomials of one degree higher for the velocity components than those used for the pressure. As a final comment, we note that the constants appearing in (2.9)–(2.10) are in general proportinal to $1/\gamma$, where γ is the stability constant appearing in (2.5).

6.2.4. Verifying the div-Stability Condition

For particular choices of $\mathbf{V}^h$ and S^h, it is usually *not* an easy matter to verify that the div-stability condition holds. To accomplish this task for families of such spaces is even more difficult. Here, we sketch three techniques for verifying the div-stability condition.

6.2.4.1. *Fortin's Method*

One seemingly attractive method of showing that the div-stability condition holds is due to Fortin, who has shown [12] that the div-stability condition (2.5) is equivalent to the existence of a linear operator Π^h from $\mathbf{H}_0^1(\Omega) \rightarrow \mathbf{V}^h$, such that given any $\mathbf{v} \in \mathbf{H}_0^1(\Omega)$,

$$b(\Pi^h \mathbf{v}, q^h) = b(\mathbf{v}, q^h) \qquad \text{for all} \quad q^h \in S^h,$$

and

$$\|\Pi^h \mathbf{v}\|_1 \leq C \|\mathbf{v}\|_1,$$

where the constant $C > 0$ may be chosen independent of h and of the particular choice of $\mathbf{v} \in \mathbf{H}_0^1(\Omega)$. Thus the task of verifying the div-stability condition (2.5) is reduced to showing the existence of the operator Π^h; unfortunately, although the latter task has been accomplished in a few specific settings, in general, it is also a difficult thing to do.

6.2.4.2. *Verfürth's Method*

Verfürth [26] has developed a method for verifying the div-stability condition (2.5) that applies to the case of *continuous* discrete pressure spaces.

Specifically, if $S^h \subset H^1(\Omega) \cap L_0^2(\Omega)$, he starts out by combining the inverse inequality, see, e.g., [9],

$$|\mathbf{v}^h|_1 \leq C_1 h^{-1} \|\mathbf{v}^h\|_0 \qquad \text{for all} \quad \mathbf{v}^h \in \mathbf{V}^h, \tag{2.12}$$

and the result

$$\sup_{0 \neq \mathbf{v}^h \in \mathbf{V}^h} \frac{b(\mathbf{v}^h, q^h)}{\|\mathbf{v}^h\|_0} \geq C_2 |q^h|_1 \qquad \text{for all} \quad q^h \in S^h, \tag{2.13}$$

to yield

$$\sup_{0 \neq \mathbf{v}^h \in \mathbf{V}^h} \frac{b(\mathbf{v}^h, q^h)}{|\mathbf{v}^h|_1} \geq \frac{C_2}{C_1} h |q^h|_1 \qquad \text{for all} \quad q^h \in S^h. \tag{2.14}$$

The inequality (2.13) can be shown to hold for many element pairs involving continuous discrete pressure fields; see, e.g., [7]. Note that (2.13) has a similar appearance to the div-stability condition (2.5), but that it involves the "wrong" norms.

Next, we combine the result, which can be found in, e.g., [15, 20]: given any $q^h \in S^h \subset L_0^2(\Omega)$, there exists a $\mathbf{w} \in \mathbf{H}_0^1(\Omega)$ such that $\operatorname{div} \mathbf{w} = q^h$ and $|\mathbf{w}|_1 \leq C_3 \|q^h\|_0$, with the approximation-theoretic assumption concerning the space $\mathbf{V}^h$: for any $\mathbf{w} \in \mathbf{H}^1(\Omega)$ there exists a $\mathbf{w}^h \in \mathbf{V}^h$ such that

$$|\mathbf{w} - \mathbf{w}^h|_k \leq C_4 h^{1-k} |\mathbf{w}|_1 \qquad \text{for} \quad k = 0, 1, \tag{2.15}$$

to yield

$$\sup_{0 \neq \mathbf{v}^h \in \mathbf{V}^h} \frac{b(\mathbf{v}^h, q^h)}{|\mathbf{v}^h|_1} \geq C_5 - C_6 h |q^h|_1 \qquad \text{for all} \quad q^h \in S^h \quad \text{with} \quad \|q^h\|_0 = 1. \tag{2.16}$$

Verfürth then shows that the div-stability condition (2.5) follows from (2.14) and (2.16) provided the constants $C_1, \ldots, C_6$ are independent of h.

Thus the main task of applying his method, once the inverse inequality (2.12) and the approximation-theoretic result (2.15) have been shown to hold for the discrete velocity space $\mathbf{V}^h$, is to show that (2.13) is valid.

6.2.4.3. *The Boland–Nicolaides Method*

A more useful method, in the sense of having wide applicability and relative ease of use, has been developed by Boland and Nicolaides [4]. One difficulty with verifying the div-stability condition (2.5) is its *global* nature; Boland and Nicolaides have shown how to *localize* the difficult part of the verification process.

Specifically, consider a subdivision of Ω into disjoint *macroelements* Ω_r,

$r = 1, \ldots, R$, each of which consists of one or a few elements in the finite element triangulation associated with $\mathbf{V}^h$ and S^h. The number of elements within a macroelement is independent of h, i.e., as we refine the mesh the macroelements are also refined so that they always contain the same number of elements. Let Γ_r denote the boundary of the macroelement Ω_r.

Now, first suppose that the div-stability condition holds for the pair $\mathbf{V}^h$ and S^h *locally* over a macroelement, i.e., there exists a constant $\hat{\gamma} > 0$, independent of h and of the particular choice of macroelement, such that

$$\sup_{\mathbf{0} \neq \mathbf{v}^h \in \mathbf{V}_r^h} \left\{ \frac{b(\mathbf{v}^h, q^h)}{\|v^h\|_1} \right\} \geq \hat{\gamma} \|q^h\|_0 \qquad \text{for all} \quad q^h \in S_r^h, \tag{2.17}$$

where

$$\mathbf{V}_r^h = \{\mathbf{v} \in \mathbf{V}^h|_{\Omega_r} | \mathbf{v} = \mathbf{0} \text{ on } \Gamma_r\} \quad \text{and} \quad S_r^h = \{q \in S^h|_{\Omega_r}\}.$$

Since $\mathbf{V}_r^h$ and S_r^h have fixed small dimension, independent of h, (2.17) may often be verified by a direct computation.

Second, suppose that the div-stability condition holds *globally* for the spaces $\tilde{\mathbf{V}}^h$ and $\tilde{S}^h$ where

$$\begin{cases} \tilde{\mathbf{V}}^h \subset \mathbf{V}^h, \\ \tilde{S}^h = \begin{Bmatrix} L_0^2(\Omega) \text{ piecewise constant functions with respect} \\ \text{to the macroelements } \Omega_r,\ r = 1, \ldots, R \end{Bmatrix}, \end{cases} \tag{2.18}$$

i.e., suppose that there exists a constant $\tilde{\gamma} > 0$, independent of h, such that

$$\sup_{\mathbf{0} \neq \mathbf{v}^h \in \tilde{\mathbf{V}}^h} \left\{ \frac{b(\mathbf{v}^h, q^h)}{\|\mathbf{v}^h\|_1} \right\} \geq \tilde{\gamma} \|q^h\|_0 \qquad \text{for all} \quad q^h \in \tilde{S}^h. \tag{2.19}$$

Summarizing the Boland–Nicolaides method, suppose we know that the pair $\mathbf{V}^h$, S^h is *locally div-stable* with constant $\hat{\gamma}$ independent of h, i.e., in the sense of (2.17). Further, suppose that the *comparison* spaces $\tilde{\mathbf{V}}^h$, $\tilde{S}^h$, which satisfy (2.18), are *globally div-stable* with constant $\tilde{\gamma}$ independent of h, i.e., in the sense of (2.19). Then the spaces $\mathbf{V}^h$, S^h are *globally div-stable* with a constant γ independent of h. Thus, through the use of the comparison spaces the div-stability of the pair $\mathbf{V}^h$, S^h need only be checked locally, i.e., over a macroelement.

This method has been successfully used, e.g., in [2, 4, 6, 15], to show the div-stability of a variety of well-known elements and some novel ones as well, both in two and in three dimensions.

6.2.5. Examples of Unstable Spaces Including the Bilinear-Constant Pair

There are different ways in which arbitrarily chosen finite element spaces may fail to satisfy the div-stability condition. Here we discuss some of these and then give specific examples, focusing on the much studied and much misunderstood bilinear velocity-constant pressure pair.

The most catastrophic type of failure is for (2.2), or equivalently (2.4), to imply that $\mathbf{u}^h = \mathbf{0}$, i.e., the only discretely solenoidal field belonging to $\mathbf{V}^h$ is the zero vector. The approximate solution is useless since, of course, $\mathbf{u}^h = \mathbf{0}$ cannot be a good approximate solution of the Navier–Stokes equations. This type of situation can usually be detected by a counting argument, i.e., the discrete divergence matrix $b(\mathbf{v}_k, q_j)$, $j = 1, \ldots, J$ and $k = 1, \ldots, K$, appearing in (2.4) has more independent rows than columns.

Less catastrophic is the situation wherein for one or a few, but not all, $q^h \in S^h$ we have that $b(\mathbf{v}^h, q^h) = 0$ for all $\mathbf{v}^h \in \mathbf{V}^h$ so that $\gamma = 0$ in (2.5). This kind of failure of the div-stability condition is usually easy to detect since it results, in practice, in the discrete divergence matrix being rank-deficient. Furthermore, if these types of pressure modes q^h are the sole reason for the invalidity of (2.5), we may often still obtain, through a filtering process, useful approximations.

A more subtle failure of the div-stability condition is the case where for at least some $q^h \in S^h$

$$C_1 h \|q^h\|_0 \leq \sup_{0 \neq \mathbf{v}^h \in \mathbf{V}^h} \left\{ \frac{b(\mathbf{v}^h, q^h)}{\|\mathbf{v}^h\|_1} \right\} \leq C_2 h \|q^h\|_0 \tag{2.20}$$

for some constants C_1 and C_2 independent of h. In this case $\gamma = O(h)$, where γ is the constant appearing in (2.5). In practice, this may result in a loss of accuracy, especially for the pressure approximations. Such instabilities are harder to detect because, of course, computations are usually carried out by using a finite value of h. In particular, no problems such as those caused by rank-deficient approximations to the continuity equation are encountered. This type of situation points out the dangers of calculating on only one grid and of not at least performing serious mesh refinement studies. It also points out the usefulness of rigorous results concerning the stability, or lack thereof, of finite element spaces.

6.2.5.1. *An Unstable Linear-Constant Pair*

A example of the first and most catastrophic instability is the following seemingly natural choice for the velocity and pressure finite element spaces. Let Ω be a square that is triangulated as in Figure 6.1. For the velocity

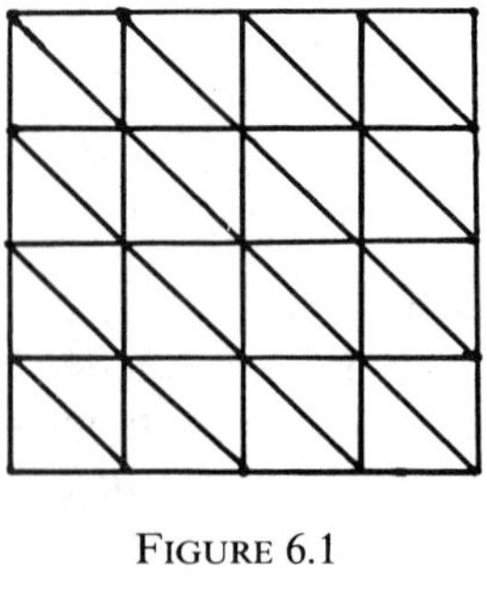

FIGURE 6.1

approximations we choose piecewise linear functions with respect to the given triangulation that are continuous over $\bar{\Omega}$ and that vanish on Γ. For the discrete pressures we choose piecewise constant functions with respect to the same triangulation and having zero mean over Ω. Clearly, $\mathbf{V}^h \subset \mathbf{H}_0^1(\Omega)$ and $S^h \subset L_0^2(\Omega)$. For this choice the only discrete velocity field $\mathbf{u}^h \in \mathbf{V}^h$ satisfying the discrete incompressibility constraint (2.2) is $\mathbf{u}^h = \mathbf{0}$, i.e., the only discretely solenoidal velocity field is the zero vector! We easily see that if there are $\mathbf{N}$ cells to a side, then the number of equations in (2.4) is $J = \dim(S^h) = 2\mathbf{N}^2 - 1$, which is greater than the number of columns $K = \dim(\mathbf{V}^h) = 2(\mathbf{N} - 1)^2$.

In the preceding example we see that the discrete incompressibility condition (2.2) imposes too many constraints in relation to the available velocity degrees of freedom. In fact, $\dim(S^h) > \dim(\mathbf{V}^h)$. In order to remedy the situation we must, at least, increase the dimension of $\mathbf{V}^h$ in relation to that of S^h.

6.2.5.2. *The Bilinear-Constant Element Pair*

We next consider the *bilinear velocity-constant pressure pair*, which is often referred to as the *$Q_1 - P_0$ element pair*. Again consider the case of Ω being a square and consider the "triangulation" of Figure 6.2. We now choose $\mathbf{V}^h$ to consist of piecewise bilinear functions with respect to this triangulation that are continuous over $\bar{\Omega}$ and that vanish on Γ. For S^h we choose piecewise constant functions over the same triangulation that have zero mean over Ω. Once again the inclusions $\mathbf{V}^h \subset \mathbf{H}_0^1(\Omega)$ and $S^h \subset L_0^2(\Omega)$ hold. The simple counting argument used for the first example does not yield any definitive information since $\dim(\mathbf{V}^h) = 2(\mathbf{N} - 1)^2$, the same as before, whereas now $\dim(S^h) = \mathbf{N}^2 - 1$.

It is well known, e.g., see [3, 12, 17, 18, 22], that this bilinear-constant element pair exhibits the disastrous "checkerboard" mode, i.e., for the particular discrete pressure field $q^h \in S^h$ that is $+1$ in the "red squares" and -1 in the "black squares" we have that $b(\mathbf{v}^h, q^h) = 0$ for all $\mathbf{v}^h \in \mathbf{V}^h$. This is an example of the second type of instability discussed previously. The single "bad" pressure mode can be easily filtered out, and therefore some have suggested that once this mode is taken care of, the bilinear-constant element pair can be safely used.

However, this is not the whole story for the bilinear-constant element pair.

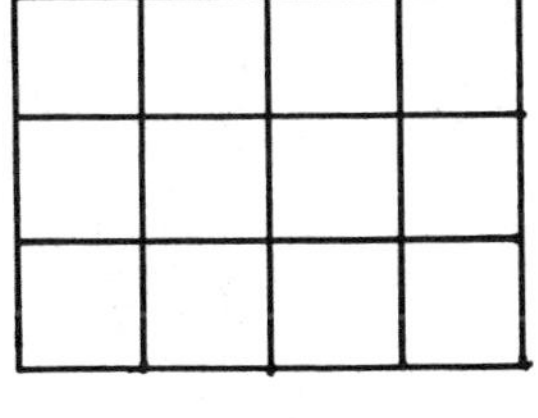

FIGURE 6.2

Boland and Nicolaides [5] have shown that there exist other pressure modes for which (2.20) is satisfied. The left-hand inequality of (2.20) was previously known [18], at least in the different context of penalty methods. Of course, the left inequality does not imply the right and certainly does not imply that for those modes the stability constant $\gamma = O(h)$. However, Boland and Nicolaides have shown that this is indeed the case. Moreover, they have shown [6] that there exist data $\mathbf{f}$ for which the pressure approximations do not converge and that it is also possible to set up problems for which the velocity approximations do not converge as well. At the least, since the constants in the error estimates (2.9)–(2.11) are proportional to γ^{-1}, there will likely be a loss of accuracy due to these pressure modes. Their conclusions are worth noting, especially in view of the fact that the bilinear-constant element pair, with the checkerboard mode filtered out, has been used on numerous occasions in "practical" computations.

6.3. Finite Element Spaces for the Primitive Variable Formulation

In this section we discuss pressure and velocity finite element spaces that have been rigorously shown to satisfy the div-stability condition. There are many such pairs known, especially for two-dimensional problems; therefore, we will restrict our attention to pairs that have proved to be of the most practical utility.

Throughout, $P_k(\mathscr{D})$ denotes the space of polynomials of degree less than or equal to k with respect to the set $\mathscr{D} \subset R^d$, and $[P_k(\mathscr{D})]^d$ dcnotes the space of d vector-valued functions each of whose components belongs to $P_k(\mathscr{D})$. Analogous definitions hold for $Q_k(\mathscr{D})$ and $[Q_k(\mathscr{D})]^d$ in the case of functions that are polynomials of degree less than or equal to k in each of the coordinate directions, e.g., $Q_1(\mathscr{D})$ denotes piecewise bilinear functions with respect to the set $\mathscr{D}$. Likewise, we define the spaces $C^k(\mathscr{D})$ and $[C^k(\mathscr{D})]^d$ of k times continuously differentiable functions with respect to the set $\mathscr{D}$.

For the most part, the following results hold for polygonal domains in R^2 and polyhedral domains in R^3. Through the use of, e.g., isoparametric elements, they will also hold for domains with curved boundaries, provided the latter satisfy the usual smoothness criteria. Furthermore, we assume that all subdivisions of Ω into finite elements that are employed in the following satisfy the standard conditions. For details concerning these issues, we may consult, e.g., [9].

6.3.1. Piecewise Linear and Bilinear Velocity Fields

We begin with some examples involving piecewise linear or bilinear velocity fields with respect to a subdivision of Ω into triangles or rectangles, respectively. In all cases the discrete velocity fields are continuous over $\bar{\Omega}$. In combination with these types of velocity finite element spaces we will consider both discontinuous piecewise constant and continuous, over $\bar{\Omega}$, piecewise linear pressure fields. Every element pair listed satisfies the div-stability condition (2.5). Moreover, provided the solution $\mathbf{u}$, p of (1.9)–(1.10) satisfies $\mathbf{u} \in \mathbf{H}^2(\Omega) \cap \mathbf{H}_0^1(\Omega)$ and $p \in H^1(\Omega) \cap L_0^2(\Omega)$, the following error estimates for the discrete solution $\mathbf{u}^h$, p^h of (2.1)–(2.2) hold uniformly in h:

$$\begin{cases} \|\mathbf{u} - \mathbf{u}^h\|_1 = O(h), \\ \|\mathbf{u} - \mathbf{u}^h\|_0 = O(h^2), \\ \|p - p^h\|_0 = O(h). \end{cases} \tag{3.1}$$

Thus, these elements yield first-order accurate pressure approximations and second-order accurate velocity approximations.

6.3.1.1. *Piecewise Constant Pressures, I*

For the linear-constant element pair mentioned in Section 6.2.5 the discrete continuity equation overconstrained the approximate velocity field. However, by employing different grids for the pressure and velocity fields, the linear-constant element pair may be made stable. For example, consider a given triangulation $\mathcal{T}_h$ of a polygonal domain Ω into triangles. Then divide each triangle in $\mathcal{T}_h$ into four triangles by joining the midsides, thus defining a refined triangulation $\mathcal{T}_{h/2}$. An example is provided in Figure 6.3.

Now define

$$\begin{cases} S^h = \left\{ q \mid q \in P_0(\Delta), \Delta \in \mathcal{T}_h; \displaystyle\int_\Omega q = 0 \right\}, \\ \mathbf{V}^h = \{\mathbf{v} \mid \mathbf{v} \in [P_1(\Delta)]^2, \Delta \in \mathcal{T}_{h/2}; \mathbf{v} \in [C^0(\bar{\Omega})]^2; \mathbf{v} = \mathbf{0} \text{ on } \Gamma\}, \end{cases} \tag{3.2}$$

so that the pressure is sought among piecewise constants with respect to the triangulation $\mathcal{T}_h$ and the velocity is sought among continuous piecewise linear fields with respect to the finer triangulation $\mathcal{T}_{h/2}$. The pair of finite element spaces defined by (3.2) is known to satisfy the div-stability condition (2.5) and thus yield optimally accurate approximations satisfying (3.1).

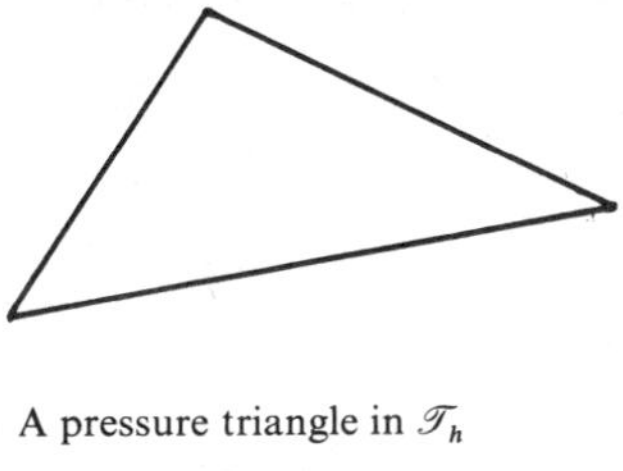

A pressure triangle in $\mathcal{T}_h$

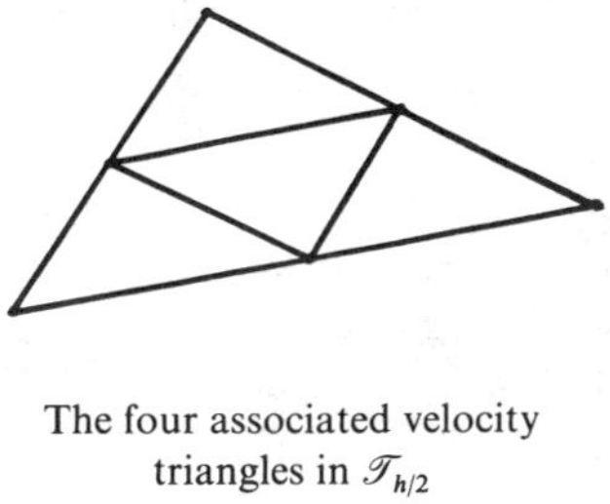

The four associated velocity triangles in $\mathcal{T}_{h/2}$

FIGURE 6.3

6.3.1.2. *Piecewise Constant Pressures, II*

For the unstable linear-constant element pair of Section 6.2.5 there was one velocity element for each pressure element; for the stable linear-constant element pair (3.2) there are four velocity triangles for each pressure triangle. Stable linear-constant element pairs wherein the ratio of discrete pressures to velocities is not so high may be defined. For example, let the velocity space $\mathbf{V}^h$ be as in (3.2); now define the pressure space S^h through the following choice of basis. For each triangle of $\mathcal{T}_h$ we define three basis functions, namely piecewise constants that are unity in the shaded areas in Figure 6.4 and zero in the unshaded areas. Of course, outside the particular triangle of $\mathcal{T}_h$, the basis functions vanish as well. This pressure space consists of three of the four possible piecewise constants associated with the four triangles in $\mathcal{T}_{h/2}$ contained within a single triangle in $\mathcal{T}_h$. Moreover, there are essentially three times as many pressure degrees of freedom for this choice of S^h as there are for the choice made in (3.2). However, this element pair is also stable, i.e., satisfies the div-stability condition (2.5) and the error estimates (3.1).

6.3.1.3. *Piecewise Linear Pressures*

We may also couple a piecewise linear velocity element with a piecewise linear pressure element and still satisfy the div-stability condition (2.5) and the estimates (3.1). Such a pair was introduced in [7], analyzed there and in [26], and is given by

$$\begin{cases} S^h = \left\{ q \mid q \in P_1(\Delta),\ \Delta \in \mathcal{T}_h;\ q \in C^0(\Omega);\ \int_\Omega q = 0 \right\}, \\ \mathbf{V}^h = \text{as in (3.2)}. \end{cases} \tag{3.3}$$

As a result of the coupling between the pressure and velocity errors we cannot take advantage of the better approximating ability of the linear pressure space. Thus, insofar as the rates of convergence, this linear–linear element pair is no better than the stable linear–constant element pairs. However, in practical calculations we have found this to be the best element combination involving linear velocity fields, better in the sense of giving more accuracy for useful values of h. Furthermore, this linear–linear element pair usually results in fewer unknowns, for the same grid, than do the linear–constant pairs. For

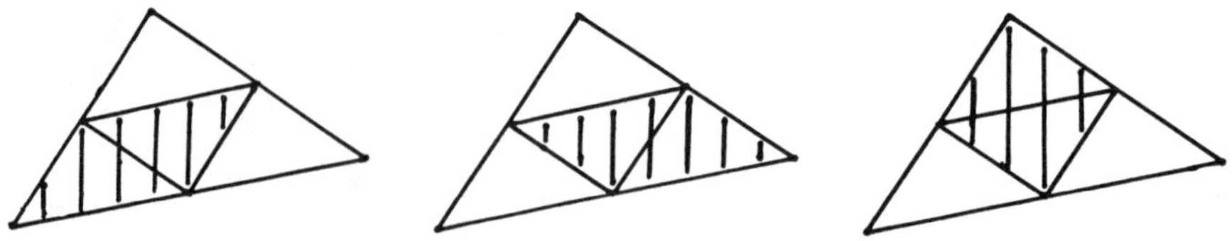

FIGURE 6.4

example, suppose the pressure triangulation $\mathscr{T}_h$ is given by Figure 6.1 (Section 6.2.5) with N intervals on each side. Thus there are $2N^2$ triangles in $\mathscr{T}_h$ and the element pair (3.2) has $2N^2 - 1$ pressure unknowns; on the other hand, the number of nodes in this triangulation is only $(N + 1)^2$ and thus the piecewise linear pressure space of (3.3) has only $(N + 1)^2 - 1$ degrees of freedom. Both element pairs have $2(2N - 1)^2$ velocity unknowns so that the linear–linear element pair (3.3) has roughly N^2 fewer degrees of freedom, for the same grid, as does the linear–constant element pair (3.2).

6.3.1.4. *Piecewise Bilinear Velocity Fields*

Entirely analogous to the triangular elements described previously are the following elements involving bilinear velocity fields with respect to rectangular elements. More general quadrilateral elements may be found from these through, e.g., isoparametric mappings.

We start with a subdividivision $\mathscr{Q}_h$ of Ω into rectangles, or more generally quadrilaterals. Subsequently we divide each rectangle into four smaller rectangles by joining the midsides, thus creating another subdivision $\mathscr{Q}_{h/2}$ of Ω into rectangles. See Figure 6.5. In all three velocity–pressure element pairs about to be described we choose the approximating velocity space to consist of piecewise bilinear functions with respect to the subdivision $\mathscr{Q}_{h/2}$ that are continuous over $\bar{\Omega}$ and vanish on Γ, i.e.,

$$\mathbf{V}^h = \{\mathbf{v} | \mathbf{v} \in [Q_1(\square)]^2, \square \in \mathscr{Q}_{h/2}; \mathbf{v} \in [C^0(\bar{\Omega})]^2; \mathbf{v} = \mathbf{0} \text{ on } \Gamma\}. \tag{3.4}$$

For the first pressure space we choose piecewise constants with respect to the larger quadrilaterals of the subdivision $\mathscr{Q}_h$ that have zero mean over Ω, i.e.,

$$S^h = \left\{q | q \in Q_0(\square), \square \in \mathscr{Q}_h; \int_\Omega q = 0\right\}.$$

As indicated in Figure 6.6, for the second pressure space we choose three of

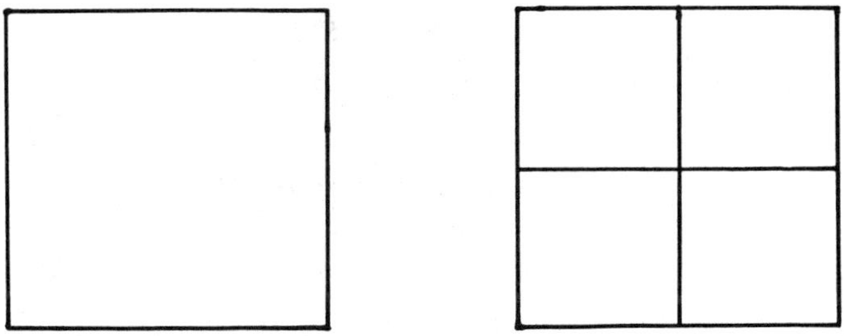

A pressure rectangle in $\mathscr{Q}_h$

The four associated velocity rectangles in $\mathscr{Q}_{h/2}$

FIGURE 6.5

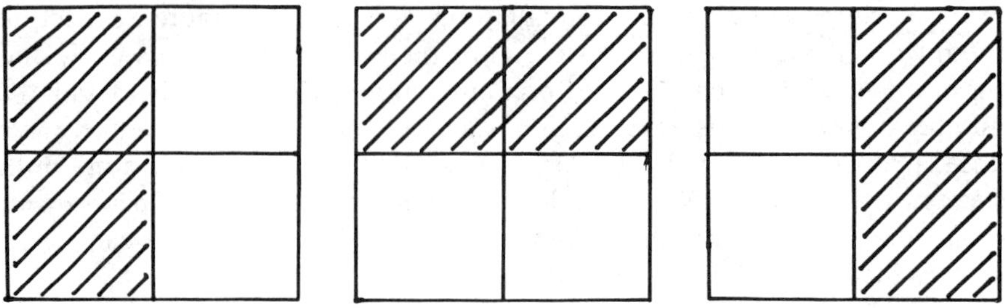

FIGURE 6.6

the four possible piecewise constants associated with the rectangles belonging to $\mathcal{Q}_{h/2}$ and that have zero mean over Ω. Finally, the third pressure space consists of piecewise bilinear functions with respect to the subdivision $\mathcal{Q}_h$ that are continuous over $\bar{\Omega}$ and have zero mean over Ω, i.e.,

$$S^h = \left\{ q | q \in Q_1(\square),\ \square \in \mathcal{Q}_h;\ q \in C^0(\bar{\Omega});\ \int_\Omega q = 0 \right\}. \tag{3.5}$$

The three velocity–pressure elements just described satisfy the div-stability condition (2.5) and the error estimates (3.1). Similar to the case for triangles and for the same reasons, the preferred element pair involving bilinear velocities is (3.4) coupled with (3.5), i.e., the bilinear velocity–bilinear pressure pair.

6.3.2. The Taylor–Hood Element Pair

We next turn to quadratic and biquadratic approximate velocity fields. Suppose we have a triangulation $\mathcal{T}_h$ of Ω. Then, the Taylor–Hood element pair [25] is defined by

$$\begin{cases} \mathbf{V}^h = \{\mathbf{v} | \mathbf{v} \in [P_2(\Delta)]^2,\ \Delta \in \mathcal{T}_h;\ \mathbf{v} \in [C(\bar{\Omega})]^2;\ \mathbf{v} = \mathbf{0} \text{ on } \Gamma\}, \\ S^h = \left\{ q | q \in P_1(\Delta),\ \Delta \in \mathcal{T}_h;\ q \in C(\bar{\Omega});\ \int_\Omega q = 0 \right\}. \end{cases} \tag{3.6}$$

Note that we are now basing $\mathbf{V}^h$ and S^h on the same grid but on different degree polynomials, in contrast to (3.3), which uses the same degree polynomials but different grids. The element pair (3.6) satisfies the div-stability condition (2.5). Furthermore, if the solution $(\mathbf{u}, p)$ of (1.9)–(1.10) has the indicated smoothness, then the following error estimates hold uniformly in h:

$$\left\{ \begin{aligned} \|\mathbf{u} - \mathbf{u}^h\|_1 &= O(h^{m-1}), \\ \|\mathbf{u} - \mathbf{u}^h\|_0 &= O(h^m), \\ \|p - p^h\|_0 &= O(h^{m-1}), \end{aligned} \right\} \quad \text{whenever} \quad \left\{ \begin{gathered} \mathbf{u} \in \mathbf{H}^m(\Omega) \cap \mathbf{H}_0^1(\Omega) \\ \text{and} \\ p \in H^{m-1}(\Omega) \cap L_0^2(\Omega) \end{gathered} \right\},$$

$$m = 2 \text{ or } 3. \tag{3.7}$$

These results have been established by many authors [4, 7, 26]. We see from (3.7) that if $\mathbf{u} \in \mathbf{H}^3(\Omega) \cap \mathbf{H}_0^1(\Omega)$ and $p \in H^2(\Omega) \cap L_0^2(\Omega)$ then, in L^2-norms, we have third-order accurate velocity approximations and second-order accurate pressure approximations. This is an improvement over any of the elements involving linear velocities.

We should note that the number of degrees of freedom, of both velocity and pressure types, associated with the use of (3.6) is identical to that associated with the use of (3.3), the most efficient linear velocity element. In fact, the structure of the discrete system resulting from a Taylor–Hood discretization is in every way identical to that resulting from the use of (3.3). Therefore, the solution times for the Taylor–Hood and the linear–linear discrete systems are roughly the same if we use the same pressure triangulation in both cases. Of course, the Taylor–Hood element pair will yield better accuracy than the linear–linear pair, provided the exact solution is sufficiently smooth.

On the other hand, on the same grid, the assembly costs of Taylor–Hood will in general be higher since we need to use higher order quadrature rules to integrate the higher degree polynomial integrands resulting from the Taylor–Hood element pair. For many solvers, the assembly time is overwhelmed by the solution time; therefore, the increased assembly cost associated with (3.6) is not a serious drawback. Of course, this is further mitigated by the fact that for the same accuracy, we may use a coarser grid for (3.6) than for (3.3).

Summarizing, provided the exact solution is sufficiently smooth, the Taylor–Hood element pair, when compared to any of the linear velocity elements, yields better accuracy for essentially the same work, or alternately, yields a desired level of accuracy for less cost.

For rectangles or quadrilaterals we have the analogous pair

$$\begin{cases} \mathbf{V}^h = \{\mathbf{v} | \mathbf{v} \in [Q_2(\square)]^2, \square \in \mathcal{Q}_h; \mathbf{v} \in [C(\bar{\Omega})]^2; \mathbf{v} = \mathbf{0} \text{ on } \Gamma\}, \\ S^h = \left\{ q | q \in Q_1(\square), \square \in \mathcal{Q}_h; q \in C(\bar{\Omega}); \int_\Omega q = 0 \right\}, \end{cases} \tag{3.8}$$

where $\mathcal{Q}_h$ denotes a subdivision of Ω into rectangles. This element pair satisfies the div-stability condition (2.5) and the error estimates (3.7).

We may well ask whether further efficiencies may be gained by using higher order elements, e.g., cubic velocities coupled with quadratic pressures. Here we need to consider the trade-off between the increased accuracy of higher order elements and the increased complexity of those elements. As in other settings, e.g., structural mechanics, we generally find that the optimum seems to be achieved by quadratic elements. Furthermore, it is questionable that in general settings the exact solution of the Navier–Stokes equations is sufficiently smooth to enable the potential better accuracy of higher order elements.

In our overall experience, we have found the best choice of velocity–pressure elements to be the Taylor–Hood element pair (3.6) or its quadrilateral counterpart (3.8).

6.3.3. Divergence-Free Elements

Ideally, we would like to choose the finite element spaces $\mathbf{V}^h$ and S^h so that the functions belonging to $\mathbf{V}^h$ are at least discretely divergence-free. Certainly, if the functions belonging to $\mathbf{V}^h$ are divergence-free then they are discretely divergence-free as well, i.e., $\operatorname{div} \mathbf{v}^h = 0$ for all $\mathbf{v}^h \in \mathbf{V}^h$ implies that $\mathbf{V}^h = \mathbf{Z}^h$. Such a case effects a great simplification since the velocity and pressure uncouple. Indeed, we need only solve

$$a(\mathbf{u}^h, \mathbf{v}^h) + c(\mathbf{u}^h, \mathbf{u}^h, \mathbf{v}^h) = (\mathbf{f}, \mathbf{v}^h) \qquad \text{for all} \quad \mathbf{v}^h \in \mathbf{V}^h,$$

for the discrete velocity field $\mathbf{u}^h$, since in this case the term $b(\mathbf{v}^h, q^h)$ in (2.1) vanishes for any $\mathbf{v}^h \in \mathbf{V}^h = \mathbf{Z}^h$. Also, since $\mathbf{Z}^h \in \mathbf{Z}$, note that in the velocity estimate (2.9), $\Theta = 0$ so that the velocity error depends only on the ability to approximate in $\mathbf{V}^h$.

Unfortunately, although some finite element pairs such that the functions in $\mathbf{V}^h$ are at least locally divergence-free are known, these have proved to be impractical, and we will not consider them here. We do mention that one obvious method of generating divergence-free discrete vector fields is to take the curl of a piecewise polynomial field, i.e., of a piecewise polynomial stream function. One problem with this approach is that if we want a conforming velocity field, i.e., $\mathbf{V}^h \subset \mathbf{H}_0^1(\Omega)$, then the discrete stream function field must be chosen to be continuously differentiable over Ω. In R^2 this, of course, necessitates the use of at least quintic stream functions over triangles, or cubic polynomials over macroelements, e.g., the Clough–Toucher element. Nonconforming velocity fields can also be generated in this manner. See [8, 10, 15] for details.

6.3.4. Three-Dimensional Elements

Compared to the two-dimensional setting, many fewer stable element pairs are known for three-dimensional problems. However, there is great interest in this subject, and therefore there has been substantial recent progress. Here we mention a few of the known stable three-dimensional elements.

In the first place, the three-dimensional analog of the Taylor–Hood element is known to be stable in three dimensions; this may be shown by the methods of Verfürth or Boland–Nicolaides. Specifically, we subdivide Ω into tetrahedrons and use continuous piecewise quadratic polynomials for the velocity and continuous piecewise linear polynomials for the pressure. The accuracy of this combination is the same as in the two-dimensional case.

Next we consider linear–constant elements. Again, subdivide Ω into tetrahedrons. For the pressure space we choose piecewise constants with respect to this initial subdivision. Now we subdivide each tetrahedron into 12 smaller tetrahedrons by first joining the centroid of the faces to the vertices and then joining the centroid of the large tetrahedron to the vertices and the centroids of the faces. For the velocity space, we choose continuous piecewise linear polynomials with respect to the smaller tetrahedrons.

Another stable linear–constant element pair is defined by first subdividing Ω into rectangular prisms, or, more generally, into distortions of such prisms. For the pressure space we choose piecewise constants over the rectangular subregions. We subdivide each rectangular prism into 24 tetrahedrons by first drawing the two diagonals of each face, then joining the centroid of the prism to the vertices and to the six intersection points of the face diagonals.

Both these linear–constant element pairs are known to be stable and yield the same accuracy results as those for the two-dimensional linear–constant pairs. See [2] for details.

6.4. Alternative Weak Forms and Boundary Conditions

In this section we examine some variants of the weak formulation (1.9)–(1.10), mostly from the viewpoint of methods by which different boundary conditions may be incorporated into a finite element method using primitive variables. We again emphasize that there are many radically different weak formulations involving $\mathbf{u}$ and p that we will not be able to consider; we are restricting ourselves to variants of the most commonly used weak formulation.

Before considering boundary conditions, we briefly consider an alternative formulation of the convection term in (1.9).

6.4.1. An Alternative Formulation of the Convection Term

For the purpose of simplifying the analysis of the approximate solution it can be useful to introduce a slightly different weak formulation wherein the trilinear form $c(\cdot, \cdot, \cdot)$ appearing in (1.9) is replaced by the skew-symmetrized form introduced by Teman [23]

$$\tilde{c}(\mathbf{w}, \mathbf{u}, \mathbf{v}) = \tfrac{1}{2}\{c(\mathbf{w}, \mathbf{u}, \mathbf{v}) - c(\mathbf{w}, \mathbf{v}, \mathbf{u})\}. \tag{4.1}$$

We may easily verify that $\tilde{c}(\mathbf{u}, \mathbf{u}, \mathbf{v}) = c(\mathbf{u}, \mathbf{u}, \mathbf{v})$ whenever $\operatorname{div} \mathbf{u} = 0$ in Ω and $\mathbf{u} \cdot \mathbf{n} = 0$ on Γ, where $\mathbf{n}$ denotes the outward normal to Γ. Therefore, from (1.2)–(1.3), it seems irrelevant whether we use (1.8) or (4.1) in a weak formulation of the Navier–Stokes equations. From an analysis point of view, the advantage of (4.1) is that $\tilde{c}(\mathbf{w}, \mathbf{u}, \mathbf{v}) = -\tilde{c}(\mathbf{w}, \mathbf{v}, \mathbf{u})$ for any $\mathbf{u}$, $\mathbf{v}$, $\mathbf{w} \in \mathbf{H}^1(\Omega)$, whereas the analogous result for (1.8) holds only when $\operatorname{div} \mathbf{w} = 0$ in Ω and one of $\mathbf{u} = \mathbf{0}$, $\mathbf{v} = \mathbf{0}$, or $\mathbf{w} \cdot \mathbf{n} = 0$ holds on Γ.

We emphazise that, insofar as the accuracy of the approximations is concerned, it makes no difference whether we use (1.8) or (4.1); we merely point out that many of the results concerning finite element approximations of solutions of (1.1)–(1.3) were first obtained through the use of (4.1). On the other hand, any implementation of (4.1) will result in more computational work than the analogous implementation of (1.8).

6.4.2. Inhomogeneous Velocity Boundary Conditions

There are many different ways to treat inhomogeneous velocity boundary conditions. In practice, the overwhelming choice is to use the boundary interpolant. We describe this method for polygonal domains $\Omega \subset R^2$; entirely analogous ideas may be used in three dimensions and for domains with curved sides, the latter through the aid of, e.g., isoparametric elements.

Consider the boundary condition

$$\mathbf{u} = \mathbf{g} \quad \text{on } \Gamma, \tag{4.2}$$

and the set

$$\mathbf{V_g} = \{\mathbf{u} \in \mathbf{H}^1(\Omega) | \mathbf{u} \text{ satisfies (4.2)}\}. \tag{4.3}$$

Note that $\mathbf{V}_0 = \mathbf{H}_0^1(\Omega)$. The weak formulation that we will discretize is as follows: seek $\mathbf{u} \in \mathbf{V_g}$ and $p \in L_0^2(\Omega)$ such that (1.9) and (1.10) hold. Note that the test function $\mathbf{v}$ still belongs to $\mathbf{H}_0^1(\Omega)$, i.e., $\mathbf{v} = \mathbf{0}$ on Γ.

In order to pose our discrete problem we choose finite element spaces $\mathbf{V}^h \subset \mathbf{H}^1(\Omega)$ and $S^h \subset L_0^2(\Omega)$. We denote by $\mathbf{V}^h|_\Gamma$ the restriction of $\mathbf{V}^h$ to the boundary Γ, i.e., $\mathbf{V}^h|_\Gamma$ consist of functions defined on Γ and that can agree with the boundary values of at least one function belonging to $\mathbf{V}^h$. The finite element functions belonging to $\mathbf{V}^h$, being, for example, piecewise polynomials, cannot in general satisfy the boundary condition (4.2); certainly, in general $\mathbf{g} \notin \mathbf{V}^h|_\Gamma$. Therefore we choose an approximation to $\mathbf{g}$, which we denote by $\mathbf{g}^h$, belonging to $\mathbf{V}^h|_\Gamma$. The most common choice for $\mathbf{g}^h$, and the one we consider here, is the interpolant of $\mathbf{g}$ in $\mathbf{V}^h|_\Gamma$.

This choice is trivial to implement; this at least partially accounts for its popularity. For example, suppose $\mathbf{V}^h$ is a Lagrange finite element space, i.e., one whose degrees of freedom are exclusively function values at points. Let $\{\mathbf{v}_k\}$, $k = 1, \ldots, \bar{K}$ denote the usual finite element basis for $\mathbf{V}^h$. Let the first K of these basis functions be associated with interior nodes $\mathbf{x}_k$ so that for $k = 1, \ldots, K$, $\mathbf{v}_k = \mathbf{0}$ for $\mathbf{x} \in \Gamma$. The remaining basis function $\{\mathbf{v}_k\}$, $k = K + 1, \ldots, \bar{K}$, are associated with nodes $\mathbf{x}_k$ lying on Γ. In practical implementations there are more efficient node numbering schemes than the one we are using; however, the latter simplifies the explanations being attempted here.

Choosing $\mathbf{g}^h$ to be the boundary interpolant of $\mathbf{g}$ is then equivalent to writing

$$\mathbf{u}^h(\mathbf{x}) = \sum_{k=1}^{K} \beta_k \mathbf{v}_k(\mathbf{x}) + \sum_{k=K+1}^{\bar{K}} \mathbf{g}(\mathbf{x}_k)\mathbf{v}_k(\mathbf{x}). \tag{4.4}$$

In (4.4), β_k, $k = 1, \ldots, K$, are the unknown coefficients to be determined; the coefficients of the basis functions associated with boundary nodes are simply set equal to $\mathbf{g}$ evaluated at the corresponding node. Note that (4.4) implies that

$$\mathbf{g}^h(\mathbf{x}) = \sum_{k=K+1}^{\bar{K}} \mathbf{g}(\mathbf{x}_k)\mathbf{v}_k(\mathbf{x}) \qquad \text{for} \quad \mathbf{x} \in \Gamma.$$

The contribution to $\mathbf{u}^h$ emanating from the second summation of (4.4) becomes part of the data of the discrete system of equations.

Once an approximation $\mathbf{g}^h$ is chosen, we may define the set

$$\mathbf{V}_{\mathbf{g}}^h = \{\mathbf{v} \in \mathbf{V}^h | \mathbf{v} = \mathbf{g}^h \text{ on } \Gamma\}.$$

Note that $\mathbf{V}_{\mathbf{0}}^h$ is the finite element subspace of $H_0^1(\Omega)$ used in conjuction with the homogeneous boundary condition (1.3); also, clearly $\mathbf{V}_{\mathbf{g}}^h \subset \mathbf{H}^1(\Omega)$ is not a subset of $\mathbf{V}_{\mathbf{g}}$. Now, the approximate problem may be defined as follows: seek $\mathbf{u}^h \in \mathbf{V}_{\mathbf{g}}^h$ and $p^h \in S^h \subset L_0^2(\Omega)$ such that (2.1)–(2.2) holds for all $\mathbf{v}^h \in \mathbf{V}_{\mathbf{0}}^h$ and $q^h \in S^h$, respectively. Again, the test functions $\mathbf{v}^h$ vanish on the boundary Γ.

The whole discussion of the div-stability condition (2.5) carries over intact to the case of the inhomogeneous boundary (4.2); in (2.5) we still use the subspace $\mathbf{V}_{\mathbf{0}}^h$ of finite element velocity fields that vanish on the boundary. Results analogous to those of Section 6.2.3 can be derived in a fairly straightforward manner with the exception of some technicalities encountered for the $\mathbf{L}^2(\Omega)$-error estimate for the velocity approximation. See [13, 14, 15] for details.

In particular, if $\mathbf{g}^h$ is chosen to be the boundary interpolant of $\mathbf{g}$ in $\mathbf{V}^h|_\Gamma$, then all the results, e.g., error estimates, concerning the finite element spaces discussed in Section 6.3 are essentially still valid for the inhomogeneous velocity boundary condition (4.2). Again, see [13, 14, 15] for details.

6.4.3. Alternate Boundary Conditions and Formulations of the Viscous Term

In this section we examine how different choices for the viscous term in (1.1) affect the natural boundary conditions of corresponding weak formulations. Some of this material can be found in [16].

Because of (1.2), when ν is constant, the viscous term in (1.1) may be written in the various equivalent forms

$$\nu \Delta \mathbf{u} \tag{4.5a}$$

$$= \operatorname{div}\{\nu((\operatorname{grad} \mathbf{u}) + (\operatorname{grad} \mathbf{u})^{\mathrm{T}})\} \tag{4.5b}$$

$$= -\nu \operatorname{curl}(\operatorname{curl} \mathbf{u}) \tag{4.5c}$$

$$= \nu(\operatorname{grad}(\operatorname{div} \mathbf{u}) - \operatorname{curl}(\operatorname{curl} \mathbf{u})). \tag{4.5d}$$

Although these different realizations are equivalent insofar as the partial differential equations are concerned, we shall see that each generates a different numerical method.

If for some reason ν is not constant or $\operatorname{div} \mathbf{u} \neq 0$, then only (4.5b) may be used. Indeed, (4.5b) is the form of the viscous term that arises naturally in the derivation of the Navier–Stokes equations from the principle of conservation of linear momentum and the Cauchy–Poisson constitutive equation. The other three forms (4.5a), (4.5c), and (4.5d) are derived from (4.5b) with the aid of (1.2) and the assumption that $\nu = $ constant. In (1.1) we have used (4.5a) only because this is the most popular choice in the literature; all of the results obtained so far hold equally well if we choose (4.5b) instead. As will be seen from the following discussion, (4.5b) is, in general, to be preferred to (4.5a).

Denote two segments of the boundary Γ by Γ_n and Γ_τ. These segments may be empty, are not necessarily disjoint, and, in fact, may be equal. Now, for fixed given functions g_n and $\mathbf{g}_\tau$, define the set

$$\mathbf{V_g} = \{\mathbf{v} \in \mathbf{H}^1 | \mathbf{v} \cdot \mathbf{n} = g_n \text{ on } \Gamma_n; \mathbf{n} \times \mathbf{v} \times \mathbf{n} = \mathbf{g}_\tau \text{ on } \Gamma_\tau\},$$

and the spaces

$$\mathbf{V}_0 = \{\mathbf{v} \in \mathbf{H}^1 | \mathbf{v} \cdot \mathbf{n} = 0 \text{ on } \Gamma_n; \mathbf{v} \times \mathbf{n} = \mathbf{0} \text{ on } \Gamma_\tau\},$$

and

$$S = L_0^2(\Omega) \quad \text{if } \Gamma_n = \Gamma, \qquad S = L^2(\Omega) \text{ otherwise},$$

where $\mathbf{v} \cdot \mathbf{n}$ denotes the component of $\mathbf{v}$ normal to the boundary Γ and $\mathbf{n} \times \mathbf{v} \times \mathbf{n} = \mathbf{v} - (\mathbf{v} \cdot \mathbf{n})\mathbf{n}$ is the projection of $\mathbf{v}$ onto the plane tangent to Γ. In the definition of $\mathbf{V_0}$ we may use $\mathbf{v} \times \mathbf{n} = 0$ because of the relation $\mathbf{v} \times \mathbf{n} = \mathbf{n} \times (\mathbf{n} \times \mathbf{v} \times \mathbf{n})$, i.e., $\mathbf{n} \times \mathbf{v} \times \mathbf{n} = \mathbf{0}$ implies that $\mathbf{n} \times \mathbf{v} = \mathbf{0}$. In R^2, $\mathbf{n} \times \mathbf{v} \times \mathbf{n} = \mathbf{v} \cdot \boldsymbol{\tau}$ where $\boldsymbol{\tau}$ is the unit tangent vector to Γ.

Suppose that we wish to specify the boundary conditions

$$\mathbf{u} \cdot \mathbf{n} = g_n \quad on\ \Gamma_n, \tag{4.6a}$$

and

$$\mathbf{n} \times \mathbf{u} \times \mathbf{n} = \mathbf{g}_\tau \quad \text{on } \Gamma_\tau, \tag{4.6b}$$

i.e., the normal velocity on Γ_n and the tangential velocity on Γ_τ, respectively. All the weak formulations that we will consider involving any of the choices in (4.5), (4.6) will be *essential boundary conditions*. Thus the trial solution functions $\mathbf{u}$ will satisfy (4.6), i.e., $\mathbf{u} \in \mathbf{V}_g$, and the test functions satisfy $\mathbf{v} \in \mathbf{V}_0$.

Consider the following weak formulation: for $i = 1, 2, 3$ or 4, seek $\mathbf{u} \in \mathbf{V_g}$ and $p \in S$ such that

$$a_i(\mathbf{u}, \mathbf{v}) + b(\mathbf{v}, p) + c(\mathbf{u}, \mathbf{u}, \mathbf{v}) = (\mathbf{f}, \mathbf{v}) + d(\mathbf{v}) \qquad \text{for all} \quad \mathbf{v} \in \mathbf{V_0}, \tag{4.7}$$

and

$$b(\mathbf{u}, q) = 0 \qquad \text{for all} \quad q \in S. \tag{4.8}$$

Here, $b(\cdot, \cdot)$ and $c(\cdot, \cdot, \cdot)$ remain as in (1.7) and (1.8), respectively, and $\mathbf{f}$ continues to denote the body force appearing in the momentum equation. The linear functional $d(\cdot)$ is given by

$$d(\mathbf{v}) = \int_{\Gamma/\Gamma_n} r\mathbf{v} \cdot \mathbf{n} + \int_{\Gamma/\Gamma_\tau} \mathbf{s} \cdot \mathbf{v} \times \mathbf{n}, \tag{4.9}$$

where the functions r and $\mathbf{s}$ are additional data for the problem. In (4.9), for example, Γ/Γ_n denotes the complement of Γ_n in Γ, i.e., $\mathbf{x} \in \Gamma/\Gamma_n$ implies that $\mathbf{x} \in \Gamma$ but $\mathbf{x} \notin \Gamma_n$. Also, since $\mathbf{v}$ is an arbitrary test function, in direction $\mathbf{v} \times \mathbf{n}$ can be taken to be vectors spanning the tangent plane to Γ.

The bilinear forms $a_i(\cdot,\cdot)$, $i = 1, \ldots, 4$, depend on the choice made in (4.5) and, corresponding to the four choices possible in (4.5), are given by

$$a_1(\mathbf{u}, \mathbf{v}) = \nu \int_\Omega \operatorname{grad} \mathbf{u} : \operatorname{grad} \mathbf{v}, \tag{4.10a}$$

$$a_2(\mathbf{u}, \mathbf{v}) = \frac{1}{2} \int_\Omega \nu(\operatorname{grad} \mathbf{u} + (\operatorname{grad} \mathbf{u})^{\mathrm{T}}) : (\operatorname{grad} \mathbf{v} + (\operatorname{grad} \mathbf{v})^{\mathrm{T}}), \tag{4.10b}$$

$$a_3(\mathbf{u}, \mathbf{v}) = \nu \int_\Omega (\operatorname{curl} \mathbf{u}) \cdot (\operatorname{curl} \mathbf{v}), \tag{4.10c}$$

and

$$a_4(\mathbf{u}, \mathbf{v}) = \nu \int_\Omega (\operatorname{curl} \mathbf{u}) \cdot (\operatorname{curl} \mathbf{v}) + (\operatorname{div} \mathbf{u})(\operatorname{div} \mathbf{v}). \tag{4.10d}$$

In the customary manner, should $\mathbf{u}$ and p be sufficiently smooth, we can, through formal integration by parts procedures, ascertain what differential equation problem the weak formulation (4.7)–(4.8) corresponds to. To begin with, we know that the boundary conditions (4.6) are satisfied since these are being required of the candidate trial functions $\mathbf{u}$. We also find that the differential equations (1.1) and (1.2) are satisfied, where in (1.1) the viscous term is replaced according to (4.5), depending on which choice is made in (4.10). Finally, we find the *natural boundary conditions* corresponding to the particular weak formulation. We will now discuss these in some detail for each possible choice in (4.10).

Corresponding to the paired choices (4.5a) and (4.10a) we have the natural boundary conditions

$$p - \nu\mathbf{n} \cdot \operatorname{grad} \mathbf{u} \cdot \mathbf{n} = r \quad \text{on } \Gamma/\Gamma_n \qquad \text{and} \qquad \nu\mathbf{n} \cdot \operatorname{grad} \mathbf{u} \times \mathbf{n} = \mathbf{s} \quad \text{on } \Gamma/\Gamma_\tau. \tag{4.11a}$$

Unfortunately, these boundary conditions have *no physical meaning*. Thus the choice (4.5a), or equivalently (4.10a), can only be used in conjunction with the boundary condition (4.6) specified on all of Γ, i.e., $\mathbf{u}$ given on $\Gamma_n = \Gamma_\tau = \Gamma$.

Next, consider the choices (4.5b) and (4.10b). The natural boundary conditions are then

$$\begin{cases} -p + \nu\mathbf{n} \cdot (\operatorname{grad} \mathbf{u} + (\operatorname{grad} \mathbf{u})^{\mathrm{T}}) \cdot \mathbf{n} = -r & \text{on } \Gamma/\Gamma_n \quad \text{and} \\ \nu\mathbf{n} \cdot (\operatorname{grad} \mathbf{u} + (\operatorname{grad} \mathbf{u})^{\mathrm{T}}) \times \mathbf{n} = -\mathbf{s} & \text{on } \Gamma/\Gamma_\tau. \end{cases} \tag{4.11b}$$

Thus $-r$ and $-\mathbf{s}$ are the normal and tangential stresses, respectively, on the boundary. Then, for the choice (4.10b), the possible combinations of boundary conditions at a point on the boundary Γ are as follows: we may specify the velocity, or we may specify the normal velocity and the tangential stress, or we may specify the tangential velocity and the normal stress. The later combinations are useful, e.g., for free surface problems or at artificial outflow boundaries. Details may be found in [16].

The third choice (4.5c), or (4.10c), yields the natural boundary conditions

$$p = r \quad \text{on } \Gamma/\Gamma_n \qquad \text{and} \qquad \boldsymbol{\omega} = \mathbf{s}/\nu \quad \text{on } \Gamma/\Gamma_\tau, \tag{4.11c}$$

so that r and $\mathbf{s}$ are the pressure p and ν times the vorticity $\boldsymbol{\omega} = \text{curl}\,\mathbf{u}$, respectively, on the boundary. The possible combinations of boundary conditions are now the following: we may specify the velocity, or we may specify the normal velocity and the vorticity, or we may specify the tangential velocity and the pressure. The pressure is often used as an outflow condition; the vorticity is useful in exterior problems when matching to an inviscid irrotational flow since it is well known that the vorticity decays to its value at infinity faster than does the velocity. Again, details may be found in [16].

Unfortunately, although the boundary conditions associated with the use of (4.10c) can be useful, in practice we cannot employ this particular formulation of the viscous term. The reason for this is that the choice (4.10c) requires the use of divergence-free finite element velocity fields in order for the form $a_3(\cdot, \cdot)$ to be coercive on $\mathbf{Z}^h$. This condition is also needed to guarantee the stability of the approximations and, for the other three cases, (4.10a), (4.10b), and (4.10d), is trivially satisfied for any choice of conforming discrete velocity space.

Fortunately, the boundary conditions (4.11c) are *approximately* the natural boundary conditions associated with the choice (4.10d). In fact, for (4.10d), we have the natural boundary conditions

$$p - \nu\,\text{div}\,\mathbf{u} = r \quad \text{on } \Gamma/\Gamma_n \qquad \text{and} \qquad \boldsymbol{\omega} = \mathbf{s}/\nu \quad \text{on } \Gamma/\Gamma_\tau. \tag{4.11d}$$

The second of these is identical to the second of (4.11c). If ν is "small," and/or if we assume the incompressibility constraint holds up to portions of the boundary where the normal velocity is *not* specified, then $(p - \nu\,\text{div}\,\mathbf{u})$ is essentially equal to p. Thus we recover, at least approximately, the first boundary condition of (4.11c).

In summary, when we have velocity and/or stress boundary conditions, we should use (4.11b) in (4.7), and when we have velocity and/or pressure and/or vorticity boundary conditions the choice (4.11d) is preferable.

The discretization of (4.7)–(4.8) follows the usual procedures once we choose the finite element spaces for the velocity and the pressure approximations. The natural boundary conditions are automatically accounted for by the inclusion of the linear functional $d(\cdot)$ in (4.7). The essential boundary conditions on the components of the velocity can be enforced in a manner analogous to that described in Section 6.4.2 for the case where the complete velocity is specified on the whole boundary. All material relating to the div-stability condition (2.5) is essentially still valid, and thus, insofar as that condition is concerned, the particular choices of finite elements discussed in Section 6.3 may still be used.

In actuality, there are very few rigorous error estimates available for boundary conditions other than the velocity. For polygonal or polyhedral domains Ω, the error estimates of Section 6.2.3 are still valid. However, for domains with *curved boundaries*, using the type of weak formulations discussed here may result in a loss of accuracy. For example, for (4.10b) with normal velocity and tangential stress boundary conditions, it was shown by Verfürth [27] that there is a loss of accuracy due to a Babuska-type paradox, i.e., the limit of solutions of problems posed on polygonal approximations to $\Omega \subset R^2$ is not the solution of the problem posed on Ω. Verfürth [28] has also shown how through the use of additional Lagrange multipliers on the boundary, a different weak formulation yields optimal accuracy.

Acknowledgments. We wish to acknowledge the support, both for the preparation of this paper and for our work on finite element methods for the Navier–Stokes equations, to the Air Force Office of Scientific Research under Grant AFOSR-81-0101 and to the Institute for Computer Applications in Science and Engineering under Contracts NAS1-17070 and NAS1-18107. We also wish to acknowledge the many fruitful interactions with colleagues and friends, most notably J. Boland, M. Cayco, G. Fix, R. Nicolaides, and J. Peterson.

References

1. A. Baiocchi and A. Capelo, *Variational and Quasivariational Inequalities*, Wiley, Chichester, 1984.
2. J. Boland and M. Cayco, "Stable three-dimensional elements for the Navier–Stokes equations," to appear.
3. A. Brooks and T. Hughes, "Streamline upwind/Petrov–Galerkin formulations for convection dominated flows," *Comput. Methods Appl. Mech. Engrg.*, **30** (1982), 227–241.
4. J. Boland and R. Nicolaides, "Stability of finite elements under divergence constraints," *SIAM J. Numer. Anal.*, **20** (1983), 722–731.
5. J. Boland and R. Nicolaides, "On the stability of bilinear-constant velocity-pressure finite elements." *Numer. Math.*, **44** (1984), 219–222.
6. J. Boland and R. Nicolaides, "Stable and semistable low-order finite elements for viscous flows." *SIAM J. Numer. Anal.*, **22** (1985), 474–492.
7. M. Bercovier and O. Pironneau, "Error estimates for finite element method solution of the Stokes problem in primitive variables," *Numer. Math.*, **33** (1979), 211–224.
8. M. Cayco, "Finite Element Methods for the Streamfunction Formulation of the Stationary Navier–Stokes Equations," Ph.D. Thesis, Carnegie–Mellon University, Pittsburgh, 1985.
9. P. Ciarlet, *The Finite Element Method for Elliptic Problems*, North-Holland, Amsterdam, 1978.
10. M. Cayco and R. Nicolaides, "Finite element methods for the streamfunction formulation of viscous incompressible flows," to appear.
11. M. Crouzeix and P. Raviart, "Conforming and non-conforming finite element methods for solving the stationary Stokes equations," *RAIRO Anal. Numér.*, **7** (1973), 33–76.

12. M. Fortin, "Analysis of the convergence of mixed finite element methods," *RAIRO Anal. Numér.*, **11** (1977), 341–354.
13. G. Fix, M. Gunzburger, and J. Peterson, "On finite element approximations of problems having inhomogeneous essential boundary conditions," *Comput. Math. Appl.*, **9** (1983), 687–700.
14. M. Gunzburger and J. Peterson, "On conforming finite element methods for the inhomogeneous stationary Navier–Stokes equations," *Numer. Math.*, **42** (1983), 173–194.
15. V. Girault and P. Raviart, *Finite Element Approximation of the Navier–Stokes Equations.* Springer-Verlag, Berlin, 1986.
16. M. Gunzburger, C. Liu, and R. Nicolaides, "Algorithmic and theoretical results on computation of incompressible viscous flows by finite element methods." *Comput. & Fluids*, **13** (1985), 361–373.
17. M. Gunzburger, R. Nicolaides, and J. Peterson, "On conforming finite element methods for incompressible viscous flow problems," *Comput. Math. Appl.*, **8** (1982), 167–179.
18. C. Johnson and J. Pitkäranta, "Analysis of some mixed finite element methods reduced integration," *Math. Comput*, **38** (1982), 375–400.
19. P. Jamet and P. Raviart, "Numerical solution of the stationary Navier–Stokes equations by finite element methods," Lecture Notes in Computer Science, Vol. 10, Springer-Verlag, Berlin, 1973.
20. O. Ladyzhenskaya, *The Mathematical Theory of Viscous Incompressible Flow*, Gordon and Breach, New York, 1969.
21. J. Oden and J. Reddy, *An Introduction to the Mathematical Theory of Finite Elements*, Wiley, New York, 1976.
22. R. Sani, P. Gresho, R. Lee, D. Griffiths, and M. Engelman, "The cause and cure(?) of the spurious pressures generated by certain FEM solutions of the incompressible Navier–Stokes equations," *Internat. J. Numer. Methods Fluids*, **1** (1981), 17–43 and 171–204.
23. R. Teman, *Navier–Stokes Equations*, North-Holland, Amsterdam, 1979.
24. F. Thomasset, *Implementation of Finite Element Methods for Navier–Stokes Equations*, Springer-Verlag, New York, 1981.
25. C. Taylor and P. Hood, "A numerical solution of the Navier–Stokes equations using the finite element method," *Comput. & Fluids*, **1** (1973), 73–100.
26. R. Verfürth, "Error estimates for mixed finite element approximation of the Stokes equations, *RAIRO Anal. Numer.* **18** (1984), 175–182.
27. R. Verfürth, "Finite element approximation of steady Navier–Stokes equations with mixed boundary conditions," *Mod. Math. Anal. Numer.*, **19** (1985), 461–475.
28. R. Verfürth, "Finite element approximation of stationary Navier–Stokes equations with slip boundary condition," to appear.

CHAPTER 7

Gradient Weighted Moving Finite Elements in Two Dimensions

Neil Carlson and Keith Miller

7.1. Introduction

The Moving Finite Element (MFE) method is especially suited to those many highly nonlinear systems which develop shocks and other sharp moving fronts. By allowing the nodes to move and concentrate automatically, and by using an implicit stiff equation solver for the resulting ordinary differential equations (ODEs), we can often solve such problems with far fewer nodes and with far larger time steps than would otherwise be the case. Recently, we have placed special emphasis on development of the *gradient weighted* MFE method (GWMFE), introduced by Miller in [7] and [8]. Experience with a general purpose one-dimensional code of Carlson and Miller over the past several years (see, for example, [8] and [10]), has shown the method to be extremely efficient and robust. In this paper we report on some preliminary trials with our recently completed two-dimensional GWMFE code which indicate that this superior efficiency and robustness carry over to multidimensions. Our trial examples involve shocks for the system of Burger's equations in two dimensions and interacting vortices for a complex nonlinear heat equations.

The MFE method was introduced by K. Miller and R. Miller [1] and by K. Miller [2]. The first two-dimensional computations were by Alexander *et al.* [3] on the Stefan problem. Djomehri [4] and Gelinas *et al.* [5] applied the method to a wide variety of partial differential equation (PDE) systems in one dimension. Djomehri and Miller [6] give a detailed outline of a proposed general purpose two-dimensional MFE systems code; this structure has served as the starting point for the two-dimensional systems code of Djomehri, Doss, Gelinas, and Miller (reported in [7, 8, 9] and elsewhere). A more extensive report on the present work will appear in [10].

There are so many alternative ideas for ways to move or adapt the nodes in finite element and finite difference methods appearing in the literature that we make no attempt to survey them here. Instead we apologize to our coworkers in this field and concentrate here only on MFE-type methods.

7.2. The Gradient Weighted MFE Method in Two Dimensions

The basic idea of two-dimensional piecewise linear MFE is quite simple. Consider a time evolving PDE $\dot{v} = L(v)$, where L is a first- or second-order nonlinear operator on the two-dimensional region Ω. Our approximant $u(t)$ is allowed to be a continuous piecewise linear function on an evolving triangular grid, with its nodal amplitudes and positions $\{a_j, x_j, y_j\}$ all treated as unknowns. We obtain ODEs for the nodal parameters by requiring that their time derivatives at each instant be chosen so as to minimize $\|\dot{u} - L(u)\|$, the L^2 norm of the PDE residual. Because $\dot{u}$ by the chain rule depends linearly on the parameter derivatives

$$\dot{u} = \sum_j \dot{a}_j \frac{\partial u}{\partial a_j} + \dot{x}_j \frac{\partial u}{\partial x_j} + \dot{y}_j \frac{\partial u}{\partial y_j} \equiv \sum_j \dot{a}_j \alpha^j + \dot{x}_j \beta^j + \dot{y}_j \gamma^j, \tag{2.1}$$

this is a linear least squares problem and the canonical equations for this minimization are that $\dot{u} - L(u)$ be orthogonal to all the basis functions α^j, β^j, γ^j. It is easy to see with a few diagrams and tiny increments in a_j, x_j, and y_j that α^j is the standard continuous piecewise linear "hat" function which is 1 at the jth node and 0 at the others, and that

$$\beta^j = -u_x \alpha^j, \qquad \gamma^j = -u_y \alpha^j, \tag{2.2}$$

where u_x and u_y are the piecewise constant partial derivatives. There are difficulties with degeneracies of the parametrization when the graph of u at any one node becomes planar, but these are taken care of by regularization techniques. If $L(u)$ contains second-order terms in divergence form such as the Laplacian, then such terms are delta functions along the triangle edges, with infinite L^2 norm; this difficulty is handled by a mollification interpretation however; see [1] and [7].

One of the major difficulties from earliest experience with MFE is that often *all* the nodes tend to run into the shocks and sharp fronts. In the past we have devised a variety of somewhat ad hoc techniques to avoid this. More recently, however, we have concentrated on the idea of gradient weighted MFE, and such methods have been extremely robust and trouble-free in one-dimensional computations over the past several years; see [7] and [8].

The problem with standard MFE is that the L^2 norm in the basic residual minimization of $\|\dot{u} - L(u)\|$ is inappropriate for those problems which develop arbitrarily steep moving fronts, because then $\dot{u}$ ($\approx -cu_x$ in one dimension, where c is the speed of the front) has a delta function type behavior. Since this in the limit has infinite L^2 norm, the method, of course, tries to crowd most of the nodes into the steepest part of the front. Notice that $[\dot{u}]_N \equiv \dot{u}(1 + u_x^2 + u_y^2)^{-1/2}$ gives the *normal* component of the motion of the graph of u, and that this is a decently bounded L^2 function of surface area (of arc length in one dimension) independently of the steepness of the moving front. Thus in the basic PDE residual minimization we switch to the L^2 norm of the normal component of the residual, integrated with respect to surface area dS. We instead choose $\{\dot{a}_j, \dot{x}_j, \dot{y}_j\}$ at each instant so as to minimize the "gradient weighted norm"

$$\||\dot{u} - L(u)\||^2 \equiv \int [\dot{u} - L(u)]_N^2 \, dS$$

$$= \int (\dot{u} - L(u))^2 (1 + u_x^2 + u_y^2)^{-1/2} \, dx \, dy. \tag{2.3}$$

The gradient weighting function $w \equiv (1 + u_x^2 + u_y^2)^{-1/2}$ thus serves to de-emphasize the steeper portions of the graph, and the nodes have a tendency to congregate instead more in the regions of sharp curvature of the graph.

The canonical equations for this minimization are now that $\dot{u} - L(u)$ be orthogonal to all the basis functions α^j, β^j, γ^j in the gradient weighted inner product

$$\frac{1}{2}\frac{\partial \psi}{\partial \dot{a}_i} = ((\dot{u} - L(u), \alpha^i)) = (\dot{u} - L(u), \alpha^i w) = 0, \tag{2.4}$$

$$\frac{1}{2}\frac{\partial \psi}{\partial \dot{x}_i} = ((\dot{u} - L(u), \beta^i)) = (\dot{u} - L(u), \beta^i w) = 0, \tag{2.5}$$

$$\frac{1}{2}\frac{\partial \psi}{\partial \dot{y}_i} = ((\dot{u} - L(u), \gamma^i)) = (\dot{u} - L(u), \gamma^i w) = 0. \tag{2.6}$$

Since w is piecewise constant, the changes required in the MFE code are not too drastic.

7.3. Regularization

When a portion of the graph of u becomes nearly "flat" (i.e., planar), there are tangential motions $\{\dot{a}_j, \dot{x}_j, \dot{y}_j\}$ of the nodal parameters which produce almost no normal motion $[\dot{u}]_N$; thus the quadratic form of the mass matrix is nearly degenerate in those tangential directions. To remove this degeneracy we penalize on each triangular element against the divergence and the curl of $[\dot{u}]_T$, where $[\dot{u}]_T$ is the tangential component of the piecewise linear motion of the graph of u induced by linear interpolation from the motions at the nodes. Thus, our ODEs for GWMFE are obtained from the variational requirement that at each instant the $\{\dot{a}_j, \dot{x}_j, \dot{y}_j\}$ be chosen so as to minimize the functional

$$\psi \equiv \int [\dot{u} - L(u)]_N^2 \, dS + c^2 \int (\operatorname{div}[\dot{u}]_T)^2 \, dS + c^2 \int (\operatorname{curl}[\dot{u}]_T)^2 \, dS, \tag{3.1}$$

where c is a tiny constant. These regularization terms can be seen to add stabilizing "forces" in the tangential directions of the same order of magnitude that would be provided naturally by the mass matrix itself if the graph of u

had a "nonflatness" or "wobble" of $\pm c$ at each node. Thus, by choosing c several times larger than the local truncation error tolerance η, we ensure that the regularization takes over to give stabilization in the tangential directions well before the mass matrix becomes degenerate in those directions. In most of our calculations, in one dimension and in two dimensions, we have a tendency to choose $\eta \approx 10^{-3}$ in normalized units, and $c \approx 2$ to 5 times η. This is the natural generalization to two dimensions of the inverse linear "segment viscosities" (see [7] and [8]) which have worked so robustly and automatically in one dimension for the past several years.

In the case of a *system* of PDEs we let our approximants $u_1(t), \ldots, u_N(t)$ for the component functions each be piecewise linear, but we prefer to require that they share a common moving grid. The functional ψ for minimization is then the sum of the preceding ψ's over the N component graphs.

7.4. Trial Results

In Figures 7.1–7.12 we show computer graphics for some initial trials (all with a 16×16 grid) of the new code GWMFE2DS (Gradient Weighted MFE for 2-D Systems). The code, highly vectorized for the CRAY, implements the second-order diagonally implicit Runge–Kutta ODE solver DIRK2 (DIRK2) (see [2]) for solving our highly stiff ODEs. The structure of this code is quite similar to that detailed in [6].

The computer graphics shown here are by the code GRAFPAK developed by Lee Busby of our group. This highly sophisticated and user friendly graphics package is an essential adjunct to our one-dimensional and two-dimensional codes.

7.4.1. Two-Dimensional Burger's Equations

We consider first the system of Burger's equations in two dimensions on the unit square Ω,

$$\begin{aligned} u_t &= -(uu_x + vu_y) + \nu\Delta u, \\ v_t &= -(uv_x + vv_y) + \nu\Delta v. \end{aligned} \tag{4.1}$$

Here (u, v) can be considered to be the velocity of a gas which is at such low density that pressure forces can be neglected. A gas molecule with initial velocity (u, v) therefore continues ahead with that same velocity until it impinges upon other molecules with different velocities attempting to occupy the same position, in which case the second-order diffusion terms, with a tiny diffusion coefficient ν, take over to diffuse the sharp gradients in velocity. Reflection boundary conditions are given on the left and right edges of Ω and Dirichlet boundary conditions on the top and bottom edges; there we inject our gas molecules with the velocities

$$u = \pm\tau_1 \sin \pi x, \qquad v = \mp 1 + \tau_2 \cos \pi x. \tag{4.2}$$

The initial conditions are linearly interpolated between the boundary values on the top and bottom edges. This test problem (by now standard for us) was considered previously in [8] and [9] for the standard two-dimensional MFE method (without gradient weighting). With $\tau_1 = \tau_2 = 0.6$ here, we have the case II considered there (except that here $v = 10^{-3}$ is twice as small as there, giving twice as thin a shock). At $t \approx 0.5$, separate shocks begin forming near the top left and bottom right corners. With time, these shocks extend farther into the midregion and twist and eventually merge into a single diagonal shock for u and for v; u however, also forms a shock on a portion of the left and right edges. We show grid views in Figures 7.1–7.2 and contour plots for u in Figures 7.3–7.4, both at the times $t = 0.5$ and $t = 0.8$. Figure 7.5 shows a ten times enlargement of the grid about the shock at the top left corner at $t = 0.8$. This took only 44 time steps of our DIRK2 implicit stiff ODE solver and 48 cpu seconds (all cpu times are on a CRAY-XMP). The shock formed very smoothly and effortlessly with no need to cut Δt greatly as it was forming.

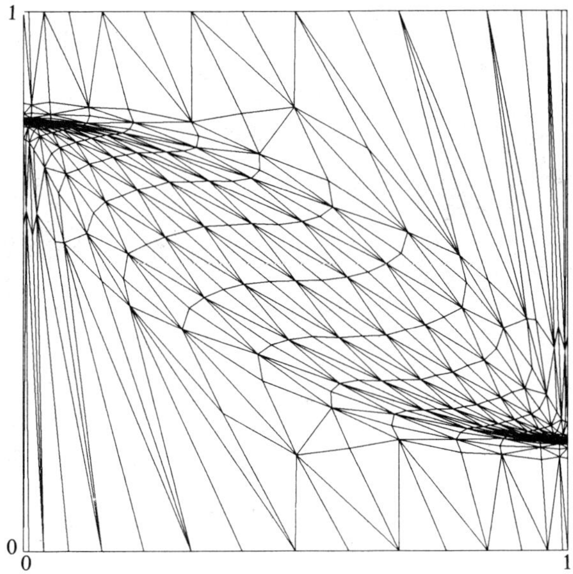

FIGURE 7.1. Burger's equations. Grid view at $t = 0.5$ of the 16×16 grid. Shocks begin forming near top left and bottom right corners.

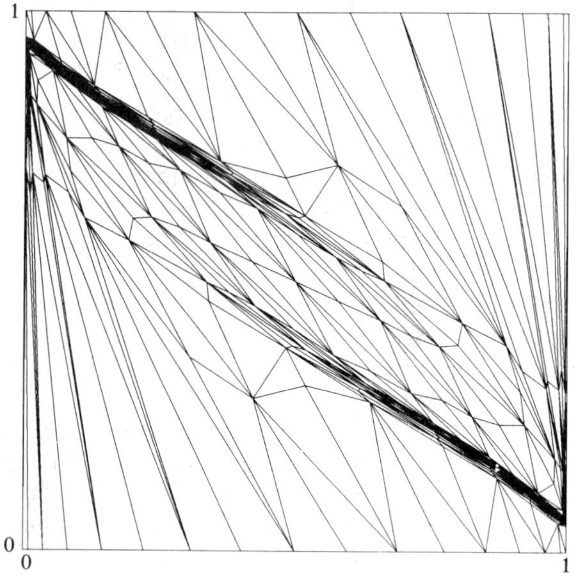

FIGURE 7.2. Burger's equations. Grid view at $t = 0.8$. Shocks have drifted toward corners and extended farther into midregion.

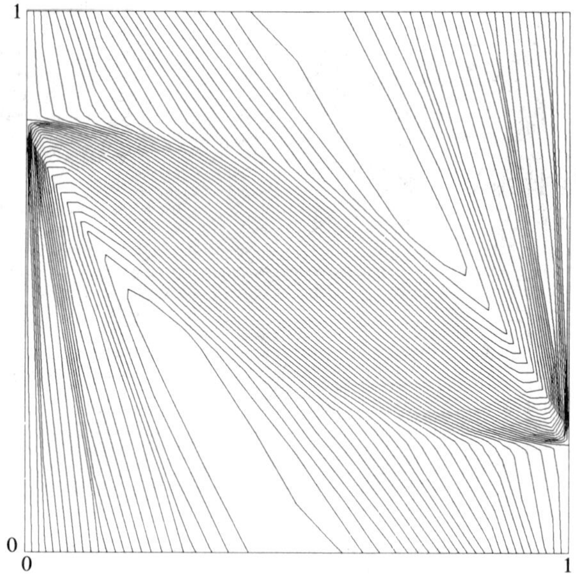

FIGURE 7.3. Burger's equations. Contour plot at $t = 0.5$ of the u component.

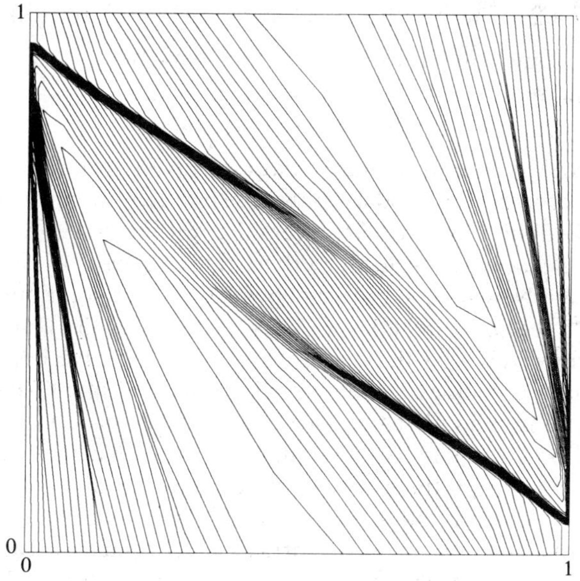

FIGURE 7.4. Burger's equations. Contour plot at $t = 0.8$ of the u component.

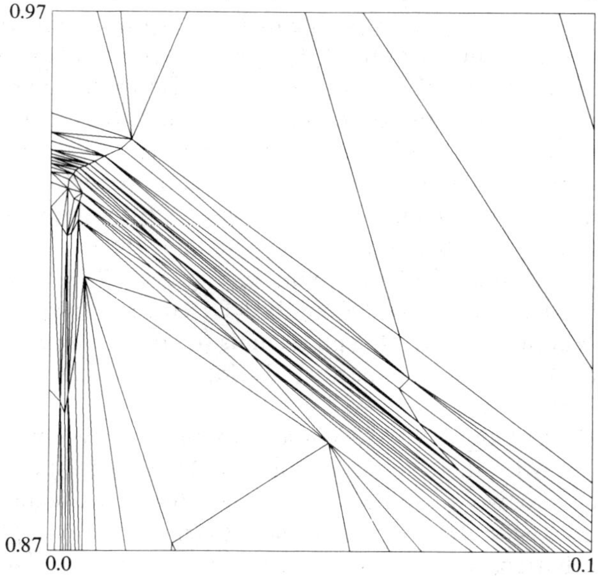

FIGURE 7.5. Burger's equations. Grid view at $t = 0.8$; ten times enlargement of the grid at the shock in upper left corner.

7.4.2. The Nonlinear Complex Heat Equation

Consider next some computations on vortex dynamics for the complex nonlinear heat equation

$$\psi_t = \Delta\psi + (1 - |\psi|^2)\psi, \tag{4.3}$$

which is a model by John Neu (mathematics, University of California, Berkeley) for some processes in modern physics. Here the vector field $\psi = u + iv$ tries to have modulus 1 everywhere, except that within any simple closed curve where the Poincaré index is nonzero there must be zeros of the field. There are certain radially symmetic time-independent vortex solutions of the form $\psi(z) = U_n(r)e^{in\theta}$, of Poincaré index $n = \pm 1, \pm 2, \ldots$. Neu [11] is studying (by asymptotics) how these vortices interact at a distance. He had conjectured that a $+1$ and a -1 index vortex would attract and annihilate each other (our computations show that they do) and that the vortices of order higher than 1 would be unstable (our computations show that they are, and Neu, from insight gained from these computations, has now proved that they are).

Figures 7.6–7.9 show our calculation of a $+3$ index vortex becoming unstable and splitting into three $+1$ vortices which then repel each other with $\approx 1/r$ velocity. The core size of a vortex is $O(1)$; our solution domain is the disc of radius 100. We begin with radially symmetric initial values with an approximate formula of Neu for the core shape function $U_3(r)$. We impose $e^{i3\theta}$ Dirichlet boundary conditions at $r = 100$. The morphogenesis of the index $+3$ vortex into three $+1$ vortices is quite interesting. The solution settles down quickly to the true radially symmetric index $+3$ vortex, but at $t \approx 4$ this vortex starts to split into five separate vortices, one -1 index zero at the origin and four $+1$ index zeros repelling themselves away from each other (in this way the exact symmetry of the MFE grid and solution about the x and y axes is maintained). But this configuration is also unstable and we see in Figure 7.6 that by $t = 13.6$ this exact numerical symmetry has also been lost; the -1 index zero at the origin has been attracted well upward and is already being attracted slightly preferentially toward the upper leftmost $+1$ index zero, which will shortly thereafter annihilate with it. By the final time $t = 138.7$ the Δt is quite large ($= 3.3$) and we see the remaining three $+1$ index vortices in the vector, u-contour, and grid plots of Figures 7.7–7.9 still smoothly repelling each other. This took 200 implicit DIRK2 time steps to solve our ODEs to high accuracy, and 148 cpu seconds (there are some sure-fire code improvements that we would expect would decrease this cpu time by factors of at least 2 or 3).

Figures 7.10–7.12 show the calculation (using x and y symmetries on a quadrant to get twice the resolution from our 16×16 grid) of a $+2$ index vortex splitting into two $+1$ index vortices. Here, to achieve better reproduceability, we begin with Neu's approximate $U_2(r)e^{i2\theta}$ formula plus a slight smooth perturbation which splits the double zero into two zeros at $z \approx \pm i10^{-3}$. The initial grid here (as with the previous case) is equally spaced on θ and equally spaced on $U_2(r)$ in the radial direction. At $t = 5000$ the Δt is 107, with 329 DIRK2 steps in 217 cpu seconds. This cpu time is actually much

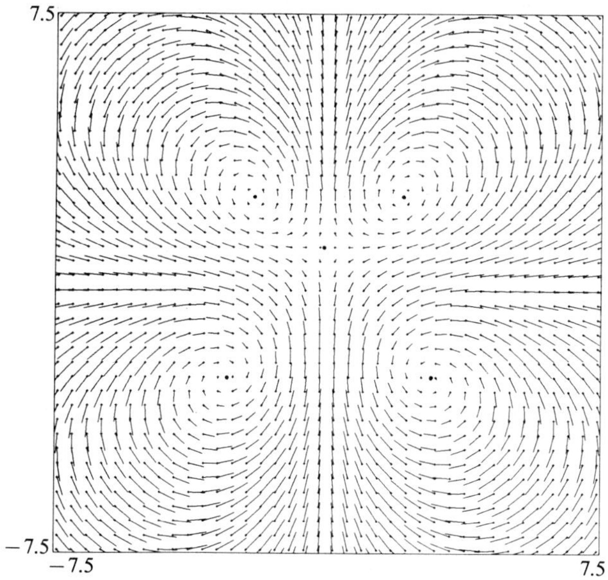

FIGURE 7.6. Index 3 vortex splitting. $13\frac{1}{3}$ times enlargement of vector field near origin at $t = 13.6$. The initial index $+3$ vortex has split into one index -1 vortex at the origin and four index $+1$ vortices. The index -1 vortex is being preferentially attracted to the upper left index $+1$ vortex and will annihilate with it shortly.

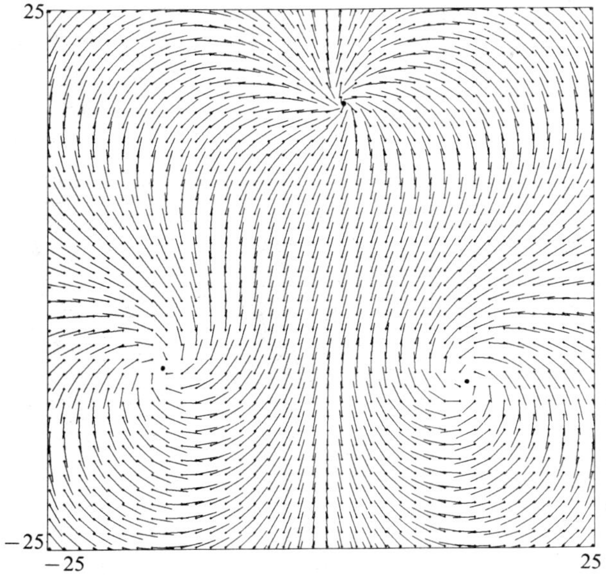

FIGURE 7.7. Index 3 vortex splitting. Four times enlargement of vector field near origin at final time $t = 138.7$. The remaining three index $+1$ vortices continue smoothly repelling each other. Δt is 3.3.

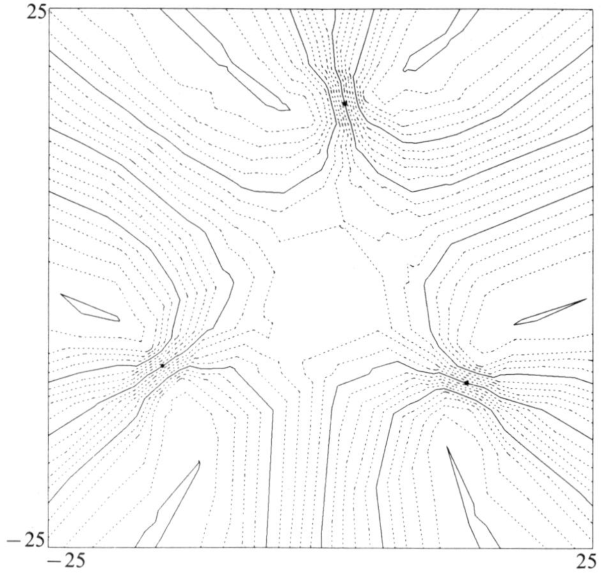

FIGURE 7.8. Index 3 vortex splitting. Four times enlargement of contour plot of the u component at final time $t = 138.7$.

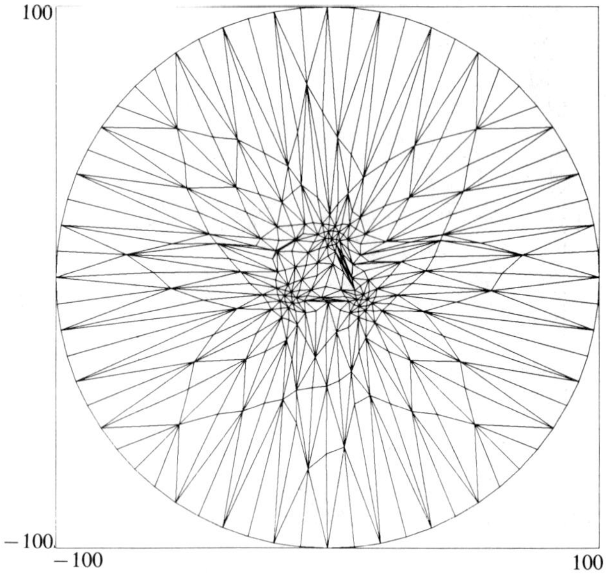

FIGURE 7.9. Index 3 vortex splitting. Grid view at final time 138.7 of the 16×16 grid.

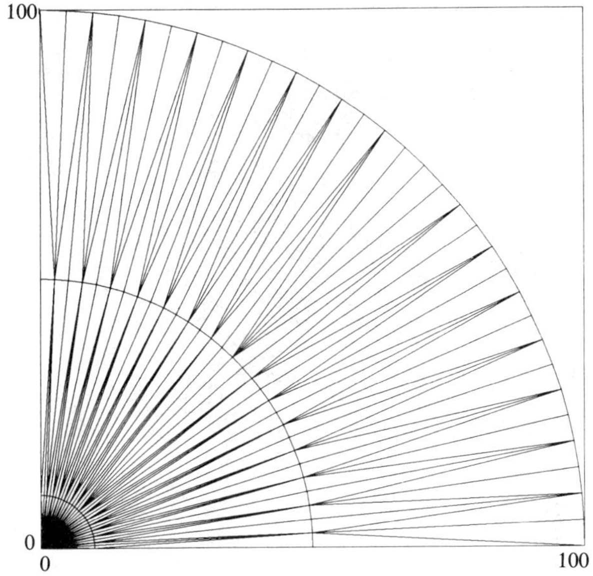

FIGURE 7.10. Index 2 vortex splitting. Grid view at initial time $t = 0$ of the 16×16 grid.

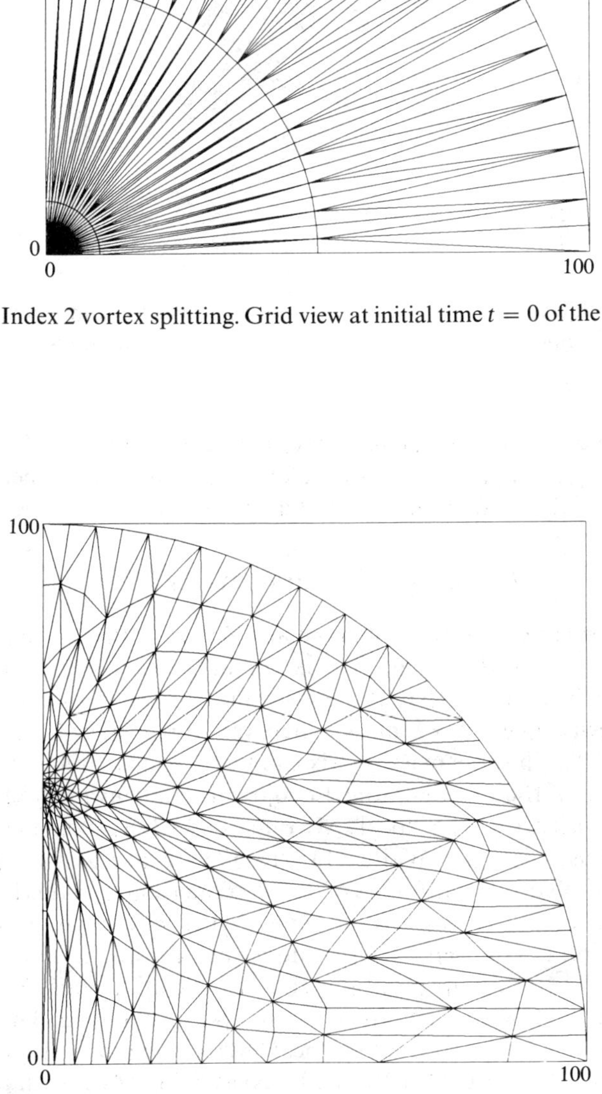

FIGURE 7.11. Index 2 vortex splitting. Grid view at final time $t = 5000$. Δt is 107.

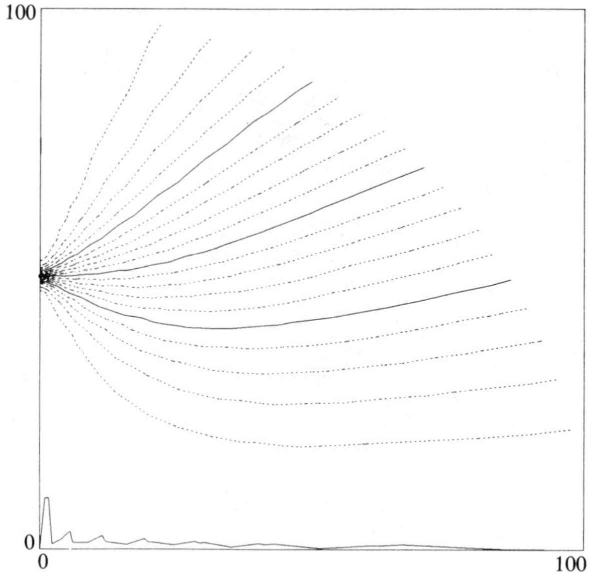

FIGURE 7.12. Index 2 vortex splitting. Contour plot of the u component at final time $t = 5000$. This index $+1$ vortex and its reflected image continue smoothly repelling each other.

greater than would usually be necessary because we were solving our ODEs to very high accuracy, this for the sake of studying the conjectured $\approx r^{-1}$ velocity of repulsion (we found instead a consistent $\approx r^{-1.3}$ type velocity).

7.5. Explicit Versions of MFE

There has been recent work by Wathen and Baines at Bristol and Reading on the structure of the MFE mass matrix; see [12, 13, 14, 15, 16, 17]. Wathen [12, 13, 14] has shown that the eigenvalues of the diagonally preconditioned piecewise linear MFE mass matrix in n-dimensions lie in the interval $[1/2, 1 + (n/2)]$. Thus the mass matrix can be inverted extremely rapidly by a few iterations of the preconditioned conjugate gradient method. Moreover, Baines introduces in [16] a "local" MFE method whose mass matrix is block diagonal. Thus Baines and Wathen [15, 16, 17] are able to use *explicit* rather than implicit solvers for the ODEs of the MFE method, and this leads to some highly efficient methods for certain purely hyperbolic (no diffusion) systems of conservation laws (when the MFE nodes are allowed to move freely with the desired wave speeds the resulting ODEs can be quite nonstiff). Miller [18] has simplified and extended Wathen's basic spectrum bound. This simplification allows him to characterize the "local" MFE methods of Baines (or more generally "local" GWMFE or FE methods) as being MFE with the L^2 norm replaced by a comparable "diagonal norm." Such methods no longer exactly minimize the PDE residual $\|\dot{u} - L(u)\|$, but they do minimize it to within a factor of two, and moreover their resulting mass matrix is block diagonal. We are convinced that the time is now ripe, with these new "local" MFE concepts, for work on *explicit* versions of MFE (for certain types of inviscid purely hyperbolic systems of conservation laws with shocks).

Acknowledgments. This work was supported by the National Science Foundation under Grant DMS 83-03267.

References

1. K. Miller and R. Miller, "Moving finite elements. Part I," *SIAM J. Numer. Anal.*, **18**, No. 6 (1981), 1019–1032.
2. K. Miller, "Moving finite elements. Part II," *SIAM J. Numer. Anal.*, **18**, No. 6 (1981), 1033–1057.
3. R. Alexander, P. Manselli, and K. Miller, "Moving finite elements for the Stefan problem in two dimensions," *Rend. Accad. Naz. Lincei* (*Rome*), **67** (1979), 57–61.
4. M.J. Djomehri, *Moving Finite Element Solution of Systems of Partial Differential Equations in One Dimension.* Ph.D. Thesis, Department of Mathematics, University of California, Berkeley, December 1982. Center for Pure Applied Mathematics, Report PAM-125, January 1983, pp. 1–83.
5. R. Gelinas, S. Doss, and K. Miller, "The moving finite element method: Applications to general equations with multiple large gradients." *J. Comput. Phys.*, **40**, No. 1 (1981), 202–249.
6. M.J. Djomehri and K. Miller, *A Moving Finite Element Code for General Systems of PDEs in 2-D*, University of California, Berkeley, October 1981, Center for Pure Applied Mathematics, Report PAM-57.
7. K. Miller, "Alternate modes to control the nodes in the moving finite element method," in *Adaptive Computational Methods for Partial Differential Equations* (I. Babuška, J. Chandra, and J. Flaherty, eds), SIAM, Philadelphia, 1983, pp. 165–182. Proceedings of a workshop held at the University of Maryland, February 14–16, 1983.
8. K. Miller, "Recent results on finite element methods with moving nodes," in *Accuracy Estimates and Adaptive Refinement in Finite Element Calculations* (I. Babuška, O.C. Zienkiewicz *et al.*, eds.), Wiley, New York, 1986, pp. 325–338.
9. M.J. Djomehri, S. Doss, R. Gelinas, and K. Miller, "Applications of the moving finite element method for systems in 2-D," submitted in July 1985 to *J. Comput. Phys.*
10. N. Carlson and K. Miller, "The gradient weighted moving finite element method in two dimensions," CPAM Report, University of California, Berkeley, in preparation; to be submitted to *SIAM J. Numer. Anal.*
11. J. Neu, "A study of vortex dynamics," CPAM Report, University of California, Berkeley, in preparation.
12. A. Wathen, "Moving Finite Elements and Oil Reservoir Modeling." Ph.D. Thesis, Mathematics, University of Reading, 1984.

13. A. Wathen, "Mesh-independent spectra in the moving finite element equations," *SIAM J. Numer. Anal.*, **23**, No. 4, (1986), 797–814.
14. A. Wathen, "Attainable Eigenvalue Bounds for the Galerkin Mass Matrix." Mathematics Report, University of Bristol, August 1985, submitted to *IMA J. Numer. Anal.*
15. A. Wathen and M. Baines, "On the structure of the moving finite element equations." *IMA J. Numer. Anal.*, **5** (1985), 161–182.
16. M. Baines, "Local Moving Finite Elements." Numerical Analysis Report, Institute for Comparative Fluid Dynamics, University of Oxford and Reading, September 1985.
17. A. Wathen, "Moving Finite Elements." Mathematics Report AM-85-05, University of Bristol, 1985.
18. K. Miller, "On the mass matrix spectrum bounds of Wathen and application to the local moving finite elements of Baines." CPAM Report, University of California, Berkeley, in preparation.

A postscript by K. Miller: I would like to include here in print a personal wish which I've been expressing wherever I've presented this material since February of 1983. It is my hope that those methods will not be used on problems having anything to do with the accelerating nuclear arms race—anything involving the design, delivery, testing, or even the interception of nuclear weapons. I am totally convinced of one fact: that (from now on out) we can assuredly obliterate our adversaries and they can assuredly obliterate us—and that nothing we can produce will ever change that fact. I feel certain that many of the new weapons developments currently being advocated create illusions and destabilization that are terribly dangerous to our long-term security, to civilization, and to this world we all love. I would prefer that my work not contribute to such illusions.

CHAPTER 8

Adaptive Finite Element Flux Corrected Transport Techniques for CFD

K. Morgan, J. Peraire, and R. Löhner

8.1. Introduction

In previous papers [1, 2] we have described an explicit finite element solution procedure for the compressible Euler and Navier–Stokes equations. The approach was a finite element equivalent of a two-step Lax–Wendroff scheme and was implemented on unstructured triangular or tetrahedral grids. An important feature of the work was the use of adaptive mesh refinement methods for the solution of steady state problems in two dimensions, using error indicators based upon interpolation theory.

In this paper, some recent developments in the extension of this approach are considered. We will describe how the basic solution procedure can be modified in a straightforward manner to produce a high resolution scheme on unstructured grids. This is accomplished by utilizing, in a finite element context, Zalesak's [3] multidimensional extension of the flux corrected transport (FCT) ideas of Boris and Book [4]. The problem of triangular mesh generation will be addressed, and the adaptive mesh approach will be widened to handle two-dimensional problems involving strongly transient phenomena. This will be implemented by allowing adaptive refinement and derefinement of the mesh as the solution proceeds. Finally, it will be demonstrated how directional refinement procedures can be incorporated for the efficient computation of steady two-dimensional flows involving significant one-dimensional features.

8.2. Basic Algorithm

The basic solution algorithm will be briefly described for the two-dimensional Navier–Stokes equations written in the conservative form

$$\frac{\partial \mathbf{U}}{\partial t} + \frac{\partial \mathbf{F}_j}{\partial x_j} = \frac{\partial \mathbf{G}_j}{\partial x_j} \qquad (j = 1, 2), \tag{2.1}$$

where $\mathbf{U}$ is the vector of conservation variables and $\mathbf{F}_j$ and $\mathbf{G}_j$ denote the advective and viscous flux vectors respectively. A time-stepping scheme for this equation can be developed, in an operator-split fashion, by treating the diffusion terms in an explicit manner and the advective terms in the Lax–Wendroff fashion [5]. The result is that

$$\mathbf{U}^{m+1} = \mathbf{U}^m - \Delta t \frac{\partial \mathbf{F}_j^m}{\partial x_j} + \frac{\Delta t^2}{2} \frac{\partial}{\partial x_j}\left[\mathbf{A}_j^m \frac{\partial \mathbf{F}_k^m}{\partial x_k}\right] + \Delta t \frac{\partial \mathbf{G}_j^m}{\partial x_j}, \tag{2.2}$$

where a superscript m denotes an evaluation at time $t = t_m$, $t_{m+1} = t_m + \Delta t$ and

$$\mathbf{A}_j = \frac{d\mathbf{F}_j}{d\mathbf{U}}. \tag{2.3}$$

The spatial domain, Ω, is discretized by using three-noded linear triangular elements, and a weighted residual [6] form of equation (2.2) is considered. The resulting integrals are evaluated exactly (in a two-step fashion by first calculating an element level approximation to $\mathbf{U}^{m+1/2}$ [7]), leading to an equation

$$\mathbf{M}\boldsymbol{\delta}\mathbf{U}^H = \mathbf{f}^m, \tag{2.4}$$

where $\mathbf{M}$ is the consistent mass matrix, $\boldsymbol{\delta}\mathbf{U}^H$ is the vector of changes in the nodal values of $\mathbf{U}$ over the time step, and the superscript H is introduced for use later. For the simulation of transient flows, this equation system can be solved iteratively and explicitly [8] and the method coupled with a domain splitting technique [9] to produce an efficient computational procedure. For steady flows, local time steps are employed and equation (2.4) is replaced by the explicit scheme

$$\mathbf{M}_l\boldsymbol{\delta}\mathbf{U}^H = \mathbf{f}^m, \tag{2.5}$$

where $\mathbf{M}_l$ is the lumped diagonal mass matrix. It should be noted that in one dimension equation (2.5) reduces to the well-known finite difference scheme of Burstein [10].

For problems involving strong shocks, the solution $\mathbf{U}^{m+1}$ is smoothed, by the application of artificial viscosity [11], before proceeding to the next time step.

8.3. FCT Extension

A more robust solution scheme, giving better resolution of flow discontinuities, can be produced by adding an FCT procedure to the preceding process. Erlebacher [12] and Parrott and Christie [13] have demonstrated how Zalesak's [3] multidimensional extension of the FCT algorithm of Boris and Book [4] can be implemented on triangular grids. The idea is to combine a high-order scheme with a low-order scheme in such a way that the high-order scheme is employed in regions where the flow variables vary smoothly, whereas the low-order scheme is favored in those regions where the variables vary abruptly. The low-order scheme should give monotonic results for the problem of interest.

The solution method of equation (2.4) will be used as a high-order scheme and a high-order solution, $\mathbf{U}^H$, after a time step can be defined by

$$\mathbf{U}^H = \mathbf{U}^m + \boldsymbol{\delta}\mathbf{U}^H. \tag{3.1}$$

Similarly, a low-order solution, $\mathbf{U}^L$, is defined by

$$\mathbf{U}^L = \mathbf{U}^m + \boldsymbol{\delta}\mathbf{U}^L, \tag{3.2}$$

where the low-order increment is calculated as

$$\boldsymbol{\delta}\mathbf{U}^L = \boldsymbol{\delta}\mathbf{U}^H + \mathbf{D}, \tag{3.3}$$

and the smoothing term $\mathbf{D}$ is given by

$$\mathbf{D} = C^L \mathbf{M}_l^{-1}(\mathbf{M} - \mathbf{M}_l)\mathbf{U}^m, \tag{3.4}$$

where C^L is a constant. This form for the smoothing is suggested by the fact that at node i on a uniform grid in one dimension

$$[\mathbf{M}_l^{-1}(\mathbf{M} - \mathbf{M}_l)\mathbf{U}^m]_i = \frac{(\mathbf{U}_{i+1}^m - 2\mathbf{U}_i^m + \mathbf{U}_{i-1}^m)}{6} \tag{3.5}$$

It should be noted that equations (3.1)–(3.3) can be rearranged to give

$$\mathbf{U}^H = \mathbf{U}^L + \mathbf{D}. \tag{3.6}$$

The new solution is computed according to

$$\mathbf{U}^{m+1} = \mathbf{U}^L + \mathbf{D}^*, \tag{3.7}$$

where $\mathbf{D}^*$ is obtained by limiting the element contributions to $\mathbf{D}$, in such a way as to attempt to ensure that $\mathbf{U}^{m+1}$ is free from extrema not found in $\mathbf{U}^m$ or $\mathbf{U}^L$ [14, 15].

The numerical performance of this FCT scheme is illustrated in Figure 8.1,

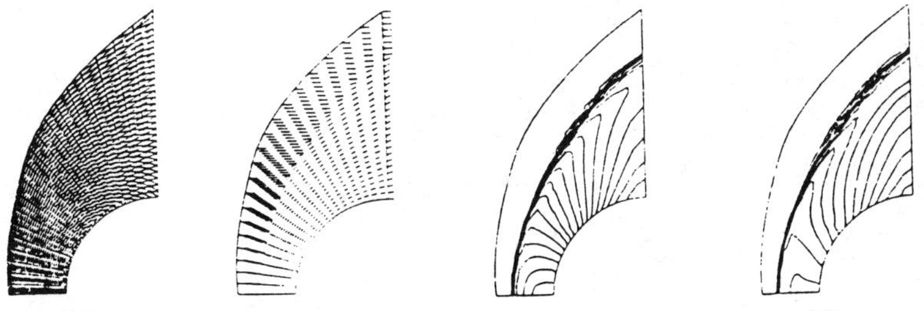

(a) (b) (c) (d)

FIGURE 8.1. Mach 8 flow past a cylinder. (a) Mesh; (b) velocity vectors; (c) pressure contours; and (d) density contours.

which shows the solutions obtained for the problem of Mach 8 flow past a cylinder.

Further generalizations of FCT that offer interesting possibilities for future investigation are possible [16].

8.4. Mesh Generation

The use of triangular elements in two dimensions means that computational domains of complex geometrical shape can be readily modeled and a variety of triangular mesh generation algorithms for planar domains are available [17, 18]. The approach to mesh generation to be outlined here begins by defining the boundaries of the solution domain in terms of Bezier polynomials and then covering this domain with a coarse "background" grid of three-noded linear triangles. This grid is normally constructed by hand, and the only geometrical requirement imposed is that the solution domain should be completely covered by this grid; i.e., the background grid is not required to approximate the geometry. At each node on the background grid, we specify the values of mesh parameters δ, s, $\boldsymbol{\alpha}$. During the mesh generation process, the local values of these parameters for the mesh being generated will be obtained by interpolation over the background grid. For the exact definition of these mesh parameters, it is useful to refer to Figure 8.2, which shows a typical generated triangle. The triangle has length $s\delta$ in the direction of $\boldsymbol{\alpha}$ and length δ in the direction at right angles to $\boldsymbol{\alpha}$. We call δ the local node spacing, s the local degree of stretching, and $\boldsymbol{\alpha}$ the local direction of stretching.

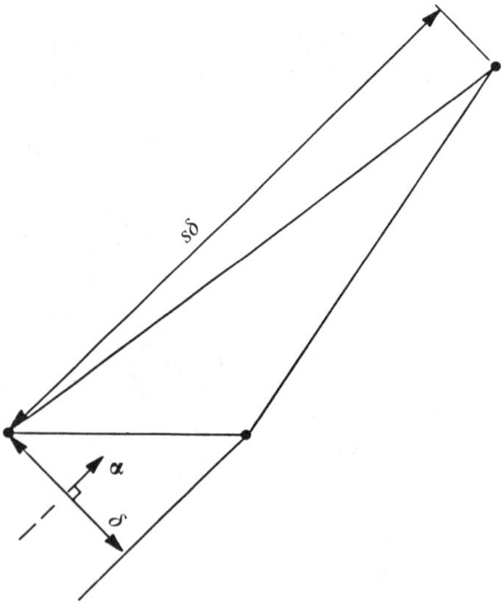

FIGURE 8.2. Definition of the mesh parameters δ, s, and $\boldsymbol{\alpha}$.

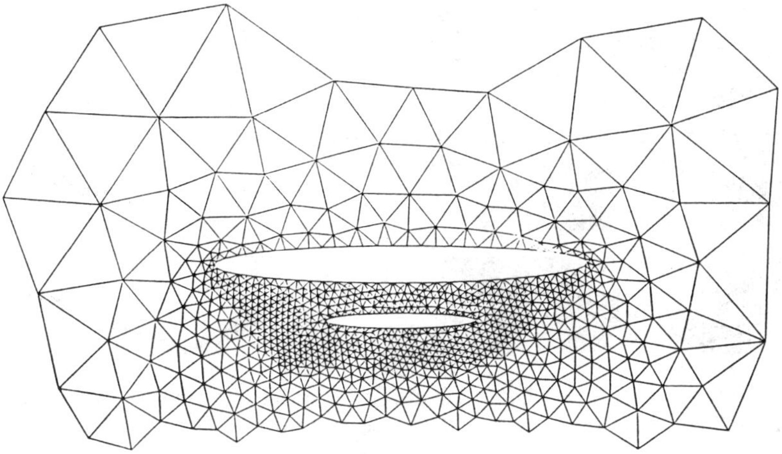

FIGURE 8.3. Detail of the initial mesh produced for the analysis of a store separation problem.

The full flexibility of the mesh generator need not be used to construct an initial mesh for a given problem, but it will be used in an adaptive mesh process to be described later. In particular, if a uniform distribution of δ is required, with no stretching, the background grid need only consist of a single element. The mesh generation process begins with the placing of the boundary nodes. The lines joining successive boundary nodes form the initial generation "front," which is the collection of sides available to construct triangles. One side in the front is chosen, and a triangle is constructed with values of δ, s, and $\boldsymbol{\alpha}$ interpolated from the background grid. The front is updated and the process is repeated until the front is empty, at which stage the whole solution domain has been discretized. Full details of the mesh generation process can be found elsewhere [19].

The performance of the mesh generator is demonstrated in Figure 8.3, which shows a detail of the initial mesh produced for the analysis of a store separation problem. This problem has been used to illustrate the full power of the generator by directly coupling it to the transient solution procedure and using it to regenerate the mesh locally as the store moves through the flow field [20].

8.5. Adaptive Mesh Strategies

Adaptive mesh strategies have a major role to play in the development of efficient solution techniques for large problems in computational fluid dynamics (CFD). The ultimate objective is the ability to solve a given problem to a prescribed accuracy with the optimum number of grid points, and, although

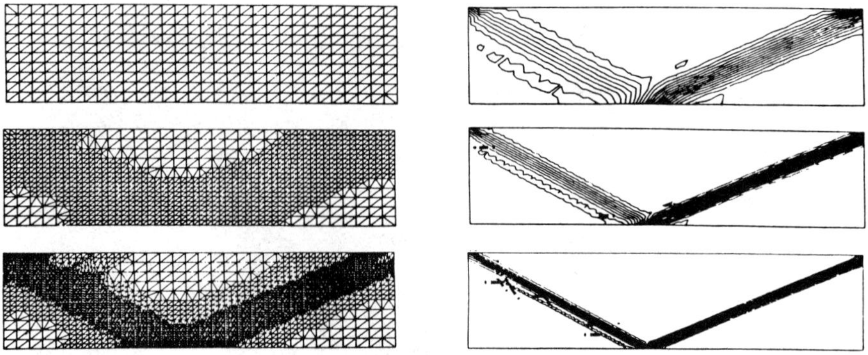

FIGURE 8.4. Regular shock reflection at a wall. Sequence of meshes produced by using mesh enrichment and the corresponding pressure contours.

this goal has not yet been met, major steps in this direction have already been made.

Mesh movement techniques can be contemplated [21, 22] but suffer from the drawback that the accuracy of the final computation can be limited by the structure and resolution of the initial grid. Mesh enrichment algorithms for steady problems [23, 24] generally advance the solution toward steady state on an initial coarse grid and then obtain an estimation of the error in each computational cell by using an error indicator. For the Euler or Navier–Stokes equation systems, the error indication is normally based upon a key variable; e.g., the density is a popular choice for the Euler equations. Indicators based upon interpolation theory can be used, with equidistribution of the error being the object of the refinement process [25]. The cells exhibiting largest error are automatically subdivided and the computation proceeds, with this process being repeated until the analyst is satisfied with the solution quality.

The mesh enrichment approach works well in practice [7], and Figure 8.4 shows the solution of the problem of regular shock reflection at a wall that has been solved in this manner. This solution was produced by using the basic solution algorithm described previously.

The extension of the mesh enrichment concepts to the solution of transient problems has been demonstrated recently [26]. Now, as the flow features of interest are moving through the solution domain, for economy of computation mesh enrichment has to be combined with the capability of derefining the mesh in regions where the error indication is small. This work has produced a highly vectorizable algorithm, with low storage requirements and with the ability to recover the original grid when the flow feature of interest has passed. The performance of the algorithm is shown in Figure 8.5, which displays the computed FCT solution for the problem of a Mach 10 shock impinging upon a half-cylinder. The solutions are depicted at three selected times during the transient.

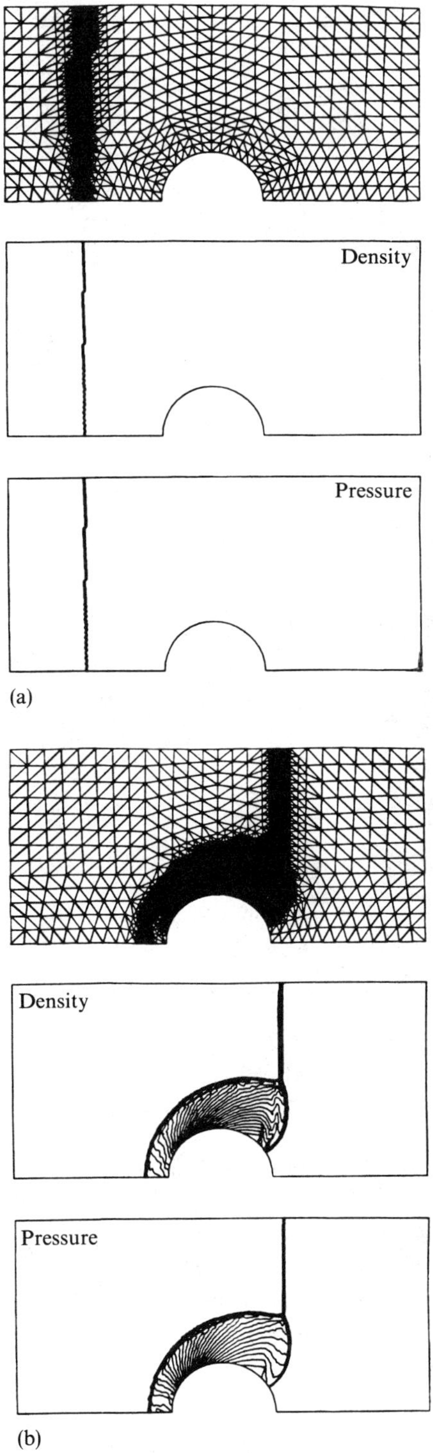

FIGURE 8.5. Shock impinging on a half-cylinder. (a) $t = 0$, 2,554 elements, 1,335 points; (b) $t = 0.3$, 9,057 elements, 4,626 points; (c) $t = 0.5$, 12,020 elements, 6,129 points.

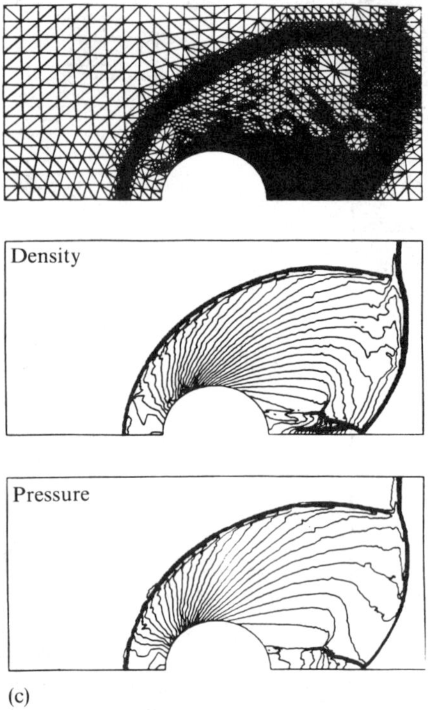

FIGURE 8.5 (*continued*)

A drawback of the mesh enrichment approach is that it provides a uniform local mesh refinement, whereas many flow features of interest are essentially one-dimensional in character. This suggests that directional refinement techniques could be computationally more efficient, and work in this area has already begun. If element error indicators are replaced by indicators along element sides [27], directional refinement for steady problems can be achieved, along with derefinement. This is illustrated in Figure 8.6, which again shows the solution obtained for the problem of regular shock reflection at a wall by using the basic solution algorithm. An alternative approach is to use the mesh generator described earlier to regenerate the mesh on the basis of information provided by the computed solution on the current mesh [19]. Figure 8.7 shows the problem of Mach 25 flow past a blunt body at an angle of attack of 20°, which has been solved by using FCT in this manner with a sequence of three grids.

Acknowledgments. The authors would like to thank the Aerothermal Loads Branch of the NASA Langley Research Center for partial support of this work under Grant No. NAGW-478 and especially A.R. Wieting and K.S. Bey for

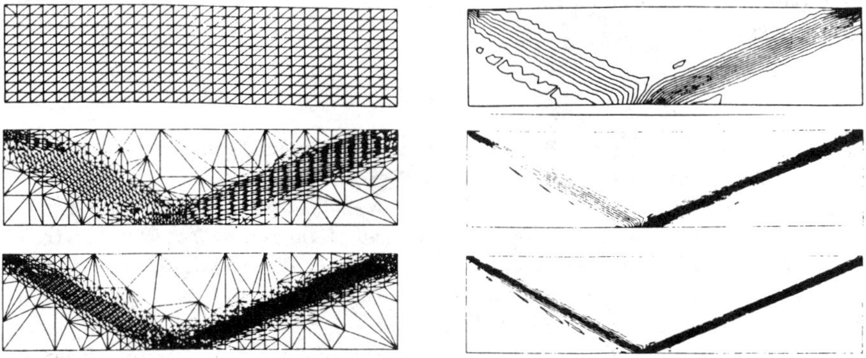

FIGURE 8.6. Regular shock reflection at a wall. Sequence of meshes produced by using directional refinement and the corresponding pressure contours.

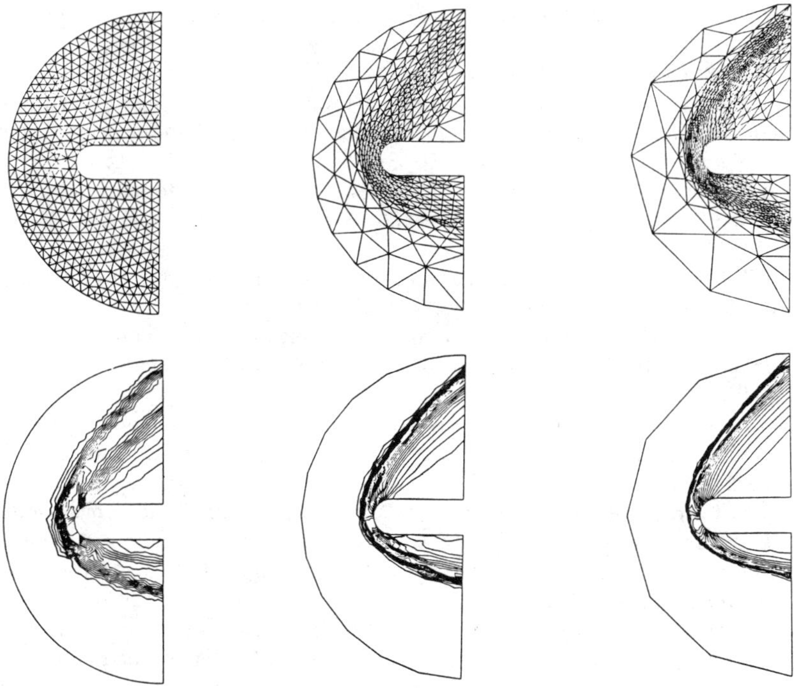

FIGURE 8.7. Mach 25 flow past a blunt body at 20° angle of attack. Sequence of meshes produced by adaptive mesh regeneration and the corresponding density contours.

their continued interest and encouragement. We would also like to acknowledge fruitful discussions with O.C. Zienkiewicz, J.P. Boris, D.L. Book, and S.T. Zalesak.

References

1. R. Löhner, K. Morgan, J. Peraire, and O.C. Zienkiewicz, "Finite element methods for high speed flows," AIAA-85-1531-CP, New York, 1985.
2. R. Löhner, K, Morgan, J. Peraire, O.C. Zienkiewicz, and L. Kong, "Finite element methods for compressible flow," in *Numerical Methods for Fluid Dynamics* (K.W. Morton and M.J. Baines, eds.), Oxford University Press, Oxford, 1986.
3. S.T. Zalesak, "Fully multidimensional flux corrected transport algorithm for fluids," *J. Comput. Phys.*, **31** (1979), 335–362.
4. J.P. Boris and D.L. Book, "Flux corrected transport I: SHASTA, a fluid transport algorithm that works," *J. Comput. Phys.*, **11** (1973), 38–69.
5. J. Donéa, "A Taylor–Galerkin method for convective transport problems," *Internat. J. Numer. Methods Engrg.*, **20** (1984), 101–119.
6. O.C. Zienkiewicz and K. Morgan, *Finite Elements and Approximation*, Wiley, New York, 1983.
7. R. Löhner, K. Morgan, and O.C. Zienkiewicz, "An adaptive finite element procedure for compressible high speed flows," *Comput. Methods Appl. Mech. Engrg.*, **51** (1985), 441–464.
8. J. Donéa and S. Giuliani, "A simple method to generate high-order accurate convection operators for explicit schemes based on linear finite elements, *Internat. J. Numer. Methods Fluids*, **1** (1981), 63–79.
9. R. Löhner, K. Morgan, and O.C. Zienkiewicz, "The use of domain splitting with an explicit hyperbolic solver," *Comput. Methods Appl. Mech. Engrg.*, **45** (1984), 313–329.
10. S.Z. Burstein, "Finite difference calculations of hydrodynamic flows containing discontinuities," *J. Comput. Phys.*, **2** (1967), 198–222.
11. R. Löhner, K. Morgan, and J. Peraire, "A simple extension to multidimensional problems of the artificial viscosity due to Lapidus," *Comm. Appl. Numer. Methods* **1** (1985), 141–147.
12. G. Erlebacher, "Solution adaptive triangular meshes with application to the simulation of plasma equilibrium," Ph.D. thesis, Columbia University, New York, 1984.
13. A.K. Parrott and M.K. Christie, "FCT applied to the 2D finite element solution of tracer transport by single phase flow in a porous medium," in *Numerical Methods for Fluid Dynamics* (K.W. Morton and M.J. Baines, eds.), Oxford University Press, Oxford, 1986.
14. R. Löhner, K. Morgan, M. Vahdati, J.P. Boris, and D.L. Book, "FEM-FCT: combining unstructed grids with high resolution," accepted for publication in *Comm. Appl. Numer. Methods* (1988).
15. K. Morgan, R. Löhner, J.R. Jones, J. Peraire, and M. Vahdati, "Finite element FCT for the Euler and Navier–Stokes equations," in Proceedings of the 6th International Symposium on Finite Element Methods in Flow Problems (INRIA, Paris, 1986), pp. 89–95.
16. R. Löhner, K. Morgan, and J. Peraire, "Further generalizations of FCT," to be submitted to *Comm. Appl. Numer. Methods* (1988).

17. J.C. Cavendish, "Automatic triangulation of arbitrary planar domains for the finite element method," *Internat. J. Numer. Methods Engrg.*, **8** (1974), 679–696.
18. S.H. Lo, "A new mesh generation scheme for arbitrary planar domains," *Internat. J. Numer. Methods Engrg.*, **21** (1985), 1403–1426.
19. J. Peraire, M. Vahdati, K. Morgan, and O.C. Zienkiewicz, "Adaptive remeshing for compressible flow computations," *J. Comput. Phys.*, **72** (1987), 1093–1109.
20. L. Formaggia, J. Peraire, and K. Morgan, "Simulation of a store separation using the finite element method," *Appl. Math. Modelling* (1988), to appear.
21. P.A. Gnoffo, "A finite volume adaptive grid algorithm applied to planetary entry flowfields," *AIAA J.*, **21** (1983), 1249–1254.
22. R. Löhner, K. Morgan, and O.C. Zienkiewicz, "Adaptive grid refinement for the compressible euler equations," in *Accuracy Estimates and Adaptivity for Finite Elements* (I. Babuska *et al.*, eds.), Wiley, New York, 1986.
23. J.F. Dannenhoffer and J.R. Baron, "Grid adaptation for the 2D Euler equations," AIAA Paper 85-0484, New York, 1985.
24. F. Angrand, V. Billey, A. Dervieux, J. Periaux, C. Pouletty, and B. Stouflet, "2D and 3D Euler flow calculations with second-order accurate Galerkin finite element method," AIAA Paper 85-1706, New York, 1985.
25. J.T. Oden, "Notes on grid optimization and adaptive methods for finite elements," University of Texas at Austin, TICOM Report, 1983.
26. R. Löhner, "An adaptive finite element scheme for transient problems in CFD," *Comput. Methods Appl. Mech. Engrg.*, **61** (1987), 323–338.
27. R. Löhner and K. Morgan, "Improved adaptive refinement strategies for finite element aerodynamic computations," AIAA-86-0499, New York, 1986.

CHAPTER 9

Remarks on Mixed and Singular Finite Elements and on the Field Boundary Element Method

Satya N. Atluri

9.1. Multifield (Mixed) Finite Element Methods

In formulating symmetric variational statements for the continuum problem in solid mechanics, we may treat each of the three fields, *viz.*, the stress tensor, the strain tensor, and the displacement vector, as an independent variable. Furthermore, if the solid is incompressible, it is possible to treat the deviatoric as well as the mean parts of the stress tensor as independent variables (see [11]). Likewise, in the analysis of incompressible, viscous, convective flows governed by the Navier–Stokes equations, each of the four fields, *viz.*, the deviatoric stress, the hydrostatic stress, velocity, and velocity strain, may be treated as an independent variable. On the other hand, in formulating Galerkin finite element methods (with similar trial and test function spaces) from these symmetric variational statements in solid mechanics, it is possible to assume trial functions, say for displacements and stresses, in each element of the solid, that do not obey, a priori, the interelement conditions of displacement–continuity and/or traction–reciprocity. In such cases, it becomes necessary to introduce additional independent fields of tractions and/or displacements at the element-boundaries, as Lagrange multipliers to enforce, a posteriori, these interelement constraints. A comprehensive summary of such finite element methods (FEM), with multiple independent fields being assumed inside each element as well as at its boundary, is given in several reports [1, 7, 11, 5, 30].

There are several reasons as to why the preceding types of mixed finite element methods are developed and thought to be advantageous over the usual single-field finite element method in compressible solid mechanics (based on assumed displacements) or the two-field method in incompressible fluid mechanics (based on assumed pressure and velocity fields). Among these are the following:

(i) In fourth-order problems, such as Kirchhoff plate bending, strict C^1 continuity of transverse displacement is demanded in a symmetric variational statement based on the displacement field alone. The so-called hybrid stress method [23] and mixed method (see [22], for instance) have been offered originally as "remedies" for the difficulties thus posed in the displacement method. However, it is now common to use alternative second-order theories for plates and shells, with additional generalized displacement fields as independent variables. For instance, in a Reissner–Mindlin-type theory, the transverse displacement as well as rotation (of the normal) are treated as independent variables. If C^0-type trial functions are assumed for both of the displacement fields, the attendant strain fields will be incomplete. Oftentimes, certain expedient "tricks" such as selective-reduced integrations are used to alleviate such problems of "membrane" and "shear" locking in such C^0 elements. Sometimes, such selective-reduced integration methods are also equivalent to "mixed" methods.

(ii) In general, in solid mechanics, the trial functions for displacements in each element (in a Galerkin finite element method) are not complete polynomials. The strain fields corresponding to these displacements are even more incomplete. This situation becomes more transparent in the so-called *isoparametric* elements with curvilinear geometry. It is well known that a single field, assumed displacement four-noded square element, can represent states of pure tension and pure shear but "locks" in pure bending (see [25]). To avoid such problems, we may start with "consistent" and "complete" trial functions as assumptions for independent strain and/or stress fields in mixed formulations, which may indeed lead to dramatic improvements [25].

(iii) Although it is popular to satisfy the incompressibility constraint in Navier–Stokes equations through a penalty reduced integration approach and assumed velocity fields, dramatic accuracies may sometimes result from mixed (i.e., assumed deviatoric stress, velocity, and pressure) formulations of fluid mechanics (see [16, 31, 32]), which may also be effective in treating the convection operator.

(iv) In inelastic solid mechanics, wherein the relation between the stress rate and strain rate depends on the current state of stress, often better accuracies are observed in multifield formulations with independent stress and displacement assumptions (see [26, 20]). Such improvements are also noted in problems of fully developed elastic perfect plasticity [12].

(v) Hybrid and mixed methods have also been very effective in treating problems of fracture with known singular fields [2, 4].

Although the preceding listings of "advantages" appear convincing at first sight, and although the mathematical basis of the convergence of mixed finite element methods appears well established [18, 13], it must be said that the development of mixed finite element methods has been notoriously plagued by problems that yet remain to be satisfactorily resolved. Some of these problems center around:

(i) the conditions under which the curvilinear elements based on mixed formulations possess the properties of observer invariance and objectivity;

(ii) the development of curvilinear mixed elements that have, embedded in them, certain cardinal states of strain and stress;
(iii) the choice of trial functions to satisfy the consistency and stability conditions that govern each *pair* of independent variables in a multifield mixed method; and
(iv) the avoidance of kinematic modes at the element level.

A general discussion of some of these problems has appeared in [29] and [30]. In the following we give a brief *generic* discussion of some of the problems with mixed methods, with specific reference to a class of mixed methods based on Reissner's variational principle in linear solid mechanics.

From Reissner's principle, modified to account for fields σ_{ij} and u_i that do not obey interelement traction reciprocity or displacement compatibility, we may derive a variety of finite element methods [1, 30]. We shall now consider two specific examples:

(i) a mixed finite element method, wherein an arbitrary symmetric σ_{ij} (but $\sigma_{ij,j} + \bar{f}_i \neq 0$) and an interelement compatible u_i are assumed in each element Ω_m; and
(ii) a hybrid finite element method, wherein an *equilibrated* σ_{ij} (i.e., $\sigma_{ij,j} + \bar{f}_i = 0$) is assumed in each Ω_m, and an interelement compatible displacement field u_i is assumed at $\partial\Omega_m$ only.

The variational basis of the preceding mixed method may be shown to be [1, 30]

$$0 = \sum_m \left\{ \int_{\Omega_m} \left[\left(u_{(i,j)} - \frac{\partial W_c}{\partial \sigma_{ij}} \right) \tau_{ij} + \sigma_{ij} v_{(i,j)} - \bar{f}_i v_i \right] dv - \int_{S_{tm}} \bar{t}_i v_i \, ds \right\}, \tag{1.1}$$

or, in abstract form:

Find $(\sigma, u) \in T \times V_{co}$ such that:

$$\begin{aligned} a(\sigma, \tau) - b(\tau, u) &= 0, \qquad \forall \tau \in T, \\ b(\sigma, v) &= \langle f, v \rangle, \qquad \forall v \in V_{co}, \end{aligned} \tag{1.2}$$

where

$$V_{co} = \{ v \in H'(\Omega_m);\ v_i^+ = v_i^- \text{ at } \rho_m;\ v_i = 0 \text{ at } S_{um}, \forall m \}. \tag{1.3}$$

Likewise, the variational basis of the "hybrid stress" method may be shown to be [1, 30]

$$0 = \sum \left\{ \int_{\Omega_m} (\ )\, dv - \int_{S_{tm}} (\ )\, ds \right\}, \tag{1.4}$$

wherein the integrands over Ω_m and S_{tm} are identical to those in (1.1), except that now the nature of functions τ_{ij} are different from those in (1.1). This can be emphasized in the abstract form:

Find $(\sigma, u) \in T_0 \times V_{co}$ such that:

$$\begin{aligned} a(\sigma, \tau) - b(\tau, u) = 0, \qquad & \forall \tau \in T_0, \\ B(\tau, v) = \langle f, v \rangle, \qquad & \forall v \in V_{co}, \end{aligned} \tag{1.5}$$

where

$$T_0 = \{\tau_{ij} \in H'(\Omega_m);\ \tau_{ij,j} = 0, (\sigma_{ij,j} + \bar{f}_i = 0) \text{ in } \Omega_m, \forall m\}, \tag{1.6}$$

and V_{co} is defined in (1.3). Inasmuch as problems (1.2) and (1.5), even though having different physical interpretations, are similar in abstract form, we shall study (1.5) henceforth in some detail. Problem (1.5) has a unique solution if the following conditions are met:

$$\sup_{\forall \sigma \in T_0} \frac{\sum_m \int_{\Omega_m} \sigma_{ij} v_{(i,j)}\, d\Omega}{\|\sigma_{ij}\|_T} \geq \beta \|v\|_V, \qquad \forall v \in V_{co}, \tag{1.7}$$

and

$$W_c(\sigma_{ij}, \sigma_{ij}) \geq \alpha \|\sigma\|_T, \qquad \forall \sigma \in \text{Ker}(B), \tag{1.8}$$

where α and β are positive, and Ker(B) is defined as

$$\text{Ker}(B) = \left\{ \sigma \in T_0, \sum_m \int_{\Omega_m} \sigma_{ij} v_{(i,j)}\, d\Omega = 0, \forall v \in V_{co} \right\}. \tag{1.9a}$$

Since $\sigma \in T_0$, (1.9) may be written as

$$\text{Ker}(B) = \left\{ \sigma \in T_0, \sum_m \int_{\partial\Omega_m} \sigma_{ij} n_j v_i = 0, \forall v \in V_{co} \right\}. \tag{1.9b}$$

Let the dimension of T_0 be m and that of V_{co} be n. As shown in Section 6 of [30] and the references cited therein, conditions (1.7) and (1.8) imply that $m > n$ and that the rank of the $n \times m$ matrix B in

$$v^t B \sigma = \sum_m \int_{\Omega_m} \sigma_{ij} v_{(i,j)}\, d\Omega \tag{1.10}$$

should be n.

The condition for existence and uniqueness of the solution may also be stated, from (1.7), as

$$\sup_{\forall \sigma \in T_0} \sum_m \int_{\Omega_m} \sigma_{ij} v_{(i,j)}\, d\Omega > 0, \qquad \forall v \in V_{co}. \tag{1.11}$$

Since $v_i = 0$ at S_{um} for $v \in V_{co}$, it can be seen that v_i appearing in (1.11) do not include any *global rigid body modes*, assuming that the original boundary conditions at S_u are such that they preclude any rigid motion of the solid as a whole.

However, even *while global rigid motions may be precluded, such rigid motion may be considered at the element level.* Thus, $v^m_{(i,j)} = 0$ for r possible rigid modes of each element displacement field, v^m. Furthermore, σ^m_{ij} is arbitrary for each element, while within each element, $\sigma^m \in T_0$. Thus, (1.11) may be written as

$$\sup_{\forall\sigma\in T_0} \sum_m \int_{\Omega_m} \sigma^m_{ij} v^{md}_{(i,j)}\, d\Omega > 0, \qquad \forall v \in V_{co}, \tag{1.12}$$

where v^{md}_i denote nonrigid deformation modes in each element. A *sufficient* condition for the validity of (1.12) is, clearly,

$$\sup_{\forall\sigma\in T_0} \sum_m \int_{\Omega_m} \sigma^m_{ij} v^{md}_{(i,j)}\, d\Omega > 0, \qquad \forall v \in V_{co}, \quad \forall m. \tag{1.13}$$

Let M_β be the number of stress modes assumed in each element Ω_m, and let N_q be the number of displacement modes for each Ω_m. Then the dimension of v^{md}_i is $(N_q - r)$. Thus, we see

$$\int_{\Omega_m} \sigma^m_{ij} v^{md}_{(i,j)}\, d\Omega = \sigma^{mt} B^{*mt} v^{md} \qquad (1 \times M_\beta)(M_\beta \times N_q - r)(N_q - r \times 1). \tag{1.14}$$

From Theorem 6.1 of [30], for (1.13) to hold, $M_\beta \geq (N_q - r)$ and the rank of B^{*m} should be $(N_q - 1)$.

Remark 1.1. If $a(\sigma, \sigma)$ in each element, $\forall\sigma \in T_0$ (denoted as $a_m(\sigma, \sigma)$), can be written as

$$a_m(\sigma, \sigma) = \underset{(1\times M_\beta)}{\sigma^{mt}} \quad \underset{(M_\beta \times M_\beta)}{H} \quad \underset{(M_\beta \times 1)}{\sigma^m},$$

then it can be shown [27] that the element stiffness matrix k_m can be written as

$$k_m = B^m H^{-1} B^{mt},$$

where B^m is defined through

$$\int_{\Omega_m} \sigma^m_{ij} v^m_{(i,j)}\, d\Omega = \sigma^{mt} B^{mt} v^m.$$

Here, v^m includes both rigid and nonrigid modes, and $\dim(v^m) = N_q$. Note that the rank of B^m is the same as that of B^{*m}. Since $a_m(\sigma, \sigma)$ is positive definite, it follows that the rank of the element stiffness matrix is $(N_q - r)$, provided that of B^m is $(N_q - r)$.

Remark 1.2. Note that both σ_{ij} and v_i in (1.14) are components in the Cartesian system x_i. The momentum balance condition involves differentiation of σ_{ij} with respect to x_j; while the strains $v_{(i,j)}$ also involve differentiation with respect to x_j. In the usual *isoparametric* element formulation, the geometrical transformation between the (nondimensional) "parent" element and that in the physical domain is $x_j = x_j(\xi^k)$ where ξ^k, usually taken to be $-1 \leq \xi^k \leq 1$, are curvilinear coordinates. In a displacement formulation, we usually assume $u_i = u_i(\xi^k)$, and in an isoparametric representation, the representation for x_j as well as u_i contains an equal number of basis functions in ξ^k. The stiffness matrix of the element in such an isoparametric displacement formulation can be shown to be *objective* or observer invariant.

That means, if k_m is the element stiffness matrix in an isoparametric displacement formulation in the x_i coordinate system, then its representation in any other Cartesian system $x' = Qx$ is given by $k'_m = Qk_mQ^{\mathrm{T}}$, where Q is orthogonal.

To maintain the objectivity of the element stiffness matrix in a mixed hybrid formulation, it has been shown [27, 25] that the stress tensor, $\boldsymbol{\sigma}$, should be assumed *in an element local coordinate system* and *not* in *a global coordinate system.*

Remark 1.3. First consider two- and three-dimensional finite elements of square and cubic (or rectangular and rectangular prism) shapes, respectively. Here, the theory of symmetric groups has been demonstrated [27, 25] to be a useful tool in choosing least order stress fields ($M_\beta = N_q - r$) that lead to the matrix B^{*m} (see (1.14)) of rank $(N_q - r)$, and a stiffness matrix k_m, which is objective and also of rank $(N_q - r)$. In this case, a Cartesian coordinate system located at the centroid of the element, and along the axes of symmetry of the element, are used. In the two cited references, both σ_{ij} and $v_{(i,j)}$ [$(\ )_{,j}$ implying $\partial(\ )/\partial x_j$] are decomposed into invariant irreducible spaces using group theory. In terms of these irreducible representations, the matrix, say $(B^*)'$, corresponding to $B(\sigma, v)$ for each element, becomes "quasi-diagonal." Thus, group theory enables us to pick σ_{ij} in each element, for a given v_i, such that the resulting element formulation is invariant and stable. It has been shown that:

(i) for a four-noded square with $N_q - r = 5$, there are two possible choices for a five-parameter equilibrated stress field;
(ii) for an eight-noded square with $N_q - r = 13$, there are twenty-one choices for a thirteen-parameter stress field;
(iii) for an eight-noded cube, there are eight choices for a stress field with $M_\beta = 18$; and
(iv) for a twenty-noded cube, there are 384 choices for a stress field with $M_\beta = 54$ all of which lead to stable and objective elements.

The "best" selection among all these choices may depend upon:

(i) the lowest eigenvalue of the matrix $(B^{*m})(B^{*mt})$, as discussed in Section 6 of [30], and
(ii) the capability of the candidate stress field to represent the cardinal states of stress of pure tension, shear, bending, and torsion in each element.

A comprehensive study of such tests is given in [24] and [25].

Remark 1.4. Consider a mixed hybrid element of a general curvilinear shape and introduce a geometric mapping of the type $x_i = x_i(\xi^k)$, with $-1 \leq \xi^k \leq 1$. Let $\mathbf{g}_k(\xi^m)$ and $\mathbf{g}^k(\xi^m)$ be the covariant and contravariant base vectors, respectively, of the curvilinear coordinates ξ^m. Let $\bar{\mathbf{g}}_k$ represent the covariant base vectors at the centroid; i.e., $\bar{\mathbf{g}}_k = \mathbf{g}_k(\xi^m = 0)$, and let $\mathbf{e}_k$ be a Cartesian system at $\xi^m = 0$. Then, it has been shown [24, 25] that requirements of invariance may be met by representing the stress tensor in the alternative forms

$$\boldsymbol{\sigma} = \sigma_{ij}(x_k)\mathbf{e}_i\mathbf{e}_j, \qquad \sigma_{ij} \in T_0, \tag{1.15a}$$

$$= \sigma_{ij}(\xi^k)\mathbf{e}_i\mathbf{e}_j, \qquad \sigma_{ij} \in T, \tag{1.15b}$$

$$= \sigma_{ij}(\xi^k)\mathbf{g}^i\mathbf{g}^j, \qquad \sigma_{ij} \in T, \tag{1.15c}$$

$$= \sigma^{ij}(\xi^k)\mathbf{g}_i\mathbf{g}_j, \qquad \sigma^{ij} \in T, \tag{1.15d}$$

$$= \sigma^{ij}(\xi^k)\bar{\mathbf{g}}_i\bar{\mathbf{g}}_j, \qquad \sigma^{ij} \in T. \tag{1.15e}$$

Other possible representations are discussed in [8]. In (1.15), T_0 is the space of *equilibrated* stresses, and T is that of differentiable (but not equilibrated) stresses. It is seen that (1.15a, b) can easily represent states of constant stress in the Cartesian coordinate system and hence can pass the so-called "constant stress" patch test [19].

Considering a state of constant stress, say, $\boldsymbol{\sigma} = C_{ij}\mathbf{e}_i\mathbf{e}_j$, where C_{ij} are constants, it is seen that representation (1.15c) can pass the patch test if $\sigma_{ij}(\xi^k)$ includes functions such that

$$\sigma_{ij}(\xi) = C_{mn}(\mathbf{e}_m \cdot \mathbf{g}_i)(\mathbf{e}_n \cdot \mathbf{g}_j) \tag{1.16a}$$

$$= C_{mn}\frac{\partial x_m}{\partial \xi^i}\frac{\partial x_n}{\partial \xi^j}. \tag{1.16b}$$

Since, in an isoparametric formulation, $(\partial x_m/\partial \xi^i)$ is a simple polynomial in ξ^k, it is possible, in general, that a polynomial representation that passes the patch test exists for $\sigma_{ij}(\xi^k)$ in (1.15c). However, the stress field will not be, in general, of the "least order."

On the other hand, (1.15d) can pass the patch test if $\sigma^{ij}(\xi^k)$ includes functions such that

$$\sigma^{ij}(\xi^k) = C_{mn}(\mathbf{e}_m \cdot \mathbf{g}^i)(\mathbf{e}_n \cdot \mathbf{g}^j) \tag{1.17a}$$

$$= C_{mn}\frac{\partial \xi^i}{\partial x_m}\frac{\partial \xi^j}{\partial x_n}. \tag{1.17b}$$

For the usual isoparametric formulation, it is seen that $\sigma^{ij}(\xi^k)$ of (1.17b) are no longer simple polynomials. Hence, representation (1.15d) with polynomial functions $\sigma^{ij}(\xi^k)$ will not, in general, pass the patch test. However, (1.15e) will pass the patch test, since, in this case,

$$\sigma^{ij}(\xi^k) = C_{mn}(\mathbf{e}_m \cdot \bar{\mathbf{g}}_i)(\mathbf{e}_n \cdot \bar{\mathbf{g}}_j) \equiv d_{mn}, \tag{1.18}$$

where d_{mn} are constants, and hence a simple (even least order) polynomial representation (including constant terms) will suffice for $\sigma^{ij}(\xi^k)$.

Remark 1.5. To formulate an isoparametric curvilinear mixed hybrid element, we may use alternative representations for stress as in (1.15a–e) and assume $v_i(\xi^k)$ to be of the same form as $x_i(\xi^k)$. Note that v_i are Cartesian components of displacement. For the alternative representations of stress as in (1.15), the bilinear form $B(\sigma, v)$, for each element, takes on the respective representation

$$B(\sigma, v) = \int_{\Omega_m} \sigma_{ij}(x_k) v_{i;m} J_{mj}^{-1} (\det J)\, d\xi^1\, d\xi^2\, d\xi^3 \tag{1.19a}$$

$$= \int_{\Omega_m} \sigma_{ij}(\xi^k) v_{i;m} J_{mj}^{-1} (\det J)\, d\xi^1\, d\xi^2\, d\xi^3 \tag{1.19b}$$

$$= \int_{\Omega_m} \sigma_{mn}(\xi^k) v_{i;k} J_{kj}^{-1} J_{mi}^{-1} J_{nj}^{-1} (\det J)\, d\xi^1\, d\xi^2\, d\xi^3 \tag{1.19c}$$

$$= \int_{\Omega_m} \sigma^{mn}(\xi^k) v_{i;n} J_{im} (\det J)\, d\xi^1\, d\xi^2\, d\xi^3 \tag{1.19d}$$

$$= \int_{\Omega_m} \sigma^{mn}(\xi^k) v_{i;k} J_{kj}^{-1} (\bar{J}_{im} \bar{J}_{jn}) \det(J)\, d\xi^1\, d\xi^2\, d\xi^3, \tag{1.19e}$$

where $(\ )_{;m}$ denotes $\partial(\ )/\partial\xi^m$, $J_{mj} = (\partial x_m/\partial\xi^j)$, and $\bar{J}_{mj} = J_{mj}(\xi^k = 0)$.

In Remark 1.3 concerning squares and cubes, a group-theoretical method that enables a choice of $\sigma_{ij}(x_k)$, for a given $v_i(x_k)$, that gives the rank $(N_q - r)$ to B^{*m} was described. For such squares and cubes, the bilinear form is computed by using (1.14). Comparing (1.14) and (1.19), it can be seen that there exists no simple way of choosing the stress as in (1.15) for curvilinear elements such that the rank of B^{*m} is determined a priori. However, it has been demonstrated in [25] that if $\sigma_{ij}(\xi_k)$ or $\sigma^{ij}(\xi^k)$ of (1.15b–e) is chosen to be of the same polynomial form (i.e., by replacing x_k by ξ^k) as that of $\sigma_{ij}(x_k)$, which is derived by using group theory for squares and cubes, then the rank of B^{*m} is maintained to be $(N_q - r)$ even for very severely distorted elements. Further, it has been clearly demonstrated [25] that the least order, invariant, isoparametric, curvilinear mixed hybrid elements are less distortion sensitive and lead to more accurate results compared to the standard displacement elements in a variety of examples.

From the preceding remarks it may be concluded that although several mixed finite elements have been developed and reported over the last twenty or so years, an understanding of their robustness and inherent properties is yet far from complete. Although some success has been achieved, through group-theoretical methods, in understanding the *local* stability (*element* kinematic modes, etc.) for elements of simple shapes, much remains to be done in understanding *local* as well as global stability conditions for general curvilinear element meshes. Further, the establishment of the convergence condition, that the constant β in (1.7) is independent of mesh, has, to the best of the author's knowledge, never been attempted even for the simplest case of plane elasticity for general meshes.

Thus, it is the author's belief that in spite of its long (20 years) history, the subject of mixed finite element methods is still in its infancy, and much remains to be done to exploit the full potential of mixed finite elements in solid as well as fluid mechanics.

9.2. Computational Fracture Mechanics and Singular Finite Elements

Comprehensive surveys of singular finite elements for analyzing problems wherein the derivative of the field variable (such as the derivative of displacement, or strain, in solid mechanics) has a singularity of order $r^{-\lambda}$, $0 < \lambda < 1$, where r is the distance from the point of singularity (or the distance from the line of singularity, measured in a plane normal to the line), were presented in [2, 4]. For three-dimensional problems of embedded or exposed elliptical cracks, a Schwartz–Neumann-type of alternating method, employing an ordinary finite element method for analyzing the uncracked solid, and an analytical solution for an embedded elliptical crack, was found to be the most efficient alternative to a singular finite element method [9].

Current problems of active research interest include fast crack propagation in viscoplastic solids with a predominant dependence of material behavior on very high strain rates. Here, moving mesh finite element procedures, as well as adaptive mesh refinements, are of importance [9].

9.3. Field Boundary Element Methods Based on Unsymmetric Variational Statements and Petrov–Galerkin Approaches

The traditional finite element method in solid mechanics is based on "symmetric" variational statements (or weak forms) in which both the trial and test functions are required to be differentiable to the same degree. Such symmetric variational statements in solid mechanics are identified to be the principles of virtual work, complementary virtual work, the principles of Reissner, and of Hu and Washizu, etc. The Galerkin finite element method, based on the use of a similar class of trial and test functions in any of these symmetric variational statements leads to the familiar symmetric banded system of equations. As is well known, this method has found wide application in a variety of nonlinear problems as well.

It is often of considerable advantage to use *unsymmetric* variational statements wherein the test functions are required to be differentiable to a higher degree than the trial functions. A Petrov–Galerkin discrete method based on the use of different classes of test and trial functions, although leading to an unsymmetric system of equations, may offer significant advantages in several linear or nonlinear problems. This Petrov–Galerkin-type unsymmetric variational method may be shown to lead to integral equation representations for the solution variables in solid mechanics (usually displacements), if the test functions are appropriately chosen. For boundary value/initial value problems in solid mechanics, it is often possible to derive *certain boundary integral representations* for displacements [6, 17, 3].[1] A boundary integral representation, when discretized, leads to the so-called boundary element method. Such boundary element methods are possible, for example, in linear, isotropic, elastostatics (see, for instance, [6]) and in problems of static bending of linear elastic isotropic plates [28, 15] in which the integral representation for displacements involves only boundary integrals of the unknown trial functions and their derivatives. On the other hand:

(i) even for linear problems wherein the fundamental solution, in infinite space, cannot be established for the entire linear differential operator;
(ii) for anisotropic/nonhomogeneous materials; and
(iii) for nonlinear problems of large deformation and material inelasticity;

the integral representation, if any, for displacements would involve not only boundary integrals but also interior integrals (i.e., integrals over domain) of the trial functions and/or their derivatives [3]. As a matter of fact, such integral representations are equivalent to the unsymmetric variational statements, or Petrov–Galerkin-type weighted residual methods, wherein the trial and test function spaces are distinctly different. A discretization of such integral equations would lead to "field-boundary" method [3].

Such *field boundary element* (FBM) methods may have significant advantages over the traditional *finite element methods* (FEM) in several instances. For instance, if the problem is linear and a pure "boundary integral" representation is possible, the dimensionality of the discretization is reduced by one. Even in nonlinear problems, the order of the discrete equations in the FBM may be significantly lower than in FEM. This significant reduction in order, even though the system of equations may be unsymmetric, is of considerable advantage in implementing algorithms for control of nonlinear dynamic response of structural systems [10]. We shall now illustrate the FBM and its advantages in the context of nonlinear dynamic analysis of shells.

The literature on the nonlinear static or dynamic analysis of shells based on the integral equation formulation is rather sparse. As is well known, because of the curvature of the shell, the in-plane displacements and the transverse displacement in a shell are inherently coupled in the kinematics of deformation as well as in the momentum balance relations for the shell. The problems imposed by the requirement of C^1 interelement continuity of transverse displacement in thin shell "finite element" analysis based on a symmetric variational formulation have also long plagued the successful development of shell finite elements. In the following, we present an FBM

[1] Certain new formulations of integral equations for deformation-gradients directly, have been reported in [35].

method for shallow shells based on an *unsymmetric* variational formulation and the use of fundamental solutions, in infinite space, to the highest order differential operator of the problem as test functions. In this method, the problem of C^1 continuity of trial functions for transverse displacement becomes totally vacuous. In some instances, in this formulation, the trial function for transverse displacement in each element may simply be piecewise constant. Since

(i) the in-plane displacements u_α are coupled to the transverse displacement w in the shell equilibrium equations and
(ii) the nonlinear deformation as well as the dynamic motion of the shell is considered;

it turns out that the integral representations for the shell displacements involve not only boundary integrals, but also domain integrals involving the trial solutions for displacements.

9.3.1. The Boundary Value/Initial Value Problem

Consider a shallow shell of an isotropic elastic material with the midsurface being described by $z = z(x_\alpha)$, $\alpha = 1, 2$. The base plane of the shell is defined by a domain Ω that is bounded by a piecewise smooth curve Γ in the Ox_1x_2 plane. The equations of large deformation for the shell may be written as

$$N_{\alpha\beta,\beta} + b_\alpha - \rho\ddot{u}_\alpha = 0 \qquad (\alpha, \beta = 1, 2), \tag{3.1a}$$

and

$$D\nabla^4 w + \frac{N_{\alpha\beta}}{R_{\alpha\beta}} - (b_3 - \rho\ddot{w}) = f_3 + (N_{\alpha\beta}w_{,\beta})_{,\alpha}, \tag{3.1b}$$

where $N_{\alpha\beta}$ are membrane forces; $(\ )_{,\beta} = \partial(\ )/\partial x_\beta$; w is the transverse deflection of the midsurface of the shell; b_i $(i = 1, 2, 3)$ are both forces; f_3 is the load normal to the shell midsurface; $D \equiv [Et^3/12(1 - \nu^2)]$; t is the thickness; E and ν are the elastic constants; ∇^4 is the biharmonic operator in the variables x_α; $\ddot{u}_\alpha$ and $\ddot{w}$ are accelerations; and

$$R_{\alpha\beta} = -\frac{1}{z_{,\alpha\beta}} \tag{3.1c}$$

are the radii of curvature of the undeformed shell. Along Γ, the boundary conditions are

$$u_\alpha = \bar{u}_\alpha \quad \text{at } \Gamma_u; \qquad N_{\alpha\beta}n_\beta = \bar{P}_\alpha \quad \text{at } \Gamma_\sigma; \qquad \Gamma = \Gamma_u \cup \Gamma_\sigma, \tag{3.2a, b}$$

where n_α are the direction cosines of the unit outward normal to Γ in the base plane. The out-of-plane boundary conditions are

$$w = \bar{w} \qquad \text{or} \qquad V_n = \bar{V}_n, \tag{3.3a}$$

$$\Psi_n = \bar{\Psi}_n \qquad \text{or} \qquad M_n = \bar{M}_n, \tag{3.3b}$$

where

$$V_n = -D\frac{\partial}{\partial n}(\nabla^2 w) + \frac{\partial}{\partial s}M_s + N_n\frac{\partial w}{\partial n} + N_{nt}\frac{\partial w}{\partial s}$$

is the reduced Kirchhoff shear force;

$\Psi_n = \dfrac{\partial w}{\partial n}$ is the rotation around the tangent to Γ;

$$M_n = -D\{(n_1^2 + \nu n_2^2)\frac{\partial^2 w}{\partial x_1^2} + 2(1-\nu)n_1 n_2\frac{\partial^2 w}{\partial x_1\,\partial x_2} + (n_2^2 + \nu n_1^2)\frac{\partial^2 w}{\partial x_2^2};$$

and n and s are directions normal and tangential, respectively, to Γ in the base plane. The nonlinear in-plane strain displacement relations are

$$\varepsilon_{\alpha\beta} = \tfrac{1}{2}\left[u_{\alpha,\beta} + u_{\beta,\alpha} + \frac{2w}{R_{\alpha\beta}} + w_{,\alpha}w_{,\beta}\right], \tag{3.4a}$$

where u_α are the in-plane displacements at the shell midsurface. The in-plane stress resultant/strain relations are

$$\begin{aligned} N_{11} &= C(\varepsilon_{11} + \nu\varepsilon_{22}); \qquad N_{22} = C(\varepsilon_{22} + \nu\varepsilon_{11}); \\ N_{12} &= C(1-\nu)\varepsilon_{12}, \end{aligned} \tag{3.4b}$$

where $C = Et/(1-\nu^2)$. The moment curvature relations are

$$\begin{aligned} M_{11} &= -D(w_{,11} + \nu w_{,22}); \qquad M_{22} = -D(w_{,22} + \nu w_{,11}); \\ M_{12} &= -D(1-\nu)w_{,12}. \end{aligned} \tag{3.5}$$

Finally, the initial conditions on the shell may be written as

$$\begin{aligned} u_\alpha(x_\beta, 0) &= u_{\alpha 0}(x_\beta) \quad \text{at} \quad t = 0, \\ \dot{u}_\alpha(x_\beta, 0) &= \dot{u}_{\alpha 0}(x_B) \quad \text{at} \quad t = 0, \\ w(x_\beta, 0) &= w_0(x_\beta);\ \dot{w}(x_\beta, 0) = \dot{w}_0(x_\beta) \quad \text{at} \quad t = 0, \end{aligned} \tag{3.6}$$

where $(\dot{\ }) \equiv d(\ )/dt$.

Here, our attention is restricted to the case of moderately large deformation.

9.3.2. Integral Equations For Shell Displacements

In an approximate analysis of the boundary/initial value problem described, let u_α and w be the assumed trial solutions. We shall consider a general weighted residual formulation and let u_α^* and w^* be the corresponding test functions. In the familiar Galerkin finite element method, the trial functions (u_α and w) and the test functions (u_α^* and w^*) belong to the same category of function spaces. In the present formulation, however, as will be seen, the test functions (u_α^* and w^*) belong to an entirely different class of function space from that of the trial functions. With this in mind, the combined weak forms of the equilibrium equations and boundary conditions for the inplane ((3.1a), (3.2a, b)) and out-of-plane ((3.1b) and (3.3a, b)) deformations, respectively, may be written (see, for instance, [30]) as

$$\int_\Omega (N_{\alpha\beta,\beta} + b_\alpha - \rho\ddot{u}_\alpha)u^*\,d\Omega = \int_{\Gamma_\sigma} (P_\alpha - \bar{P}_\alpha)u_\alpha^*\,d\Gamma + \int_{\Gamma_u} (\bar{u}_\alpha - u_\alpha)P_\alpha^*(u_\sigma^*)\,d\Gamma, \tag{3.7}$$

and

$$\int_\Omega \left\{D\nabla^4 w + \frac{N_{\alpha\beta}}{R_{\alpha\beta}} - b_3 + \bar{\rho}\ddot{w} - f_3 - (N_{\alpha\beta}w_{,\beta})_{,\alpha}\right\} w^*\,d\Omega$$
$$= \int_{\Gamma_v} (\bar{v}_n - v_n)w^*\,d\Gamma + \int_{\Gamma_M} (\bar{M}_n - M_n)\psi_n^*\,d\Gamma + \int_{\Gamma_\psi} (\psi_n - \bar{\psi}_n)M_n^*\,d\Gamma$$
$$+ \int_{\Gamma_w} (w - \bar{w})V_n^*\,d\Gamma. \tag{3.8}$$

To make a specific choice for the test functions that results in convenient integral representations for the shell displacements u_α and w, we rewrite the in-plane equilibrium equations in a slightly different form, as follows. From the relations between $(N_{\alpha\beta})$ and $(u_{\alpha,w})$ as given in (3.4a, b), we may write

$$N_{\alpha\beta} = N'_{\alpha\beta} + C\kappa_{\alpha\beta}w + N_{\alpha\beta}^{(n)}, \tag{3.9}$$

where

$$N'_{11} = C(u_{1,1} + \nu u_{2,2}); \qquad N'_{22} = C(u_{2,2} + \nu u_{1,1});$$
$$N'_{12} = \tfrac{1}{2}C(1 - \nu)(u_{1,2} + u_{2,1})$$

or

$$N'_{\alpha\beta} = C_{\alpha\beta\gamma\sigma}u_{\gamma,\sigma}; \tag{3.10}$$

$$\kappa_{11} = \frac{1}{R_{11}} + \frac{\nu}{R_{22}}; \qquad \kappa_{22} = \frac{1}{R_{22}} + \frac{\nu}{R_{11}}; \qquad \kappa_{12} = \frac{1-\nu}{R_{12}}, \tag{3.11}$$

and the nonlinear parts

$$N_{11}^{(n)} = \frac{C}{2}\left[\left(\frac{\partial w}{\partial x_1}\right)^2 + \nu\left(\frac{\partial w}{\partial x_2}\right)^2\right]; \qquad N_{22}^{(n)} = \frac{C}{2}\left[\left(\frac{\partial w}{\partial x_2}\right)^2 + \nu\left(\frac{\partial w}{\partial x_1}\right)^2\right];$$
$$N_{12}^{(n)} = \frac{C}{2}(1 - \nu)\frac{\partial w}{\partial x_1}\frac{\partial w}{\partial x_2}. \tag{3.12}$$

Use of (3.9) in (3.7) and the application of the divergence theorem results in

$$\int_\Gamma N'_{\alpha\beta}n_\beta u_\alpha^*\,d\Gamma - \int_\Omega N'_{\alpha\beta}u^*_{\alpha,\beta}\,d\Omega + \int_\Omega C(\kappa_{\alpha\beta}w)_{,\beta}u_\alpha^*\,d\Omega + \int_\Omega N^{(n)}_{\alpha\beta,\beta}u_\alpha^*\,d\Omega$$
$$+ \int_\Omega (b_\alpha - \rho\ddot{u}_\alpha)u_\alpha^*\,d\Omega = \int_{\Gamma_\sigma} (P_\alpha - \bar{P}_\alpha)u_\alpha^*\,d\Gamma + \int_{\Gamma_u} (\bar{u}_\alpha - u_\alpha)P_\alpha^*(u_\sigma^*)\,d\Gamma. \tag{3.13}$$

Since the material is linear elastic and isotropic, we have

$$N'_{\alpha\beta}u^*_{\alpha,\beta} = C_{\alpha\beta\gamma\sigma}u_{\gamma,\sigma}u^*_{\alpha,\beta} \equiv N^{*\prime}_{\gamma\sigma}(u^*_\mu)u_{\gamma,\sigma}, \tag{3.14}$$

where the definition of $N'^*_{\gamma\sigma}$ is apparent. Now note that

$$P_\alpha = N_{\alpha\beta}n_\beta, \tag{3.15a}$$

or

$$P_\alpha = N'_{\alpha\beta}n_\beta + C\kappa_{\alpha\beta}wn_\beta + N^{(n)}_{\alpha\beta}n_\beta. \tag{3.15b}$$

Using (3.14), (3.15a, b) in (3.13) and applying the divergence theorem, it is easy to obtain

$$\int_\Omega [N'^*_{\alpha\beta}(u^*_\alpha)]_{,\beta}u_\alpha \, d\Omega - \int_\Omega (b_\alpha - \rho\ddot{u}_\alpha)u^*_\alpha \, d\Omega + \int_\Gamma \hat{P}_\alpha u^*_\alpha \, d\Gamma - \int_\Gamma P^*_\alpha \hat{u}_\alpha \, d\Gamma$$
$$- \int_\Omega C\kappa_{\alpha\beta}wu^*_{\alpha,\beta} \, d\Omega - \int_\Omega N^{(n)}_{\alpha\beta}u^*_{\alpha,\beta} \, d\Omega = 0, \tag{3.16a}$$

where

$$\hat{P}_\alpha = \bar{P}_\alpha \quad \text{at } \Gamma_\sigma; \qquad \text{and} \qquad \hat{P}_\alpha = P_\alpha \quad \text{at } \Gamma_u, \tag{3.16b}$$

and

$$\hat{u}_\alpha = \bar{u}_i \quad \text{at } \Gamma_u; \qquad \text{and} \qquad \hat{u}_\alpha = u_\alpha \quad \text{at } \Gamma_\sigma. \tag{3.16c}$$

Now, we choose u^*_β to be the "fundamental solution" in infinite space, of the equation

$$[N'^*_{\alpha\beta}(u^*_\sigma)]_{,\beta} + \delta(x_\mu - \xi_\mu)\delta_{\alpha\theta}e_\theta = 0, \tag{3.17}$$

where $\delta(x_\mu - \xi_\mu)$ is the Dirac delta function at $x_\mu = \xi_\mu$, $\delta_{\alpha\theta}$ is the Kronecker delta, and e_θ denotes that the direction of the application of the point load is along the x_θ direction. The "fundamental solution" of (3.17) will be denoted as $u^*_{(\theta)\alpha}$; where $u^*_{(\theta)\alpha}$ is the displacement along the x_α direction in a plane infinite body at any point x_μ, due to a unit load along the x_θ direction, applied at the location $x_\mu = \xi_\mu$. Likewise, $P^{*\prime}_{(\theta)\alpha}(x_\mu, \xi_\mu)$ will be considered to be the traction along the x_α direction on an oriented surface at x_μ, with a unit normal n_σ, due to a unit load along x_θ at the location ξ_μ. These solutions are well known and may be written as

$$u^*_{(\theta)\alpha}(x_\mu, \xi_\mu) = \frac{1}{8\Pi G}\left[(v - 3)\ln \rho\delta_{\theta\alpha} + (1 + v)\frac{\partial\rho}{\partial x_\theta}\frac{\partial\rho}{\partial x_\alpha}\right], \tag{3.18a}$$

and

$$P^{*\prime}_{(\theta)\alpha}(x_\mu, \xi_\mu) = -\frac{t}{4\Pi\rho}\left\{\frac{\partial\rho}{\partial n_\sigma}\left[(1 - v)\delta_{\theta\alpha} + 2(1 + v)\frac{\partial\rho}{\partial x_\theta}\frac{\partial\rho}{\partial x_\alpha}\right]\right\}$$
$$- (1 - v)\left(n_\theta\frac{\partial\rho}{\partial x_\alpha} - n_\alpha\frac{\partial\rho}{\partial x_\theta}\right), \tag{3.18b}$$

where $\rho = |x_\mu - \xi_\mu|$ is the radius vector from x_μ to ξ_μ;

$$G = \frac{Et}{[2(1+\nu)]}.$$

Because of the property of integrals involving Dirac functions, we have

$$\int_\Omega (N'^{*}_{\alpha\beta})_{,\beta} u_\alpha \, d\Omega \equiv -\int_\Omega \delta(x_\mu - \xi_\mu)\delta_{\alpha\theta} e_\theta u_\alpha(x_\mu) \, d\Omega \equiv -u_\theta(\xi_\mu). \tag{3.19}$$

Using (3.18) and (3.19) in (3.16a), we have

$$\begin{aligned}\gamma u_\theta(\xi_\mu) = &\int_\Omega [b_\alpha(x_\mu) - \rho \ddot{u}_\alpha(x_\mu)] u^*_{(\theta)\alpha}(x_\mu, \xi_\mu) \, d\Omega + \int_\Gamma \hat{P}_\alpha(x_\mu) u^*_{(\theta)\alpha}(x_\mu, \xi_\mu) \, d\Gamma \\ &- \int_\Gamma \hat{u}_\alpha(x_\mu) P^*_{(\theta)\alpha}(x_\mu, \xi_\mu) \, d\Gamma - \int_\Omega C\kappa_{\alpha\beta} w(x_\mu) u^*_{(\theta)\alpha,\beta}(x_\mu, \xi_\mu) \, d\Omega \\ &- \int_\Omega N^{(n)}_{\alpha\beta}(x_\mu) u^*_{(\theta)\alpha,\beta}(x_\mu, \xi_\mu) \, d\Omega.\end{aligned} \tag{3.20}$$

It can be shown that whereas the coefficient γ in the left-hand side of (3.20) is unity when ξ_μ is in the interior of Ω, the value of γ is 0.5 when ξ_μ falls on the "smooth" boundary Γ [6]. Equation (3.20) is the sought-after integral equation for u_α in a shallow shell.

We now choose the test function $w^*(x_\mu)$ to be the "fundamental solution" in an *infinite plate* corresponding to a unit point load at the location ξ_μ. Thus, w^* corresponds to the solution of the linear equation

$$D\nabla^4 w^* = \delta(x_\mu - \xi_\mu) \tag{3.21}$$

in an *infinite* domain in the base plane of the shallow shell. It is well known that the solution for w^* is given by

$$w^*(x_\mu, \xi_\mu) \equiv \frac{1}{8\pi}\rho^2 \ln \rho, \tag{3.22}$$

where $\rho = |x_\mu - \xi_\mu|$.

Using (3.22) and (3.9) in (3.8) and employing repeated integrations by part in the resulting equation, we easily obtain the integral equation:

$$\begin{aligned}&\gamma_w D w(\xi_\mu) \\ &= \int_\Gamma \hat{V}_n(x_\mu) w^*(x_\mu, \xi_\mu) \, d\Gamma - \int_\Gamma \hat{M}_n(x_\mu) \psi^*_n(x_\mu, \xi_\mu) \, d\Gamma \\ &\quad + \int_\Gamma \hat{\psi}_n(x_\mu) M^*_n(x_\mu, \xi_\mu) \, d\Gamma - \int_\Gamma \hat{w}(x_\mu) V^*_n(x_\mu, \xi_\mu) \, d\Gamma \\ &\quad - \int_\Gamma C\kappa_{\alpha\beta} n_\beta \hat{u}_\alpha(x_\mu) w^*(x_\mu, \xi_\mu) \, d\Gamma + \int_\Omega C[\kappa_{\alpha\beta} w^*(x_\mu, \xi_\mu)]_{,\beta} u_\alpha(x_\mu) \, d\Omega\end{aligned}$$

$$- \int_{\Omega} C \frac{\kappa_{\alpha\beta}}{R_{\alpha\beta}} w(x_\mu) w^*(x_\mu, \xi_\mu)\, d\Omega - \int_{\Omega} \frac{N_{\alpha\beta}^{(n)}}{R_{\alpha\beta}} (x_\mu) w^*(x_\mu, \xi_\mu)\, d\Omega$$

$$+ \int_{\Omega} [b_3 - \rho \ddot{w} + f_3 + (N_{\alpha\beta} w_{,\beta})_{,\alpha}](x_\mu) w^*(x_\mu, \xi_\mu)\, d\Omega$$

$$+ \sum_{1}^{k} [\langle M_t \rangle w^* - \langle M_t^* \rangle w], \tag{3.23}$$

where

$$V_n = -D \frac{\partial}{\partial n}(\nabla^2 w) + \frac{\partial}{\partial s} M_s + N_n \frac{\partial w}{\partial n} + N_{nt} \frac{\partial w}{\partial s},$$

$$M_n = M_{11} n_1^2 + 2M_{12} n_1 n_2 + M_{22} n_2^2,$$

$$M_t = (M_{22} - M_{11}) n_1 n_2 + M_{12}(n_1^2 - n_2^2),$$

$$\gamma_w = 1 \text{ for } \xi_\mu \in \Omega; \quad \gamma_w = \tfrac{1}{2} \text{ for } \xi_\mu \in \Gamma \text{ (smooth)}.$$

In (3.23) the terms with the superposed symbol "^" should be taken to imply the respective prescribed values, if any, at Γ; otherwise, they are to be treated as the unknown solution variables. Also, the symbol $\langle(\ )\rangle$ denotes the jump in the quantity ( ) at a corner at Γ, in the direction of the increasing arc length along Γ; and the summation (1 to k) extends to all the k such corners.

Since $(\partial w/\partial n)$ is also an independent variable at Γ, an integral relation for $(\partial w/\partial n)$ should also be derived. To this purpose, consider a second fundamental solution

$$w_v^* = \frac{1}{2\pi} \rho \ln \rho \cos \phi,$$

where ϕ is the angle between the outward normal to Γ and the radius ρ (see [28]). The resulting integral equation is

$$\gamma_v D \frac{\partial W}{\partial n}$$

$$= \int_{\Gamma} \hat{V}_n(x_\mu) w_v^*(x_\mu, \xi_\mu)\, d\Gamma - \int_{\Gamma} \hat{M}_n(x_\mu) \psi_v^*(x_\mu, \xi_\mu)\, d\Gamma$$

$$+ \int_{\Gamma} \hat{\psi}_n(x_\mu) M_v^*(x_\mu, \xi_\mu)\, d\Gamma - \int_{\Gamma} [\hat{w}(x_\mu) - \hat{w}(\xi_\mu)] V_v^*(x_\mu, \xi_\mu)\, d\Gamma$$

$$- \int_{\Gamma} C \kappa_{\alpha\beta} n_\beta \hat{u}_\alpha(x_\mu) w_v^*(x_\mu, \xi_\mu)\, d\Gamma + \int_{\Omega} C[\kappa_{\alpha\beta} w_v^*(x_\mu, \xi_\mu)]_{,\beta} u_\alpha(x_\mu)\, d\Omega$$

$$- \int_{\Omega} C \frac{\kappa_{\alpha\beta}}{R_{\alpha\beta}} w(x_\mu) w_v^*(x_\mu, \xi_\mu)\, d\Omega - \int_{\Omega} \frac{N_{\alpha\beta}^{(n)}}{R_{\alpha\beta}} (x_\mu) w_v^*(x_\mu, \xi_\mu)\, d\Omega$$

$$+ \int_{\Omega} [b_3 - \rho \ddot{w} + f_3 + (N_{\alpha\beta} w_{,\beta})_{,\alpha}](x_\mu) w^*(x_\mu, \xi_\mu)\, d\Omega$$

$$+ \sum_{1}^{k} [\langle M_t \rangle w_v^* - \langle M_{vt}^* \rangle w]. \tag{3.24}$$

9.3.2.1. *Comments*

In summary, (3.20), (3.23), and (3.24) represent the complete set of integral equations for u_α, w, and $\partial w/\partial n$. An examination of these equations reveals the following features:

(i) For *given* body forces b_α, the integral relation for u_θ (3.20) involves the trial functions u_θ *only* at the boundary Γ. On the other hand, as a result of the curvature induced coupling of the trial functions (u_α and w) in the shallow shell problem, the integral relations for u_θ contain a *domain integral* (over Ω) *involving the trial function for* w. If the in-plane inertia forces ($\rho\ddot{u}_\alpha$) appear in the problem, then the integral relation for u_α involves a domain integral (over Ω) of $\ddot{u}_\alpha$ as well.

(ii) Again, because of the curvature induced coupling of the in-plane and out-of-plane displacements in the shallow shell presently considered, the integral equations for w and $\partial w/\partial n$, (3.23) and (3.24), respectively, contain *domain integrals* (*over* Ω) *involving trial functions for both* w *and* u_α. Also, in a transient dynamic analysis, the term ($\ddot{w}$) appears inside a domain integral.

(iii) In the nonlinear problem, the nonlinear terms $N^{(n)}_{\alpha\beta}$ and $(N_{\alpha\beta}w_{,\beta})_{,\alpha}$ involving trial functions for both w and u_α inevitably bring the domain integrals (over Ω) into the equations.

(iv) For reasons (i) to (iii), unlike the classical homogeneous isotropic elastostatics [6] wherein a discretization of the relevant integral equations requires a use of basis functions for the displacements at the *boundary* alone, the present shallow shell formulation requires the assumption of basis functions for the *trial solutions* u_α *and* w, *at the boundary* Γ *as well as in the domain* Ω. Thus, the present solution methodology may, strictly speaking, be classified as a field boundary element method based on a direct discretization of *integral equations.*

(v) Suppose now that in (3.20), (3.23), and (3.24), we let ξ_μ tend to a point on the boundary, i.e., $\xi_\mu \in \Gamma$. Thus, we obtain three integral relations for the *boundary* values of u_α, w, and $\partial w/\partial n$. An examination of these relations reveals that, in order to *discretize* these equations, we need to assume only *very simple* trial functions u_α, w, $\partial w/\partial n$ not only at the boundary but also in the interior of Ω. For instance, Ω may be discretized into a number of finite elements and Γ into a number of boundary elements. As ξ_μ tends to Γ, the integral relations (3.20), (3.23), and (3.24) clearly show that w and u_α need only be *piecewise* linear functions in each finite element in Ω, in any discretization process. In contrast, it is recalled that in the Galerkin finite element method, u_α need be C^0 continuous and w be C^1 continuous in each element. The difficulties with such a finite element approach are too well documented in literature to warrant further comment here.

(vi) At each point on the boundary, two of the in-plane variables u_α ($\alpha = 1, 2$), P_α ($\alpha = 1, 2$) are specified; and the other two are unknown. Likewise, two of the out-of-plane variables, V_n, M_n, ψ_n, and w, are specified; and the other two are unknown. At each point in Ω, as seen from (3.20), (3.23), and (3.24), the three displacements, u_α and w, are unknown. Thus, if (3.20), (3.23), and (3.24) were discretized, through finite as well as boundary elements and the nodal values of the variables are appropriately understood to be either specified or unknown, we would easily obtain exactly as many equations as the number of unknowns, so that the problem may be considered as well posed.

Note that because of the appearance of the nonlinear terms in the obtained integral equations (3.20), (3.23), and (3.24), the discretized interior/boundary element equations cannot be solved directly; that is, an incremental approach is necessary. This will be discussed in detail in Section 9.3.3.

9.3.3. Incremental Approach and Solution Strategy

In the incremental approach, the load and the prescribed boundary conditions are applied in small but finite increments, and in a time-dependent problem these increments correspond to the time steps.

Consider that the shell is at the end of the Kth load increment. All the quantities that have been known during the previous K steps of analysis are denoted by a superscript K. The displacement increments are denoted as

$$\Delta u_\alpha = u_\alpha^{k+1} - u_\alpha^k; \qquad \Delta w = w^{k+1} - w^k. \tag{3.25}$$

With these notations the integral equations in incremental form may be written as

$$\begin{aligned}
\gamma(u_\theta^k + \Delta u_\theta) = {} & \int_\Omega [b_\alpha^k + \Delta b_\alpha - \rho(\ddot{u}_\alpha^k + \Delta\ddot{u}_\alpha)] u_{(\theta)\alpha}^* \, d\Omega \\
& + \int_\Gamma (\hat{P}_\alpha + \Delta\hat{P}_\alpha) u_{(\theta)\alpha}^* \, d\Gamma - \int_\Gamma (\hat{u}_\alpha^k + \Delta\hat{u}_\alpha) P_{(\theta)\alpha} \, d\Gamma \\
& - \int_\Omega C\kappa_{\alpha\beta}(w^k + \Delta w) U_{(\theta)\alpha}^* \, d\Omega - \int_\Omega (N_{\alpha\beta}^{(n)K} + \Delta N_{\alpha\beta}^{(n)} \\
& + \text{higher order terms}) u_{(\theta)\alpha,\beta}^* \, d\Omega,
\end{aligned} \tag{3.26a}$$

$$\begin{aligned}
\gamma_w D(w^k + \Delta w) = {} & \int_\Gamma (\hat{V}_n^k + \Delta\hat{V}_n) w^* \, d\Gamma - \int_\Gamma (\hat{M}_n^k + \Delta\hat{M}_n) \Psi_n^* \, d\Gamma \\
& + \int_\Gamma (\hat{\Psi}_n^k + \Delta\hat{\Psi}_n) M_n^* \, d\Gamma - \int_\Gamma (\hat{w}^k + \Delta\hat{w}) V_n^* \, d\Gamma \\
& - \int_\Omega \left[\frac{N_{\alpha\beta}'^k}{R_{\alpha\beta}} + \frac{\Delta N_{\alpha\beta}'}{R_{\alpha\beta}} + C\frac{\kappa_{\alpha\beta}}{R_{\alpha\beta}}(w^k + \Delta w) \right] w^* \, d\Omega \\
& - \int_\Omega \left[\frac{N_{\alpha\beta}^{(n)k}}{R_{\alpha\beta}} + \frac{\Delta N_{\alpha\beta}^{(n)}}{R_{\alpha\beta}} + \text{higher order terms} \right] w^* \, d\Omega
\end{aligned}$$

$$+ \int_\Omega [b_3^k + \Delta b_3 - \rho(\ddot{w}^k + \Delta\ddot{w}) + f_3^k + \Delta f_3] w^* \, d\Omega$$

$$+ \int_\Omega [(N_{\alpha\beta}^k + \Delta N_{\alpha\beta})(w^k + \Delta w)_{,\beta}]_{,\alpha} w^* \, d\Omega$$

$$+ \sum_1^k [\langle M_t^k + \Delta M_t \rangle w^* - \langle M_t^* \rangle (w^k + \Delta w)], \qquad (3.26b)$$

where the increments of the nonlinear parts of in-plane forces are

$$\begin{aligned} \Delta N_{11}^{(n)} &= C[w_{,1}^k \Delta w_{,1} + \nu w_{,2}^k \Delta w_{,2}], \\ \Delta N_{22}^{(n)} &= C[w_{,2}^k \Delta w_{,2} + \nu w_1^k \Delta w_{,1}], \\ \Delta N_{12}^{(n)} &= C\frac{1-\nu}{2}[w_{,1}^k \Delta w_{,2} + w_{,2}^k \Delta w_{,1}], \end{aligned} \qquad (3.27)$$

and

$$\Delta N_{\alpha\beta} = \Delta N'_{\alpha\beta} + C\kappa_{\alpha\beta}\Delta w + \Delta N_{\alpha\beta}^{(n)} + \text{higher order terms},$$

where the definition of the increments of the linear part is apparent.

Here, the incremental form of (3.24) is similar to (3.26b), and its treatment follows the same routine.

In (3.26a, b), the higher order terms involve the products of the incremental displacements. In solving these unknown incremental displacements, those higher order terms are ignored. In those incremental equations, the terms with the superscript K should have satisfied the equilibrium conditions at the end of the Kth load increment, but the equilibrium conditions are in fact not exactly satisfied because of the absence of the higher order terms. Therefore, those terms are written in the incremental equations as the equilibrium constraints.

Note that in (3.26b), the nonlinear term $(N_{\alpha\beta}^k + \Delta N_{\alpha\beta})(w^k + \Delta w)_{,\beta}$ can be written as

$$(N_{\alpha\beta}^k + \Delta N_{\alpha\beta})(w^k + \Delta w)_{,\beta} = N_{\alpha\beta}^k w_{,\beta}^k + N_{\alpha\beta}^k \Delta w_{,\beta} + \Delta N_{\alpha\beta} w_{,\beta}^k + \text{higher order terms}. \qquad (3.28)$$

Ignoring the higher order terms and examining (3.27), (3.28), we may see that those nonlinear terms are linearized with respect to the displacement increments. Using (3.27), (3.28) in (3.26a, b) and applying the divergence theorem, we may obtain the final integral equations in terms of unknown displacement increments

$$\gamma(u_\theta^k + \Delta u_\theta) = \int_\Omega [b_\alpha^k + \Delta b_\alpha - \rho(\ddot{u}_\alpha^k + \Delta\ddot{u}_\alpha)] u_{(\theta)\alpha}^* \, d\Omega$$

$$+ \int_\Gamma (\hat{P}_\alpha^k + \Delta\hat{P}_\alpha) u_{(\theta)\alpha}^* \, d\Gamma - \int_\Gamma (\hat{u}_\alpha^k + \Delta\hat{u}_\alpha) P_{(\theta)\alpha}^* \, d\Gamma$$

$$- \int_\Omega C\kappa_{\alpha\beta}(w^k + \Delta w)\mathrm{u}^*_{(\theta)\alpha,\beta}\, d\Omega + \int_\Gamma N'^*_{(\theta)\alpha\beta} w^k_{,\alpha} n_\beta \Delta\hat{w}\, d\Gamma$$

$$- \int_\Omega N'^*_{(\theta)\alpha\beta} w^k_{,\alpha\beta} \Delta w\, d\Omega + w^k_{,\theta}|_{\xi_\mu}, \tag{3.29a}$$

$$\gamma_w D(w^k + \Delta w) = \int_\Gamma (\hat{V}^k_n + \Delta\hat{V}_n) w^*\, d\Gamma - \int_\Gamma (\hat{M}^k_n + \Delta\hat{M}_n)\Psi^*\, d\Gamma_n$$

$$+ \int_\Gamma (\hat{\Psi}^k_n + \Delta\hat{\Psi}_n) M^*_n\, d\Gamma - \int_\Gamma (\hat{w}^k + \Delta\hat{w}) V^*_n\, d\Gamma$$

$$- \int_\Gamma [C\kappa_{\alpha\beta} n_\beta w^* + A_\alpha]\Delta\hat{u}_\alpha\, d\Gamma$$

$$+ \int_\Omega [C(\kappa_{\alpha\beta} w^*)_{,\beta} + B_\alpha]\Delta u_\alpha\, d\Omega$$

$$- \int_\Gamma [(C\kappa_{\alpha\beta} w^k_{,\alpha} w^* + N^k_{\alpha\beta} w^*_{,\alpha}) n_\beta + A_w]\Delta\hat{w}\, d\Gamma$$

$$+ \int_\Omega \left[C(\kappa_{\alpha\beta} w^k_{,\alpha} w^*)_{,\beta} + N^k_{\alpha\beta} w^*_{,\alpha\beta} - C\frac{\kappa_{\alpha\beta}}{R_{\alpha\beta}} w^* - B_w \right] \Delta w\, d\Omega$$

$$+ \int_\Omega \left[(N^k_{\alpha\beta} w^k_{,\alpha})_{,\beta} - \frac{N'^k_{\alpha\beta}}{R_{\alpha\beta}} - C\frac{\kappa_{\alpha\beta}}{R_{\alpha\beta}} w^k + b^k_3 + \Delta b_3 \right.$$

$$\left. - \rho(\ddot{w}^k + \Delta\ddot{w}) + f^k_3 + \Delta f_3 \right] w^*\, d\Omega$$

$$+ \Sigma^k_1 [\langle M^k_t + \Delta M_t \rangle w^* - \langle M^*_t \rangle (w^k + \Delta w)], \tag{3.29b}$$

where, A, B may be given as follows:

$$A_1 = C\left[(w^k_{,1} w^*_{,1} + \nu w^k_{,2} w^*_{,2}) n_1 + \frac{1-\nu}{2} w^k_{,1} w^*_{,2} + w^k_{,2} w^*_{,1}) n_2 \right], \tag{3.30a}$$

$$B_1 = C\left\{ (w^k_{,1} w^*_{,1})_{,1} + \nu(w^k_{,2} w^*_{,2})_{,2} + \frac{1-\nu}{2}[(w^k_{,1} w^*_{,1})_{,2} + (w^k_{,2} w^*_{,1})_{,1}] \right\}, \tag{3.30b}$$

A_2, B_2 can be similarly written by cycling the subscripts

$$A_w = C\left\{ (w^k_{,1})^2 w^*_{,1} n_1 + (w^k_{,2})^2 w^*_{,2} n_2 + \nu w^k_{,1} w^k_{,2} (w^*_{,1} n_2 + w^*_{,2} n_1) \right.$$

$$\left. + \frac{1-\nu}{2}[(w^k_{,1})^2 w^*_{,2} n_2 + (w^k_{,2})^2 w^*_{,1} n_1 + w^k_{,1} w^k_{,2}(w^*_{,1} n_2 + w^*_{,2} n_1)] \right\}, \tag{3.30c}$$

$$B_w = C\left\{2(w^k_{,1}w^k_{,11}w^*_{,1} + w^k_{,2}w^k_{,22}w^*_{,2}) + (w^k_{,1})^2w^*_{,11} + (w^k_{,2})^2w^*_{,22}\right.$$

$$+ \frac{1+\nu}{2}(w^k_{,1}w^k_{,22}w^*_{,1} + w^k_{,2}w^k_{,11}w^*_{,2} + 2w^k_{,1}w^k_{,2}w^*_{,12})$$

$$+ \frac{3-\nu}{2}(w^k_{,1}w^*_{,2} + w^k_{,2}w^*_{,1})w^k_{,12} + \frac{1-\nu}{2}[(w^k_{,1})^2w^*_{,22} + (w^k_{,2})w^{2*}_{,11}]$$

$$\left. - \kappa_{ij}w^*_{,i}w^N_{,j}\right\}. \qquad (3.30d)$$

We discretize the domain (Ω) as well as the boundary Γ by using some appropriate interpolation functions for those unknown displacement increments. By carrying out the indicated integrations, we may obtain the field/boundary element equations with the displacement increments as unknowns. For each load increment, these equations have to be solved iteratively because the equilibrium conditions are only approximately satisfied because of the absence of the higher order terms.

The full Newton–Raphson algorithm is used to obtain the solution. This involves domain integrations for constructing the coefficient matrix and reduction of the matrix in each iteration. Since, in the present boundary/interior element method, the coefficient matrix is not as large as in the usual finite element approach, the reduction of this matrix is not as critical, and moreover it is found that the solution converges very rapidly.

In the present numerical implementation, at the beginning of each load increment, the equilibrium is checked first by subtracting the integrals of the other terms with the superscript K from the total load integration; the residual load vector is used to solve the displacement increments; the obtained displacement increments are then used to update the coefficient matrix and check the equilibrium again, and so on.

Several examples of postbuckling behavior as well as of nonlinear transient dynamic response are presented in [33, 34].

Acknowledgments. These remarks are based on research being performed under the support of AFOSR and ONR. It is a pleasure to acknowledge this support.

References

1. S.N. Atluri, "On hybrid finite element models in solid mechanics," in *Advances in Computer Methods for Partial Differential Equations* (R. Vishnevetsky, ed.), AICA, Rutgers University, New York, 1975, pp. 346–356.
2. S.N. Atluri, "Higher-order, special, and singular finite elements," in *State of the Art Survey of Finite Element Technology* (A.K. Noor and W. Pilkey, eds.), ASME, New York, 1983.

3. S.N. Atluri, "Computational solid mechanics (finite elements and boundary elements): Present status and future directions," in *Proceedings of the 4th International Conference on Applied Numerical Modeling*, Keynote paper (H.M. Hsia *et al.*, eds.), pp. 19–37.
4. S.N. Atluri, *Computational Methods in the Mechanics of Fracture*, North-Holland, Amsterdam, 1986.
5. S.N. Atluri, R.H. Gallagher, and O.C. Zienkiewicz (eds.), *Hybrid and Mixed Finite Element Methods*, Wiley, New York, 1983.
6. S.N. Atluri and J.J. Grannell, "Boundary element (BEM) and combination of BEM-FEM," Report GIT-ESM-SA-78-16, Center for the Advancement of Computational Mechanics, Georgia Institute of Technology, 1978.
7. S.N. Atluri and H. Murakawa, "On hybrid finite element models in nonlinear solid mechanics," in *Finite Elements in Nonlinear Mechanics* (P.G. Bergan *et al.*, eds.), Tapir, Norway, 1977, pp. 3–44.
8. S.N. Atluri, H. Murakawa, and C. Bratianu, "Use of stress functions and asymptotic solutions in FEM analysis of continua," in *New Concepts in Finite Element Analysis*, AMD, Vol. 44, ASME, New York, 1981, pp. 11–28.
9. S.N. Atluri and T. Nishioka, "Numerical studies in dynamic fracture mechanics," *Internat. J. Fracture*, **27** (1985), 245–261.
10. S.N. Atluri, P.E. O'Donoghue, and J.D. Zhang, "Analysis and control of finite deformations of plates and shells: Formulations and interior/boundary element algorithms," in *Finite Element Analysis of Shells: Formulations & Algorithms*, (T.J. Hughes, ed.), Pineridge Press, Swansea, 1985 (In Print).
11. S.N. Atluri, P. Tong, and H. Murakawa, "Recent studies on hybrid and mixed finite element methods in mechanics," in *Hybrid and Mixed Finite Element Methods* (S.N. Atluri *et al.*, eds.), Wiley, New York, 1983, pp. 51–71.
12. S.W. Attaway, "A stress based finite element method for computational elasto-plastic analysis using an endochronic theory of plasticity," Ph.D. thesis Georgia Institute of Technology, 1986.
13. I. Babuska, "The finite element method with Lagrange multipliers," *Numer. Math.*, **20** (1973), 179–192.
14. G. Bezine, "A mixed boundary integral finite element approach to plate vibration problems," *Mech. Res. Comm.*, **7** (1980), 141–150.
15. G. Bezine, "A boundary integral equation method for plate flexure with conditions inside the domain," *Internat. J. Numer. Methods Engrg.*, **17** (1981), 1647–1657.
16. C. Bratianu and S.N. Atluri, "A hybrid finite element method for stokes flow, I: Formulation and numerical studies," *Comput. Methods Appl. Mech. Engrg.*, **36** (1983), 23–38.
17. C.A. Brebbia and S. Walker, *Boundary Element Techniques in Engineering*, Newnes–Butterworth Press, United Kingdom, 1980.
18. F. Brezzi, "On the existence, uniqueness, and approximation of saddle-point problems arising from Lagrange multipliers," *RAIRO* 8-R2 (1974), 129–151.
19. B.M. Irons and S. Ahmad, *Techniques of Finite Element Analysis*, Ellis Harwood, Chickester, England, 1980.
20. H. Murakawa and S.N. Atluri, "Finite elasticity solutions using hybrid finite elements based on a complementary energy principle," *J. Appl. Mech.*, **45** (1978), 539–547.
21. T. Nishioka and S.N. Atluri, "Analytical solution for embedded elliptical cracks,

and finite element altering method for elliptical surface cracks, subjected to arbitrary loadings," *Engrg. Fracture Mech.*, **17** (1982), 247–268.

22. A.K. Noor and J.M. Peters, "Mixed models and reduced selective integration displacement models for vibration analysis of shells," in *Hybrid and Mixed Finite Element Methods* (S.N. Atluri, R.H. Gallagher, and O.C. Zienkiewicz, eds.), Wiley, New York, 1983, pp. 537–564.
23. T.H.H. Pian and P. Tong, "Basis of finite element methods for solid continua, *Internat. J. Numer. Methods Engrg.*, **1** (1969), 3–28.
24. E.F. Punch, "Stable invariant, least order isoparametric mixed hybrid stress element: Linear elastic continua, and finitely deformed plates and shells" Ph.D. thesis, Georgia Institute of Technology, 1983.
25. E.F. Punch, and S.N. Atluri, "Development and testing of stable, invariant, isoparametric curvilinear 2- and 3-D hybrid stress elements," *Comput. Methods Appl. Mech. Engrg.*, **47** (1984), 331–356.
26. K.W. Reed and S.N. Atluri, "Analysis of large quasistatic deformations of inelastic bodies by a new hybrid stress finite element algorithm," *Comput. Methods Appl. Mech. Engrg.*, **39** (1983), 245–295.
27. R. Rubinstein, E.F. Punch, and S.N. Atluri, "An analysis of, and remedies for, kinematic modes in hybrid stress finite elements: Selection of stable, invariant stress fields," *Comput. Methods Appl. Mech. Engrg.*, **38** (1983), 63–92.
28. M. Stern, "General boundary integral formulation for the numerical solution of plate bending problems," *Internat. J. Solids and Structures*, **15** (1979), 769–782.
29. W-M Xue and S.N. Atluri, "Existence and stability, and discrete BB and Rank conditions, for general mixed hybrid finite elements in elasticity" in *Hybrid and Mixed Finite Element Methods* (K.L. Spilker and K.W. Reed eds.) AMD, Vol. 73, ASME, New York, 1985, pp. 91–112.
30. W-M Xue, L.A. Karlovitz, and S.N. Atluri, "On the existence and stability conditions for mixed hybrid finite element solutions based on Reissner's variational principle," *Internat. J. Solids and Structures*, **21**, No. 1 (1985), 97–116.
31. C-T Yang, and S.N. Atluri, "An assumed deviatoric stress–pressure velocity mixed finite element method for unsteady, convective, incompressible viscous flow, I: Theoretical development," *Internat. J. Numer. Methods Fluids*, **3** (1983), 377–398.
32. L-A Ying and S.N. Atluri, "A hybrid finite element method for stokes flow, II: Stability and convergence studies," *Comput. Methods Appl. Mech. Engrg.*, **36** (1983), 39–60.
33. J.D. Zhang and S.N. Atluri, "A boundary/interior element method for quasi static and transient response analysis of shallow shells," *Computers & Structures*, **24** (1986), 213–224.
34. J.D. Zhang and S.N. Atluri, "Post buckling analysis of shallow shells by the field-boundary element method" *Internat. J. Numer. Methods Engrg.*, (1987).
35. H. Okada, H. Rajiyah, and S.N. Atluri, "Non-hypersingular integral-representations for velocity (displacement) gradients in elastic/plastic solids (small or finite deformations)," *Comput. Mech.* (1987) (In Print).

CHAPTER 10

The p and h-p Versions of the Finite Element Method: The State of the Art

I. Babuška

10.1. Introduction

There are three versions of the finite element method. The classical *h version*, which achieves the accuracy by refining the mesh while using low degrees p of the elements, usually $p = 1, 2$. The *p version* keeps the mesh fixed, and the accuracy is achieved by increasing the degree p. The *h-p version* combines both approaches.

The p and h-p versions are new developments. The p version was implemented at Washington University in St. Louis in an experimental code called COMET-X in the middle of 1970. The essential feature of the code was that hierarchic elements were implemented. These type of elements was first considered in [65] in conjunction with joining finite elements of different polynomial degrees. Hierarchic C^0 elements were then described by various authors, e.g., [36, 42, 58, 66]. P convergence was first demonstrated in [57], and a cohesive description of the p version was given in [49].

The first theoretical analysis of the p version was given in [21] and in [20]. The performance of the h-p version was first studied theoretically in [4]. For theoretical analysis of the p version in three dimensions, we refer to [24] and [25]. Additional recent results are mentioned later. For the implementational aspects of the p version, we refer to [50–52].

The p and h-p versions for two-dimensional problems were implemented in the commercial system PROBE by Noetic Tech., St. Louis with the first release in 1985, and the second in 1986 (computations in the present paper are made by PROBE). At present PROBE is also being developed for three-dimensional problems. The three-dimensional finite element code FIESTA having some p version capabilities was developed at the Istituto Sperimentale Modelli e Strutture (ISMES) in Bergamo, Italy, and since early 1980 this program has been available in the U.S.A. A new three-dimensional program STRIPE is also being developed by the Aeronautical Research Institute of Sweden (Flygtekniska Försöksanstalten FFA). The p and h-p versions are used in the industry today.

Although the p and h-p versions of the finite element method are relatively new developments, many basic results are available. The aim of the present paper is to give a survey of the basic available results and directions for further development.

10.2. Finite Element Method and the Approximation Problem

Let $B(u, v)$ be a bilinear form defined on $H_1 \times H_2$, where H_1 and H_2 are reflexive Banach spaces equipped with the norm $\|\cdot\|_1$ and $\|\cdot\|_2$, respectively, and $F \in H_2^*$; i.e., let F be a linear functional on H_2.

By the problem $\{B, F\}$, we denote the problem to find $u_0 \in H_1$ so that

$$B(u_0, v) = F(v) \tag{2.1}$$

holds for all $v \in H_2$.

If the bilinear form $B(u, v)$ is continuous and satisfies the so-called *inf–sup condition* (see [3], Ch. 5]), then the problem $\{B, F\}$ has a unique solution.

Let now $S_1 \subset H_1$, $S_2 \subset H_2$. Then the finite element problem $\{B, F, S_1, S_2\}$ is to find the finite element solution $u_{S_1} \in S_1$ such that

$$B(u_{S_1}, v) = F(v) = B(u_0, v) \tag{2.2}$$

holds for all $v \in S_2$.

If the bilinear form $B(u, v)$ satisfies the inf–sup condition on $S_1 \times S_2$, then u_{S_1} exists and is unique, and

$$\|u_{S_1} - u_0\|_1 \leq C(S_1, S_2)Z(u_0, H_1, S_1), \tag{2.3}$$

where

$$Z(u_0, H_1, S_1) = \inf_{\omega \in S_1} \|u_0 - \omega\|_1. \tag{2.4}$$

For details, see [3, Ch. 6] and [1]. We will assume that

$$C(S_1, S_2) \leq D, \tag{2.5}$$

where D is independent of S_1, S_2, and hence the norm of the error $e = u_{S_1} - u_0$ of the finite element solution is completely given by $Z(u_0, H_1, S_1)$.

Remark 2.1. We do not necessarily need (2.5) to hold. Nevertheless, assumption (2.5) simplifies our conclusions.

Remark 2.2. The condition (2.5) is satisfied, for example, if $H_1 = H_2$, $B(u, v) = B(v, u)$, $S_1 = S_2$, and

$$B(u, u) \geq \gamma \|u\|_1^2, \qquad \gamma > 0.$$

This condition is satisfied for self-adjoint positive definite problems as in the theory of elasticity, etc.

The exact solution u_0 is, of course, not known. Nevertheless, we will assume that it is a priori known that $u_0 \in \mathscr{K} \subset H_1$, where $\mathscr{K}$ is a certain set called the *solution set* that is compact in H_1. We define

$$Z(\mathscr{K}, H_1, S_1) = \sup_{u \in \mathscr{K}} Z(u, H_1, S_1), \tag{2.6}$$

which characterizes the error under the assumption that we know only that the solution $u_0 \in \mathscr{K}$.

Remark 2.3. A typical example is $H_1 = H^1(\Omega)$ and

$$\mathscr{K} = \{u \mid \|u\|_{H^k(\Omega)} \leq 1, k > 1\}.$$

This choice then leads to the classical error estimate of the error of the finite element solution (*h* version):

$$\|e\|_{H^1} \leq Ch^{\mu} \|u_0\|_{H^k}.$$

See, e.g., [3, Ch. 4].

There are many results available concerning the characterization of $Z(\mathscr{K}, H_1, S_1)$, the best selection of S_1 of dimension n, etc. For an excellent abstract treatment and survey of available results, we refer to [45].

The space S_1 in (2.6) is fully characterized by the finite element method, its *h*, *p*, or *h-p* versions. The set $\mathscr{K}$ is characterized by the class of problems to be solved. Hence, the performance of the finite element method in relation to the solution set $\mathscr{K}$ is described by $Z(\mathscr{K}, H_1, S_1)$, which will be the center of our interest. Of course, other aspects are also essential for the assessment of the performance of the finite element method. They will also be described in this survey.

10.3. The Model Problem

The performance of the method strongly depends on the class of problems for which it is used. As indicated in Section 10.2, the performance is directly related to the solution set $\mathscr{K}$ under consideration. We will concentrate here especially on the class of problems that are characterized by the piecewise analytic input data.

Let $\Omega \subset R^2$ be a bounded domain and its boundary $\partial\Omega$ be a piecewise analytic curve $\Gamma = \bigcup_{i=1}^{n} \Gamma_i$ where Γ_i are (closed) arcs with the endpoints $A_i, A_{i+1}, i = 1, \ldots, n$ $(A_{n+1} = A_1)$. An example and the notation are shown in Figure 3.1. By A_i, $i = 1, 2, \ldots, n$, we denote the vertices of Ω and by ω_i, $i = 1, 2, \ldots, n$, the internal angles. We will not exclude the case when the internal angle $\omega = 2\pi$. This case is very important in practice (cracks) when two arcs (fully or partially) coincide.

Let $\Gamma_D = \bigcup_{i \in \mathscr{D}} \Gamma_i$ and $\Gamma_N = \Gamma - \Gamma_D$ be the Dirichlet and the Neumann boundaries, respectively. We shall be interested in solving the problem

$$-\Delta u + u = f \quad \text{on } \Omega, \tag{3.1a}$$

$$u = h \quad \text{on } \Gamma_D, \tag{3.1b}$$

$$\frac{\partial u}{\partial n} = g \quad \text{on } \Gamma_N. \tag{3.1c}$$

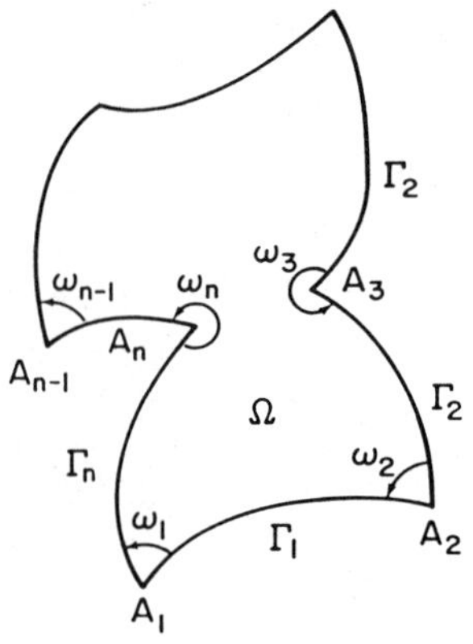

FIGURE 3.1. The scheme of a domain with piecewise analytic boundary.

We will cast problem (3.1) (for $h = 0$) into the form of a $\{B, F\}$ problem. To this end, let $H = H_1 = H_2 = H_D^1(\Omega) = \{u \in H^1(\Omega) | u = 0 \text{ on } \Gamma_D\}$, where by $H^1(\Omega)$ we denote the usual Sobolev space of functions with the square integrable first derivatives. Let

$$B(u, v) = \int_\Omega \left(\frac{\partial u}{\partial x} \frac{\partial v}{\partial x} + \frac{\partial u}{\partial y} \frac{\partial v}{\partial y} + uv \right) dx\, dy, \tag{3.2a}$$

and

$$F(v) - \int_\Omega fv\, dx\, dy + \int_{\Gamma_N} gv\, ds. \tag{3.2b}$$

If $h \neq 0$, then as usual we write $u = z + w$ with $w \in H^1(\Omega)$, $w = h$ on Γ_D, and $z \in H_D^1(\Omega)$ being the solution of the $\{B, F_1\}$ problem with F_1 being properly adjusted F.

The model problem $\{B, F\}$ satisfies the conditions listed in Section 10.2 provided that f, h, and g satisfy some mild conditions as, for example, $f \in L_2(\Omega)$, $g \in L_2(\Gamma_N)$, and $h \in H^1(\Gamma_i)$, $i \in \mathscr{D}$ and h continuous on Γ_D. See also [7–9].

The illustrative numerical computations presented in the next sections are related to the two-dimensional elasticity problem, i.e., for the (strongly elliptic) system of two partial differential equations of second order instead of the simple model problem mentioned previously. The elasticity problem has properties very similar to those of the introduced model problem but has greater practical importance.

The finite element solution (for $h = 0$) is characterized by the selection $S_1 = S_2 = S \subset H_D^1$, and all conditions including condition (2.5) in Section 10.2 are satisfied. If $h \neq 0$ and h is not a trace of a function in S, then we replace h by h', which is a trace of a function in S, and consider the additional error caused by this replacing. See [7–9, 18, 19].

10.3.1. Characterization of the Solution

Set $\mathscr{K}$ relates to the available information about input data, i.e., the information about Γ, f, g, h. We will assume that Γ is piecewise analytic, f is analytic on $\bar{\Omega}$, and g, h are analytic on $\bar{\Gamma}_i$. This assumption, namely that the data are piecewise analytic, is practically always satisfied in the problems of structural mechanics.

Remark 3.1. In our illustrative computations we will also present the results that are outside the frame described, namely that g is a Dirac function (concentrated load). Such an example is well tailored to our illustrations but needs more refined theoretical analysis that will not be addressed here.

Although our main emphasis will be on the problem with piecewise analytic input data, we will also mention the results for the more usual solution set $\mathscr{K}$ as, for example, $\mathscr{K} = \{u \mid \|u\|_{H^k(\Omega)} \leq 1\}$, etc.

Usual assumptions in the regularity theory of the differential equations of elliptic type are based on the theory of Sobolev spaces of finite order, i.e., $f \in H^k(\Omega)$, $g \in H^l(\Gamma)$, etc., and often the boundary of the domain is assumed to be smooth (i.e., not piecewise smooth). Such assumptions are not sufficiently realistic in applications. Either they are too restrictive (smooth domain Ω) or too general ($f \in H^k(\Omega)$).

To simplify our exposition further, we will assume that Ω *is a polygon*. We will make some remarks about the general case.

Remark 3.2. We will also, as an illustrative example, deal with a one-dimensional analog of our model problem, namely the problem

$$-u'' = f, \qquad x \in (0, 1) = I, \tag{3.3a}$$

$$u(0) = u(1) = 0, \tag{3.3b}$$

with f such that the exact solution $u_0(x)$ is

$$u_0(x) = (x - \xi)_+^\alpha + a + bx, \qquad \alpha > \tfrac{1}{2}, \tag{3.4}$$

where

$$(x - \xi)_+ = \begin{cases} x - \xi & \text{for} \quad x \geq \xi, \\ 0 & \text{for} \quad x < \xi, \end{cases}$$

and a, b are such that (3.3b) is satisfied. Obviously $u_0 \in H^1(I)$.

10.4. Characterization of the Solution Set

As stated earlier, the solution set $\mathscr{K}$ describes the solutions of the class of problem to be solved. The performance of the method is then directly related to this set.

Let $\boldsymbol{\beta} = (\beta_1, \ldots, \beta_n)$ be an n-tuple of real numbers $0 < \beta_i < 1$, $1 \le i \le n$. For any integer $k > 0$ we shall write $\beta + k = (\beta_1 + k, \beta_2 + k, \ldots, \beta_n + k)$. By $r_i(x)$, $i = 1, \ldots, n$, we shall denote the Euclidean distance between $x \in \Omega$ and the points $B_i \notin \Omega$, $i = 1, \ldots, n$. We then denote

$$\Phi_{\beta+k}(x) = \prod_{i=1}^{n} r_i^{\beta_i}(x) \qquad \text{and} \qquad \Phi_{\beta+k}(x) = \prod_{i=1}^{n} r_i^{\beta_i+k}(x).$$

The points B_i could be located at the boundary of Ω, e.g., in the vertices A_i or outside.

Now define

$$\mathscr{K}_1 = \{u \in H_D^1(\Omega) | \left(\int_\Omega |D^\alpha u|^2 \Phi_{\beta+k-2}^2(x)\, dx\, dy \right)^{1/2} \le Cd^{k-2}(k-2)!,$$

$$k = 2, 3, \ldots, |\alpha| = k, d > 1, d \text{ independent of } k\}.$$

As usual, we denote $\alpha = (\alpha_1, \alpha_2)$. $|\alpha| = \alpha_1 + \alpha_2$, $\alpha_i \ge 0$, $i = 1, 2$, integers and $D^\alpha u = \partial^{|\alpha|} u / \partial_x^{\alpha_1} \partial_y^{\alpha_2}$. The functions belonging to $\mathscr{K}_1$ are analytic on $\bar{\Omega} - \bigcup_{i=1}^n B_i$. If $B_i \in \partial\Omega$, then they have singular behavior in the neighborhood of B_i, and the character of the singularity is given by β_i and d.

It has been shown in [6] that if the domain Ω is a polygon, $B_i = A_i$ (i.e., B_i are the vertices) and functions f, g, h are analytic on $\bar{\Omega}$ and $\bar{\Gamma}_i$, respectively, then the solution of problem (3.1) belongs to $\mathscr{K}_1$ for properly chosen constants β, C, d. The case $B_i \notin \bar{\Omega}$ characterizes the solutions with the singularities outside Ω, e.g., when the domain has circular arcs and $h = 0$. This case also describes the case when the natural domain of the analyticity of the solution contains $\bar{\Omega}$ well.

The set $\mathscr{K}_1$ obviously belongs to the family of countably normed spaces. For the generalization to the domains with curved boundaries we refer to [7, 8].

We refer to [10] for the analysis of the regularity of the solution of problems in three dimensions.

Let us now introduce the more standard family of solution sets

$$\mathscr{K}_2 = \{u \in H_D^1(\Omega) | \|u\|_{H^{k_1}(\Omega)} \le C, k_1 > 1\},$$

$$\mathscr{K}_3 = \{u \in H^1(\Omega) | \|u\|_{H^{k_2}(\Omega)} \le C, k_2 > \tfrac{3}{2}\},$$

$$\mathscr{K}_4 = \{u \in H_D^1(\Omega) | u = r_i^{\alpha_i} |\log r_i|^{\gamma_i} \varphi_i(\theta_i) \chi_i(r_i)\},$$

where (r_i, θ_i) are the polar coordinates with the origin in the vertex A_i, $\alpha_i > 0$ noninteger, $\varphi_i(\theta_i)$ is analytic, and $\chi_i(r_i)$ is a C^∞ cutoff function.

The motivation of the solution set $\mathscr{K}_j$, $j = 2, 3, 4$, is that the solution u of (3.1) can be written in the form

$$u = u_1 + u_2 + u_3, \tag{4.1}$$

where

$$u_1 \in \mathscr{K}_2, \qquad u_2 \in \mathscr{K}_3,$$

and

$$u_3 = \sum C_{i,j} u_i^{[j]}, \qquad u_i^{[j]} \subset \mathscr{K}_4.$$

Functions u_1 and u_3 satisfy the homogeneous Dirichlet (essential) conditions; u_2 relates to the nonhomogeneous Dirichlet conditions. The restriction $k_2 > \frac{3}{2}$ has been made for simplicity only and can be replaced by $k_2 > 1$. For the theory leading to the form (4.1) we refer to [30, 38, 39].

Remark 4.1. We restricted ourselves to problem (3.1) only. The practically important case of nonhomogeneous materials that is described by the equation

$$\frac{\partial}{\partial x} a \frac{\partial u}{\partial x} + \frac{\partial}{\partial y} a \frac{\partial u}{\partial y} = f$$

with a being a piecewise constant on the domains bounded by the piecewise analytic curves can be handled in a similar way. An analogous situation also occurs when dealing with the problem of elasticity.

10.5. The Finite Element Space S

We now introduce the finite element spaces we will deal with later.

For reasons of simplicity of exposition, we will restrict our choices, but our numerical example will also present more general cases.

Let $\mathscr{M} = \{\mathscr{T}\}$ be a family of meshes $\mathscr{T} = \{\tau_i\}$, where $\tau_i \subset \Omega$ is an open triangle, called an *element*. Let $h_{\tau_i} = \operatorname{diam} \tau_i$, $h(\mathscr{T}) = \max_{\tau_i \in \mathscr{T}} h_{\tau_i}$, and ρ_{τ_i} be the diameter of the largest ball contained in τ_i. We will assume that $\mathscr{M}$ is such that for any $\tau_i \in \mathscr{T} \in \mathscr{M}$

$$\frac{h_{\tau_i}}{\rho_{\tau_i}} \le \beta, \tag{5.1}$$

i.e., that all triangles satisfy the *minimal angle condition*. Further, we shall assume that $\bar{\Omega} = \bigcup_{\tau_i \in \mathscr{T}} \bar{\tau}_i$ and that any pair $\bar{\tau}_i$, $\bar{\tau}_j \in \mathscr{T}$, $i \neq j$, has either an entire side or a vertex in common, or their intersection is empty. The number $M(\mathscr{T}) < \infty$ of elements $\tau \in \mathscr{T}$ will be called the *cardinality of* $\mathscr{T}$.

Remark 5.1. We have restricted ourselves to triangles only. The results we will present are valid more generally, e.g., for rectangles, curvilinear triangles and rectangles.

In addition to a general mesh, we will introduce two special families of meshes, the *quasi-uniform* and the *geometric mesh*.

(a) The family $\mathscr{M}_U^\gamma$, $1 \le \gamma < \infty$, of quasi-uniform meshes: There exists constant γ such that

$$h(\mathscr{T}) = \max_{\tau_i} h_{\tau_i} \le \gamma \min_{\tau_i} h_{\tau_i}$$

holds for any $\tau \in \mathscr{M}_U^\gamma$.

(b) The family $\mathcal{M}_G^{q,k}(B_1, \ldots, B_n)$, $B_j \notin \Omega$, $j = 1, \ldots, n$, $0 < q < 1$, $1 \leq k < \infty$, of geometric meshes:
Let $\tau_i \in \mathcal{M}_G^{q,k}$, $\bar{\tau}_i \cap \bigcup_{j=1}^n B_i = \emptyset$; $\kappa(i) = \min_{1 \leq j \leq n} d(B_j, \tau_j)$ then

$$k^{-1} \leq \frac{\kappa(i)}{h(\tau_i)} \leq k,$$

and if $\bar{\tau}_i \cap \bigcup_{j=1}^n B_j = \emptyset$ then

$$k^{-1} q \leq h(\tau_i) \leq kq$$

By $d(B_j, \tau_i)$ we denote d the Euclidean distance between B_j and τ_i.

Obviously, if $\tau \in \mathcal{M}_U^\gamma$, then $M(\mathcal{T}) \approx h^{-2}(\mathcal{T})$ and the cardinality $M(\mathcal{T})$ can be arbitrarily large. If $B_j \notin \bar{\Omega}$, $j = 1, \ldots, n$, then for any $\mathcal{M}_G^{q,k}$ we have $M(\mathcal{T}) \leq Q$, where Q depends on Ω, q, k. If $B_j \in \partial\Omega$, then the cardinality can be arbitrarily large.

Finally, we denote $S(\mathcal{T}, p) = \{u \in H_D^1(\Omega) | u|_{\tau_i}$ is a polynomial of degree $\leq p$ for any $\tau_i \in \mathcal{T}\}$. By $N(S, \mathcal{T}, p) = \dim S(\mathcal{T}, p)$ we denote the *number of degrees of freedom.*

If $\mathcal{T} \in \mathcal{M}_U^\gamma$, then $N(S, \mathcal{T}, p) \approx p^2(h(\mathcal{T}))^{-2}$. If $\tau \in \mathcal{M}_G^{q,k}$, $B_i \notin \bar{\Omega}$, then $N(S, \mathcal{T}, p) \approx p^2$ and if $\mathcal{T} \in \mathcal{M}_G^{q,k}$ with $B_i \in \partial\Omega$, then $N(S, \mathcal{T}, p) \approx p^2/q$.

Remark 5.2. We have assumed only triangular elements. For rectangular elements, instead of polynomials of the total degree p, we use polynomials of degree p in every variable. In the case of the curvilinear elements we use the standard mapped polynomials.

Remark 5.3. We have mentioned only the case when the degree p is uniform, i.e., is the same for all elements. The available theory also covers the nonuniform, selective choice of the degrees p.

Remark 5.4. The meaning of the finite element subspaces is especially clear in the one-dimensional setting. For example, in the case of the geometric mesh with $A = 0$, then nodal x_i points are

$$x_i = q^{M-i}, \qquad i = 0, 1, \ldots, M,$$

where $q < 1$ is called the *ratio of the mesh.*

10.6. The Basic Approximation Theorems

We will mention here some basic approximations for $H_1 = H^1(\Omega)$; i.e., we will study the magnitude of $Z(\mathcal{K}_j, H^1(\Omega), S(\mathcal{T}, p))$. A numerical illustration will be given in Section 10.7.

First, let us consider any fixed mesh $\mathcal{T}_0$ and consider the p version.

Theorem 1

$$Z(\mathscr{K}_2, H_D^1(\Omega), S(\mathscr{T}_0, p)) \le Cp^{-(k_1-1)} \approx CN^{-(k_1-1)/2} \tag{6.1}$$

holds for any mesh $\mathscr{T}_0$ (satisfying (5.1)). For the proof, see [16].

Theorem 2

$$Z(\mathscr{K}_4, H_D^1(\Omega), S(\mathscr{T}_0, p)) \le C|\log p|^\gamma p^{-2\alpha} \approx C|\log N|N^{-\alpha},$$
$$\alpha = \min \alpha_i, \quad \gamma = \max \gamma_i. \tag{6.2}$$

For the proof, see [16].

Theorem 3. *Let $\mathscr{K} = \mathscr{K}_1$ and $B_i \notin \bar{\Omega}$, $i = 1, \ldots, n$. Then*

$$Z(\mathscr{K}_1, H_D^1(\Omega), S(\mathscr{T}_0, p)) \le Ce^{-\alpha p} \approx Ce^{-\alpha N^{1/2}}, \tag{6.3}$$

where α depends on B_i, $i = 1, \ldots, n$, but is independent of p. For the proof, see [33].

So far we have assumed that the essential boundary conditions are homogeneous. Let us now discuss the case $\mathscr{K}_3$ when the Dirichlet conditions are not homogeneous. Let $\tilde{\Gamma} \subset \Gamma_D$ be any side of an element $\tau \in \mathscr{T}$ with the endpoints P_i, $i = 1, 2$ and let φ be defined on $\tilde{\Gamma}$ so that:

(a) φ is a polynomial of degree p.
(b) $\varphi(P_i) = u(P_i)$, $i = 1, 2$, $u \in \mathscr{K}_3$ (because $u \in \mathscr{K}_3$, $u \in H^{k_2}(\Omega)$, $k_2 > \frac{3}{2}$, $u(P_i)$ is well defined).
(c) $\int_{\tilde{\Gamma}} \varphi'\psi'\, ds = \int_{\tilde{\Gamma}} u'\psi'\, ds$ for any ψ that is a polynomial of degree p and $\psi(P_i) = 0$, $i = 1, 2$.

Now we replace the boundary condition on Γ_D by φ, i.e., $u_S = \varphi$, to get Theorem 4.

Theorem 4

$$Z(\mathscr{K}_3, H^1(\Omega), S(\tau_0, p)) \le Cp^{-(k_2-1)} \approx CN^{-(k_2-1)/2}. \tag{6.4}$$

For the proof, see [16]. *See also* [8, 18, 19].

Remark 6.1. In the one-dimensional case we can prove more exact theorems. Let us consider problem (3.2) with solution (3.3) and the case of one element only. Then we have Theorem 5.

Theorem 5. *Let $\hat{H}_0^1(I) = \{u' \in L_2(I), u(0) = u(1) = 0\}$ and $\|u\|_{\hat{H}_0^1(I)} = \|u'\|_{L_2(I)}$. Then*

(a) *If $\xi = 0$, then*

$$Z(u_0, \hat{H}_0^1(I), S(p)) = C_0(\alpha)\frac{1}{p^{2\alpha-1}}\left(1 + O\left(\frac{1}{p}\right)\right), \tag{6.5a}$$

with

$$C_0(\alpha) = \frac{\alpha\Gamma(\alpha)^2 |\sin \pi\alpha|}{\pi\sqrt{2\alpha - 1}}. \tag{6.5b}$$

(b) *If* $\xi < 0$, *then*

$$Z(u_0, \hat{H}_0^1(I), S(p)) = C_1(\alpha)\left(\frac{1-r^2}{2r}\right)^{\alpha-1} \frac{r^p}{p^\alpha}\left(1 + O\left(\frac{1}{p^\sigma}\right)\right), \tag{6.6a}$$

where

$$\sigma > 0, \qquad r = \frac{\sqrt{1-\xi} - \sqrt{-\xi}}{\sqrt{1-\xi} + \sqrt{-\xi}}, \qquad C_1(\alpha) = \frac{\alpha\Gamma(\alpha)|\sin \pi\alpha|}{\sqrt{\pi}\,2^{\alpha-1/2}}, \tag{6.6b}$$

the form $O(1/p^\sigma)$ *is uniform with respect to* $\xi \le -\varepsilon$, $\varepsilon > 0$, *and we have also the estimation*: *if* $0 < r^2 \le 1 - (1/p)$, *then*

$$Z(u_0, \hat{H}_0^1(I), S(p)) \approx \frac{1}{\sqrt{1+r^2}} \frac{r^{p+1-\alpha}}{p^\alpha}\left[\frac{1}{p^{\alpha-1/2}} + (1-r^2)^{\alpha-1/2}\right], \tag{6.6c}$$

if $1 - 1/p < r^2 < 1$, *then*

$$Z(u_0, \hat{H}_0^1(I), S(p)) \approx \frac{r^{p+1-\alpha}}{p^{\alpha-1/2}}\left[\frac{1}{p^{\alpha-1/2}} + (1-r)^{\alpha-1/2}\right], \tag{6.6d}$$

where equivalency constants depend only on α.

(c) *If* $0 < \xi < 1$, *then there exists a constant* $C > 0$ *depending only on* α *such that*

$$Z(u_0, \hat{H}_0^1(I), S(p)) \le \begin{cases} C(\delta/p)^{\alpha-1/2}, & 0 < \theta < \pi - 1/p, \\ C(1/p)^{2\alpha-1}, & \pi - 1/p \le \theta < \pi, \end{cases} \tag{6.6e}$$

where $\delta = \min(\theta, r - \theta)$, $\theta = \arccos(|\xi - 1|)$. On the other hand, if $\varepsilon \le \theta \le \pi - \varepsilon$, $\varepsilon > 0$, *then there is a constant* $C(\alpha, \varepsilon) > 0$ *such that*

$$Z(u_0, \hat{H}_0^2(I), S(p)) \ge C\left(\frac{\theta}{p}\right)^{\alpha-1/2}. \tag{6.6f}$$

For the proof, see [31].

If the singularity is outside the domain, the accuracy increases dramatically with p. In Figure 6.1 we see the value of Z for $\alpha = 1.5$ in dependence on ξ and p.

Let us now consider the h-p version on the quasi-uniform mesh $\mathcal{T}$.

Theorem 6. *Let* $\mathcal{K} = \mathcal{K}_2$ *(resp.* $\mathcal{K}_3$*). Then*

$$Z(\mathcal{K}, H_D^1(\Omega)(\text{resp. } H^1(\Omega)), S(\mathcal{T}, p)) \le Ch^{\eta_i - 1}(\mathcal{T})p^{(k_i-1)} \approx CN^{-(\eta_i-1)/2}p^{-(k_i-\eta_i)},$$

$$\eta_i = \min(p+1, k_i), \tag{6.7}$$

The constant C in (6.7) *is independent of h and p. For the proof, see* [17].

The estimate for $\mathcal{K}_3$ *has to be understood in the same way as in Theorem* 4.

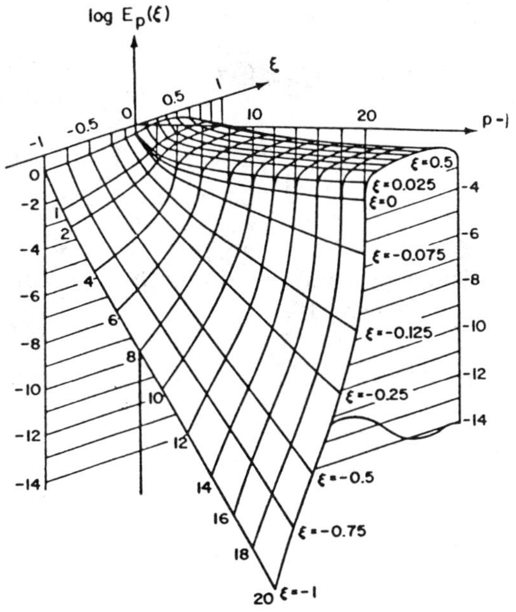

FIGURE 6.1. The error of the p version with one element.

Theorem 7. *Let* $\mathscr{K} = \mathscr{K}_4$. *Then*

$$Z(\mathscr{K}_4, H_D^1(\Omega), S(\mathscr{T}, p)) \leq Cg(h, p, \gamma) \min(h^\alpha, h^{\min(\alpha, p-\alpha)} p^{-2\alpha}), \tag{6.8}$$

where

$$g(h, p, \gamma) = \max(|\log h|^\gamma, |\log p|^\gamma).$$

For the proof, see [17].

Finally, let us address the case of the h-p version on the geometric mesh.

Theorem 8. *Let* $\mathscr{K} = \mathscr{K}_1$ *and let* B_i *be vertices of* Ω. *Then there exists a sequence of meshes and the degrees p (dependent on the mesh) such that*

$$Z(\mathscr{K}_1, H_D^1(\Omega), S(\tau, p)) \leq Ce^{-\alpha\sqrt[3]{N}}, \tag{6.9}$$

where α *is independent of* N.

For the proof, see [33]. See also [7, 8].

Remark 6.2. In the one-dimensional setting much more can be said. We will discuss the case $\xi = 0$.

Let us first ask the question about the lower bound of the error $Z(u_0, \hat{H}_0^1(I), S(\mathscr{T}, \mathbf{p}))$ among all meshes and all degree distributions. The answer is given in Theorem 9.

Theorem 9

$$Z(u_0, \hat{H}_0^1(I), S(\mathscr{T}, \mathbf{p})) \geq C(\alpha) \frac{1}{\sqrt{N^{\alpha-1/2}}} q_0^{\sqrt{(\alpha-1/2)N}}, \tag{6.10}$$

where

$$q_0 = (\sqrt{2} - 1)^2.$$

For the proof, see [31].

Let us now consider the geometric mesh with the ratio q and various degree distributions. We have Theorem 10.

Theorem 10. *As $N \to \infty$ the optimal degree distribution tends to be linear with the slope*

$$s_0 = (\alpha - \tfrac{1}{2}) \frac{\ln q}{\ln r}, \tag{6.11a}$$

with

$$r = \frac{1 - \sqrt{q}}{1 + \sqrt{q}}. \tag{6.11b}$$

This means precisely that for each $i = 1, 2, \ldots,$

$$\lim (p_{M_N - i}^{(M_N)} - p_{M_N - i - 1}^{(M_N)}) = s_0.$$

For the proof, see [31].

For the error estimates with linear degree vector of slopes S we have the following:

Theorem 11. *For the geometric mesh with the ratio q combined with a linear degree vector of the slopes s we have:*

(a) *if $s > s_0$, then*

$$Z(u_0, \hat{H}_0^1(I), S(\mathscr{T}, \mathbf{p})) \leq C(\alpha, q, s) q^{(\alpha-1/2)\sqrt{2N/s}}; \tag{6.12a}$$

(b) *if $s < s_0$, then*

$$Z(u_0, \hat{H}_0^1(I), S(\mathscr{T}, \mathbf{p})) \leq C(\alpha, q, s) r^{\sqrt{2Ns}}; \tag{6.12b}$$

(c) *if $s = s_0$, then*

$$Z(u_0, \hat{H}_0^1(I), S(\mathscr{T}, p)) \leq C(\alpha, q) e^{-\sqrt{(\alpha-1/2)N}\sqrt{2 \ln q \ln r}}, \tag{6.12c}$$

where $r = (1 - \sqrt{q})/(1 + \sqrt{q})$ and $s_0 = (\alpha - \frac{1}{2})(\ln q / \ln r)$ is the optimal slope in the sense that the exponential rate is maximal. Furthermore, the optimal geometric mesh and linear degree vector combination is given by

$$\begin{aligned} q_{op} &= (\sqrt{2} - 1)^2, \\ s_{op} &= 2\alpha - 1. \end{aligned} \tag{6.13}$$

In this case

$$Z(u_0, \hat{H}_0^1(I), S(\mathcal{T}, p)) \leq C(\alpha)[(\sqrt{2}-1)^2]^{\sqrt{(\alpha-1/2)N}}. \tag{6.14}$$

In (6.12)–(6.16) the equivalence constants depend on (α, q, s), and α, respectively.

For the proof, see [31].

Remark 6.3. For the optimal combination of geometric mesh and linear degree vector, the estimate can be written as

$$Z(u_0, \hat{H}_0^1(I), S(\mathcal{T}, p)) \approx e^{-1.762\sqrt{(\alpha-1/2)N}}, \tag{6.15}$$

and we have seen in Theorem 9 that this exponential rate of convergence is the best possible one.

Theorem 12. *For the geometric mesh with the ratio q combined with uniformly distributed degree p, the relation between the optimal choice of p and the number of element M in the mesh is asymptotically linear; i.e.,*

$$p \approx s_0 M \qquad (\text{as } M \to \infty),$$

with s_0 the same as in Theorem 10.

For the proof, see [31].
The analog of Theorem 11 is Theorem 12.

Theorem 13. *For the geometric mesh with ratio q and uniformly distributed degree p related to the number of elements M by $p = sM$, we have:*

(a) *if $s > s_0$, then*

$$Z(u_0, \hat{H}_0^1(I), S(\mathcal{T}, p)) \approx C(\alpha, q, S)\frac{q^{(\alpha-1/2)\sqrt{N/s}}}{(\sqrt{sN})^{2\alpha-1}}; \tag{6.16a}$$

(b) *if $s < s_0$, then*

$$Z(u_0, \hat{H}_0^1(I), S(\mathcal{T}, p)) \approx C(\alpha, q)\frac{r^{\sqrt{sN}}}{(\sqrt{sN})^{\alpha}}; \tag{6.16b}$$

(c) *if $s = s_0$, then we get an optimal rate of convergence*

$$Z(u_0, \hat{H}_0^1(I), S(\mathcal{T}, p)) \approx C(\alpha, q)\frac{e^{-\sqrt{(\alpha-1/2)N}\sqrt{\ln r \ln q}}}{(\sqrt{N})^{\sigma}}, \tag{6.16c}$$

where

$$s_0 = \frac{(\alpha - \frac{1}{2})\ln q}{\ln r}, \qquad r = \frac{1-\sqrt{q}}{1+q}, \qquad \sigma = \min(2\alpha - 1, \alpha). \tag{6.16d}$$

The optimal combination is also given by

$$q = q_{op} = (\sqrt{2} - 1)^2, \tag{6.16e}$$

$$e = s_{op} = 2\alpha - 1,$$

and for optimal combination we get

$$Z(u_0, H_0^1(I), S(\mathscr{T}, p)) \approx C(\alpha) \frac{[(\sqrt{2} - 1)^2]^{\sqrt{(\alpha - 1/2)N/2}}}{(\sqrt{N})^s}. \tag{6.16f}$$

For the proof, see [31].

Remark 6.4. We can also interpret the optimal *h-p* version with uniform *p* as the envelope of the *h* version with fixed *p*. In this case, for $N \to \infty$ the mesh tends to be geometric with the ratio $q \approx e^{-4/e^2} = 0.5820$, and the relation between the degree *p* and the number of elements *M* tends to be linear with $p \approx (4/e^2)(\alpha - \frac{1}{2})M = 0.5413(\alpha - \frac{1}{2})M$. For more detailed analysis, we refer once more to [31].

Remark 6.5. We have discussed only the model problem (3.1). The results hold for more general equations. For the higher order equations, see [32, 48]. In the case of systems of second-order or higher order differential equations, there is a much broader scope of essential boundary conditions. For their treatment we refer to [8, 18, 19].

Let us now make some comments on the theorems we have discussed.

(a) Comparing thc performance of the finite element method with quasi-uniform mesh with respect to the number of degrees of freedom, then the *p* version (with few elements) performs better than the *h* version. In the case that the solution has singularity of the type occurring in the corner of the domain, the rate of the *p* version is twice that of the *h* version. For the *h* version there is a classical theorem mentioned in the basic books about the finite element method

$$\inf_{\omega \in S(\tau, p)} \|u - \omega\|_{H^1(\Omega)} \le C(p) h^\eta \|u\|_{H^k(\Omega)}, \tag{6.17a}$$

$$\eta = \min(p, k - 1), \tag{6.17b}$$

where $C(p)$ is not specified (more precisely, the proof indicates that $C(p) \to \infty$ as $p \to \infty$). This sometimes leads to the (false) statement that (6.17b) indicates that for a singular solution it is improper to use higher order elements.

(b) The *h-p* version leads to the exponential rate of convergence when the input data are piecewise analytic. In the one-dimensional case, the ratio $q = (\sqrt{2} - 1)^2$ is the optimal one independent of the strength of the singularity. In the two-dimensional case, the ratio of the same magnitude seems to be optimal, although detailed theoretical analysis is not available yet. For practical reasons $q = 0.15$ is recommended.

(c) Although we have mentioned only the simple model problem, the results hold much more generally. We mention especially the elasticity problem.

(d) Very important problems arise in relation to the "locking" problem as in the elasticity problems with near incompressibility (Poisson ratio $\nu \approx \frac{1}{2}$). It has been shown [61] that the p (and h-p) version is not influenced by the locking problem and reliably solves elasticity problems of nearly incompressible materials without any difficulty. See also [20].

10.7. Numerical Illustrations

In this section we will present numerical illustrations related to the theorems we mentioned in the previous section.

EXAMPLE 7.1. Let us consider the plane strain elasticity problem when Ω is the L-shaped domain shown in Figure 7.1. Let us assume that on $\partial\Omega$ tractions are prescribed; i.e., $\Gamma_D = \emptyset$. We will assume that the solution of this problem is the displacement vector $u = (u_1, u_2)$ where

$$
\begin{aligned}
u_1 &= \frac{1}{2G} r^\alpha[(\kappa - Q(\alpha + 1)) \cos \alpha\theta - \alpha \cos(\alpha - 2)\theta], \\
u_2 &= \frac{1}{2G} r^\alpha[(\kappa + Q(\alpha + 1)) \sin \alpha\theta - \alpha \cos(\alpha - 2)\theta],
\end{aligned}
\tag{7.1}
$$

where

$$
\alpha = 0.544483737, \qquad Q = 0.543075579.
$$

G is the modulus of rigidity and $\kappa = 3 - 4\nu$, where $\nu = 0.3$. The sides OA and OE are traction-free. The solution has a typical singularity at 0 and is the first mode of the stress intensity factor solution. Instead of the norm $\|\cdot\|_{H^1(\Omega)}$

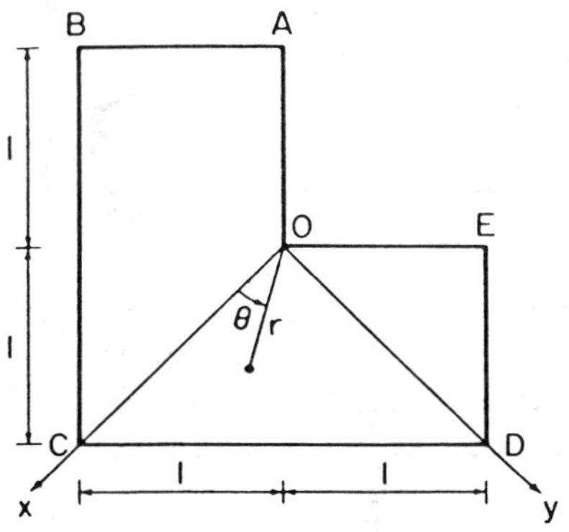

FIGURE 7.1. L-shaped domain.

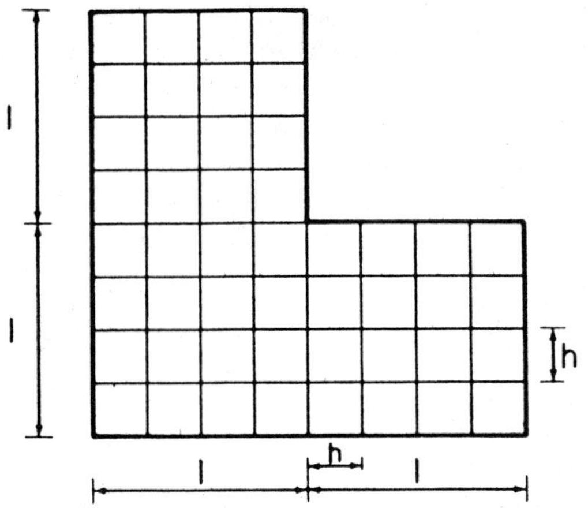

FIGURE 7.2. The scheme of the uniform mesh.

we will be interested in the energy norm $\|\cdot\|_E$, which is equivalent to the norm $\|\cdot\|_{H^1(\Omega)}$. We then define the relative error $\|e\|_{E,R} = \|e\|_E/\|u\|_E$.

First, we will consider the case of the uniform mesh with square elements shown in Figure 7.2.

The solution u can be interpreted as a member of the set $\mathscr{K}_1$, with $B_1 = 0$, $\mathscr{K}_2$ with $k_2 = \alpha + 1 - \varepsilon$, $\varepsilon > 0$, arbitrary or $\mathscr{K}_4$ with $\gamma = 0$. Interpreting the solution u as the number of the solution set $\mathscr{K}_4$, the estimate (6.8) gives

$$\|u - u_S\|_E \leq C \min\left[h^\alpha, \frac{h^{\min(\alpha, p-\alpha)}}{p^{2\alpha}}\right],$$

where C depends on α but is independent of h and p.

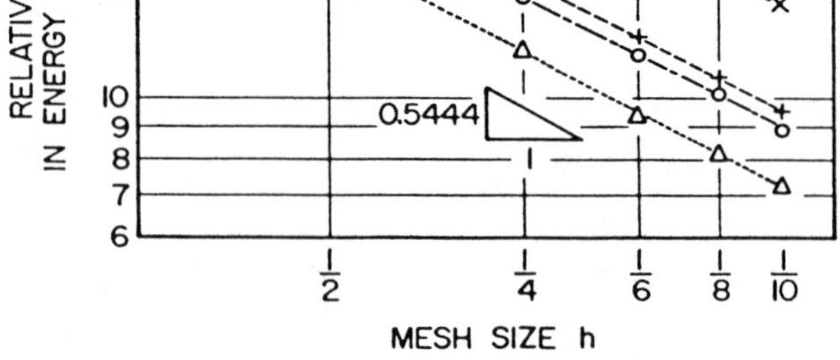

FIGURE 7.3. The relative error measured in the energy norm in dependence on h.

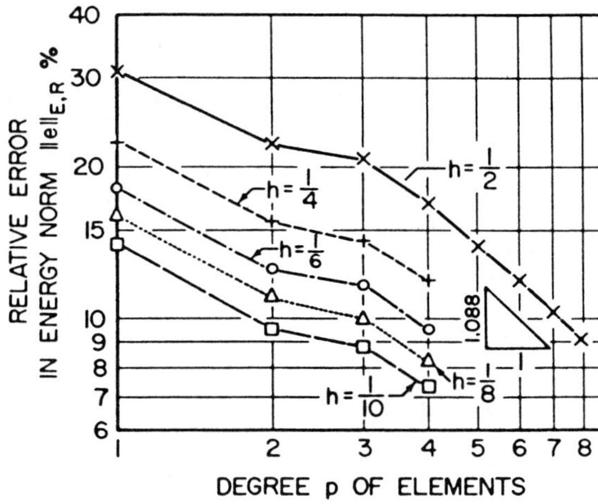

FIGURE 7.4. The relative error measured in the energy norm in dependence on p.

Figure 7.3 shows the relative error $\|e\|_{E,R}$ (for different p) in dependence on h (in $\log\|e\|_{E,R} \times |\log h|$ scale). We also show the slope h^{α} in the figure. We see that with respect to h, the error is also in the asymptotic range for moderate p and h. Figure 7.4 shows the error in dependence on p and different h. Because of the size of the computations the error is given for $p > 4$ only for $h = \frac{1}{2}$, (for $p = 4$ and $h = \frac{1}{10}$ the number of degrees of freedom $N = 5119$). The slope $p^{-2\alpha}$ $(2\alpha = 1.088)$ is apparent only for $p \geq 3$. Figure 7.5 shows the error in dependence on the number of degrees of freedom N for various p. Also the performance of the p version for $h = \frac{1}{2}$ is shown in Figure 7.5. We see that the p version is more effective than the h version and that the

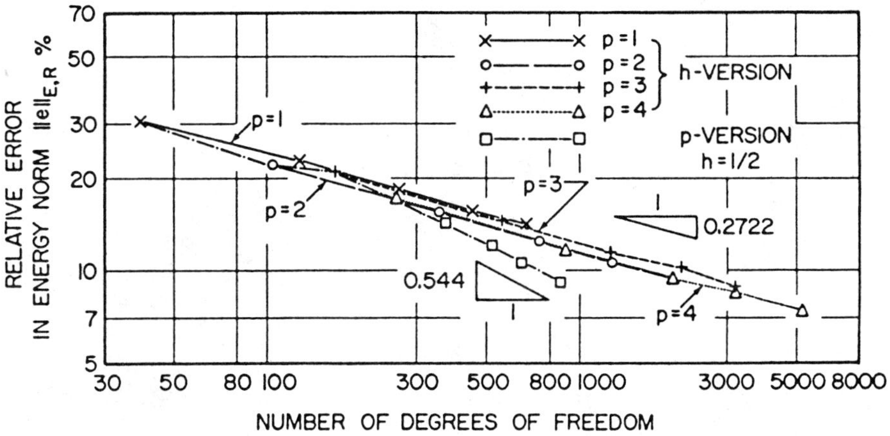

FIGURE 7.5. The relative error measured in the energy norm in dependence on v.

TABLE 7.1. The error and the values of the numerical constant in the one-dimensional case.

	$\alpha = 0.7$		$\alpha = 3.5$	
p	E_p	R_p^A	E_p	R_p^A
1	$4.743E-1$	0.9877	1.021	0.2032
2	$3.627E-1$	0.9967	$3.402E-1$	4.335
3	$3.090E-1$	0.9985	$3.093E-2$	4.488
4	$2.756E-1$	0.9992	$2.379E-3$	1.940
5	$2.522E-1$	0.9995	$4.760E-4$	1.480
6	$2.344E-1$	0.9996	$1.400E-4$	1.300
7	$2.204E-1$	0.9997	$5.154E-5$	1.208
8	$2.090E-1$	0.9998	$2.210E-5$	1.153
9	$1.994E-1$	0.9998	$1.057E-5$	1.118
10	$1.912E-1$	0.9999	$5.495E-6$	1.094
11	$1.840E-1$	0.9999	$3.053E-6$	1.077
12	$1.777E-1$	0.9999	$1.790E-6$	1.064
13	$1.722E-1$	1.000	$8.999E-7$	1.054
14	$1.671E-1$	1.000	$6.978E-7$	1.046
15	$1.626E-1$	1.000	$4.585E-7$	1.040

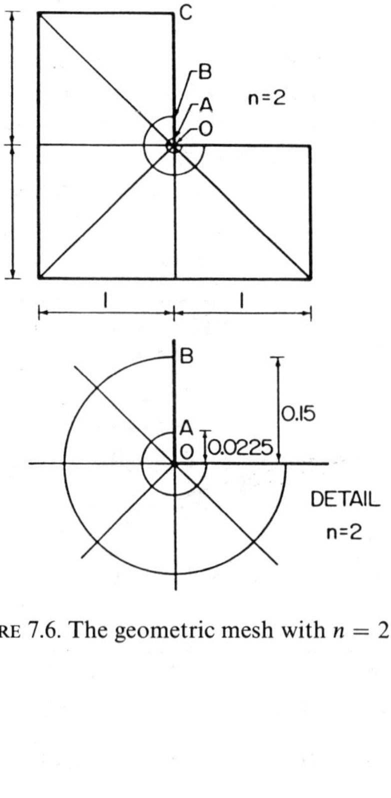

FIGURE 7.6. The geometric mesh with $n = 2$ layers.

theoretical asymptotic slope (shown in the figure) is achieved for moderate accuracy and N.

Remark 7.1. Let us illustrate the estimate given in Theorem 5. Introduce the numerical constant R_p^A

$$R_p^A = \frac{E_p}{E_p^A},$$

where E_p^A is the right-hand side of (6.5) and $E_p = \|e\|_{\hat{H}_0^1(I)}$.

Table 7.1 shows the error E_p and the numerical constant R_p^A for $\alpha = 0.7$, 3.5 and $\xi = 0$. We see that for α smaller $R_p^A \to 1$ more quickly than for α larger (i.e., smoother function).

For more numerical results we refer to [31].

So far we have addressed the performance of the uniform mesh. Let us now discuss the performance of the finite element method on the geometric mesh. Figure 7.6 shows the geometric mesh with $n = 2$ layers and the ratio 0.15.

Figure 7.7 shows the performance of the p version for various number n of layers in $\log \|e\|_{E,R} \times \log N$ scale. Figure 7.7 also shows for comparison the performance of the method on the uniform meshes shown in Figure 7.2. We see that for every number of layers the rate is $N^{-0.544}$ where N is sufficiently large (dependent on n) and that the error has the reverse S slope form. In the first phase the accuracy is exponential (curved down); in the second phase the accuracy is algebraic. Further, we see that the optimal number of layers and the degree p depend on the required accuracy. We will return to this question later in the discussion of the expert system for the h-p version (see Section 10.11). Figure 7.8 shows the relative error in the $\log \|e\|_{E,2} \times N^{1/3}$ scales. We see that the envelope of these curves that characterize the performance of the h-p version is nearly a straight line, as has been theoretically indicated by Theorem 8.

The envelope together with the optimal pair (n, p) is shown in Figure 7.9, together with the approximate straight line.

Remark 7.2. We addressed only the source problem (3.1). Nevertheless, the analysis of the regularity of the eigenfunctions and performance of the p and h-p versions were studied in [11].

Remark 7.3. We addressed here only the two-dimensional problem. The various results and computational experience for solving three-dimensional problems will not be given here.

The p and h-p versions have various properties that are essentially different from those of the h version. We have already mentioned the robustness of the p version with respect to the Poisson ratio (the nearly incompressible material). Let us now briefly discuss the "pollution" problem, which describes the influence of the locally unsmooth solution. For the h version the well-

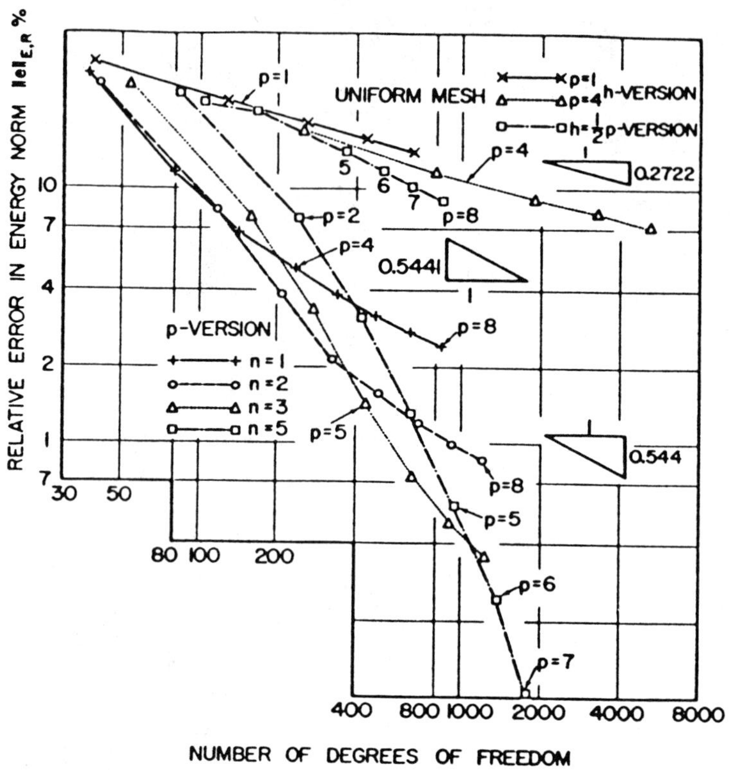

FIGURE 7.7. The performance of the p version on various meshes in the scale $\log\|e\|_{E,R} \times \log N$.

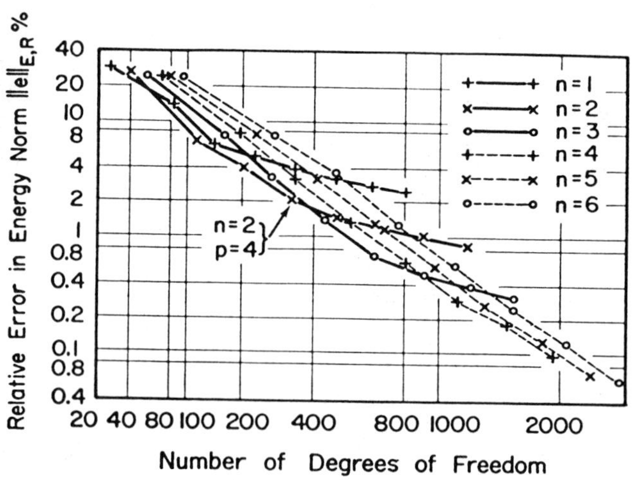

FIGURE 7.8. The performance of the p version on meshes with various number of layers n in the scale $\log\|e\|_{E,R} \times N^{1/3}$.

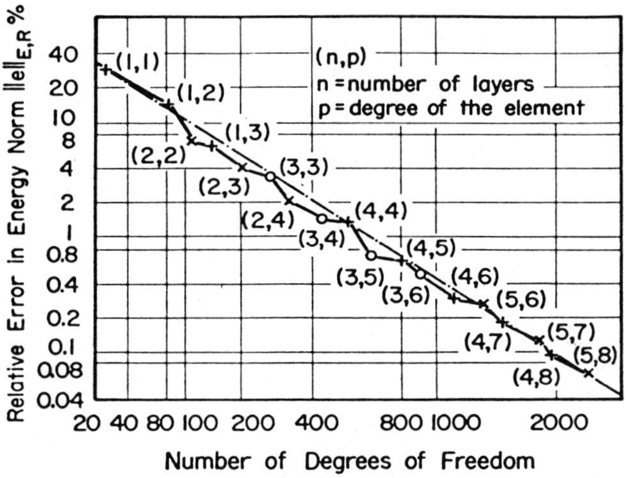

FIGURE 7.9. The performance of the h-p version.

known interior estimate shows that the influence of the local unsmoothness of the solution is also local. For the p version the effect is not so local as for the h version. To illustrate this effect, let us solve the plane strain elasticity problem ($\nu = 0.3$) loaded by the concentrated load (Dirac function) as shown in Figure 7.10. On the sides $BCDC'B'$ the tractions are prescribed so that the exact solution is the classical Boussinesq solution on the half-plane (hence the $B'AB$ side is traction-free). The energy of the solution is infinite, and hence the usual theory has to be generalized. We will be interested (because of the obvious symmetry) in the finite element solution and its accuracy on $\hat{\Omega}$ (shaded in Figure 7.10). As before, we will measure the accuracy in the energy norm. The meshes used for the computation are shown in Figure 7.11 (only half of the domain is shown). The domain $\hat{\Omega}$ is once more shaded. The minimal (pollution-free) error on $\hat{\Omega}$ is the error of the finite element method when the exact tractions are prescribed on $\partial\hat{\Omega}$.

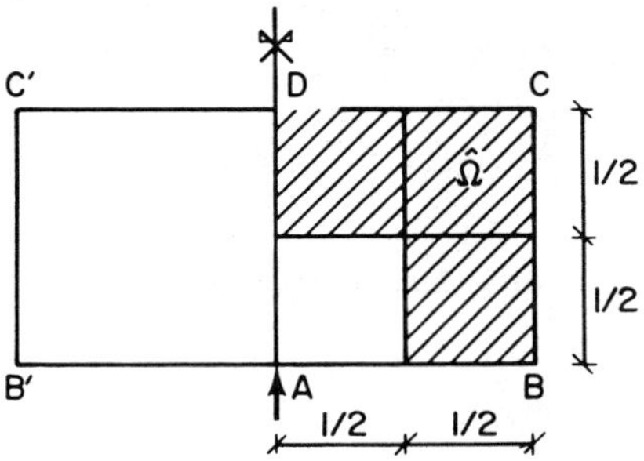

FIGURE 7.10. The scheme of the problem of a concentrated load.

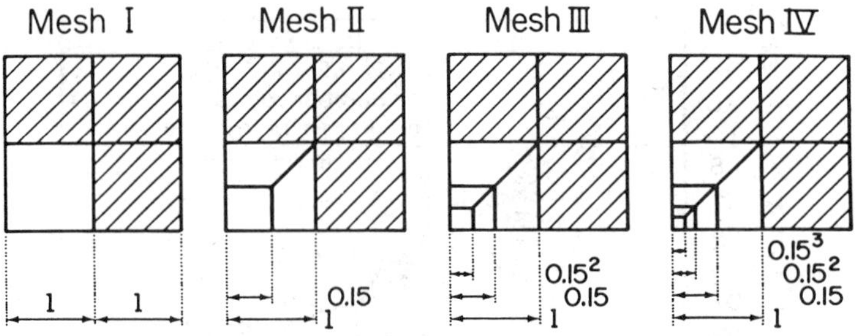

FIGURE 7.11. The meshes used for the computation.

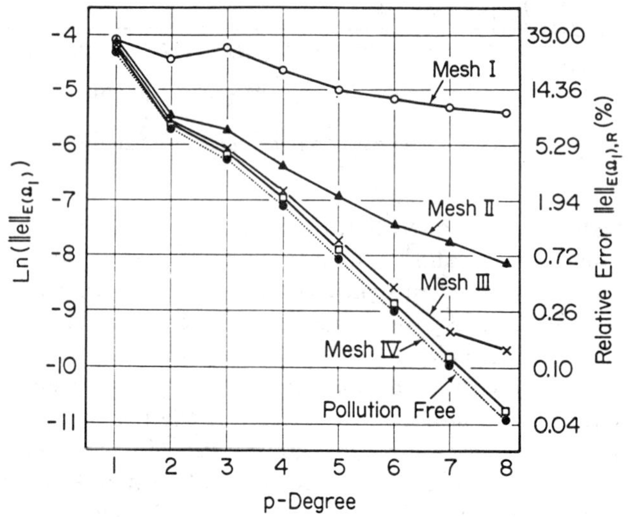

FIGURE 7.12. The error $\|e\|_{E(\hat{\Omega})}$ of the p version and the "pollution-free" error.

Figure 7.12 shows the error $\|e\|_{E(\hat{\Omega})}$ for the meshes I–IV and the "pollution-free" error. We see that the p version for any mesh shown does not converge although for three layers the error is in the range of engineering accuracy. For more details refer to [13]. A similar situation also occurs for weaker singularities and a high accuracy, e.g., in stresses. The accuracy has to be achieved in practice by "filtering" out the singularity influence by using few layers depending on the strength of the singularity.

10.8. Some Implementational Aspects

The h and h-p versions of the finite element method give great flexibility in the change of degrees of the elements either uniformly or element by element. This has to be respected in the architecture of the program. One of the essential features is the use of the hierarchical shape functions. In the most simple setting of a square master element shown in Figure 8.1 the hierarchical shape

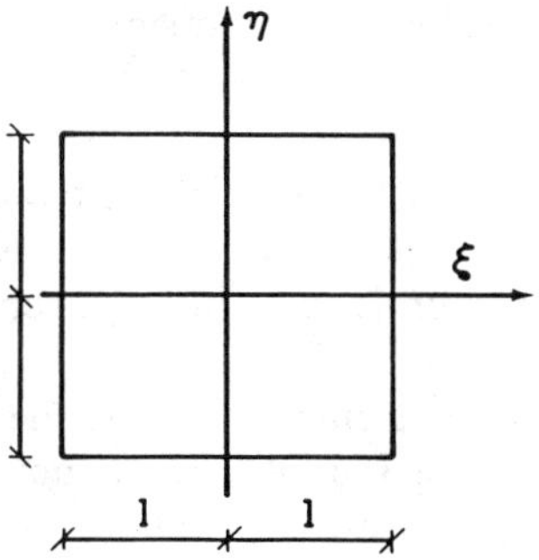

FIGURE 8.1. The scheme of the master element.

functions are as follows:

We have three kinds of shape functions:

(a) the modal shape functions $\overset{0}{N}_i(\xi, \eta)$;
(b) the side shape functions $\overset{1}{N}_i(\xi, \eta)$; and
(c) the internal shape functions $\overset{2}{N}_i(\xi, \eta)$.

The preceding index describes the dimension of the part of the element associated with the shape functions.

(a) The nodal shape functions $\overset{0}{N}_i(\xi, \eta)$, $i = 1, \ldots, 4$, are the usual bilinear functions associated with the four nodes.
(b) The side shape functions $\overset{1}{N}_i(\xi, \eta)$, $i = 1, \ldots, 4(p-1)$. Let

$$P_j(\xi) = \sqrt{\frac{2j+1}{2}} \int_{-1}^{\xi} l_j(x)\, dx,$$

where l_j, $j \geq 1$, is the Legendre polynomial. The side shape functions associated with the side $-1 < \xi < 1$, $\eta = -1$, are

$$\overset{1}{N}_j(\xi, \eta) = P_j(\xi)\frac{1-\eta}{2}, \qquad j = 1, \ldots, p-1,$$

and hence we have $4(p-1)$ side shape functions.
(c) The internal shape functions $\overset{2}{N}_i(\xi, \eta)$, $i = 1, \ldots, (p-1)^2$. The internal shape functions are

$$\overset{2}{N}_{j,k}(\xi, \eta) = P_j(\xi)P_k(\eta), \qquad j \geq 1, \quad k \geq 1.$$

The number of the internal shape functions can be reduced, as in the case of the serendipity elements. In this case the total number Q of all of the shape functions is

$$Q = 4p \qquad \text{for} \quad p < 4,$$

$$Q = 4p + \max(0, \tfrac{1}{2}(p-2)(p-3)) \qquad \text{for} \quad p \geq 4.$$

as used in the PROBE program. See [50]. PROBE uses $1 \leq p \leq 8$.

Remark 8.1. The above-mentioned elements of PROBE belong to the class of serendipity elements.

The shape functions are called *hierarchical* because the increase of p leads only to the addition of new shape functions. The use of the integrals of Legendre polynomials has essential advantages in connection with the round-off error control.

The hierarchical type of this element leads to the construction of the local stiffness matrix by bordering technique when p increases. The same structure can be achieved for the global stiffness matrix. This character then allows the flexibility to change the degree of elements and its use for the quality control and adaptive approaches.

So far we have mentioned only the square elements. A similar hierarchical structure can also be used for the triangular elements.

The p version (and h-p version) utilizes large elements. Hence it is important to deal properly with the curved boundaries. It seems to be advantageous (as implemented in PROBE) that the boundary is described exactly and not approximated, e.g., by polynomials. This is achieved by using blending function mappings. This approach, of course, leads to the loss of the important rigid body motion property of the isoparametric elements. Nevertheless, if the boundaries are smooth, e.g., piecewise analytic, then for higher p (say $p = 3$–4), the effect of this loss is in practice negligible as shown theoretically and by experience with PROBE. In addition, the quality control also indicates this possible effect when present (see also Section 10.10).

The architecture of the p version finite element program has essential differences from the h version. We mention here the types of meshes that are used that influence the mesh generator and many other aspects. We will not elaborate more.

Remark 8.2. The selection of the shape functions could be heavily influenced by various factors as, for example, when considering iterative methods for solving the system of equations. This is especially important in three-dimensional problems.

10.9. The Assessment of the Complexity and the Computational Cost

The cost of using any finite element code consists of the computer cost and the human cost. The relative part of the human cost growths due to the progress in the hardware development and is often the major part of the cost of the computational analysis. Hence, the assessment of the complexity and total cost of the method is very subtle. Some of the aspects will also be discussed in section 10.12, where the use of the h-p version and the industry experience with the code PROBE will be addressed.

The computer cost of running the code consists of the cost of the basic computation itself, the cost of the assessment of the quality of the results, and the cost of various "bells" and "whistles" and the "niceties" for the user that any commercial code has to have. The computational cost also depends on the robustness of the mesh design criterion, the selection of other parameters leading to the desired accuracy, and the user's ability to make the most useful choice of these parameters. The computer cost is the cost of numerical operations and the input/output cost, which often is the major cost. This makes any comparison a difficult task. It seems that one of the best ways is to compare various codes in an industrial environment (see also Section 10.12).

In general, we have to relate the cost of the achieved (a priori given by the user) accuracy and the assessment of the reliability of the computed data (see also Sections 10.10 and 10.11). In previous sections we measured the effectiveness of the method by the relation between the accuracy measured in the energy norm and the number of degrees of freedom. Of course, the number of degrees of freedom does not reflect some aspects of computational effort, such as the computation of the local stiffness matrices and the influence of the sparsity of the global stiffness matrix, etc. The only count of arithmetical operations, although important, does not provide a sufficiently realistic basis for determining the computational cost. Hence, very likely the most reliable way is to write a program that implements the method and to analyze its performance. To this end, we used part of the program PROBE adjusted to the given purpose. Although the results depend on the programming, the computer, the compiler, etc., an assessment based on such an analysis is the most reliable technique.

The major parts of the finite element computations are the following:

(a) input, treatment of the geometry, controls of the input, etc.;
(b) computation of the local stiffness matrices and load vectors;
(c) assembly and elimination procedures;
(d) postprocessing including accuracy assessment.

We will address here only parts (b) and (c). To analyze this question, let us consider the mesh shown in Figure 9.1 with $r \le m$.

We mention that this mesh is topologically very close to the mesh shown

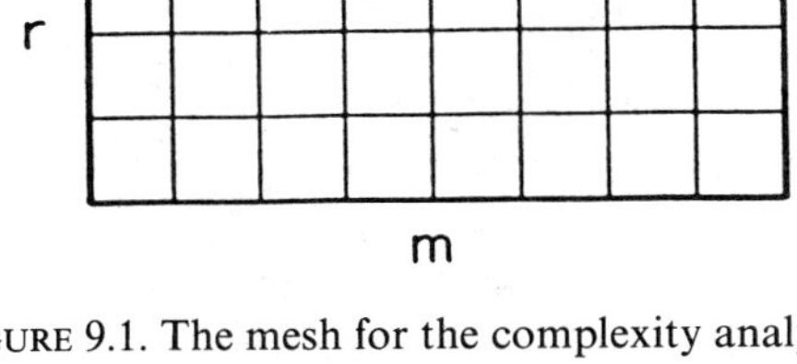

FIGURE 9.1. The mesh for the complexity analysis.

in Figure 7.6 ($r = \min(6, n + 1)$, $m = \max(6, n + 1)$) or in Figure 7.11. The cost of parts (b) and (c) of the computations dealing with the examples treated in Section 10.7 (Figures 7.5, 7.11) is essentially the same as when dealing with the mesh of Figure 9.1 for properly chosen r and m.

Let us now count the number of operations for the two major parts (b) and (c) mentioned for the elasticity problem (two unknown functions):

(a) *Computation of the local stiffness matrix.* It consists of

(1) Computation of the values of the shape function and the Jacobian in one Gaussian point.

(2) Computation of the values in $\approx p^2$ Gaussian points and the summation for obtaining one of the entries of the local stiffness matrix.

(3) Computations of all entries $\approx p^4$ of the local stiffness matrix.

(4) Computations of mn local stiffness matrix.

The crude operation count indicates that we can expect $s \approx mnp^7$ operations. Nevertheless, this count is deceptive, as will be seen later.

The number of the shape functions is for $p \geq 4$

$$Q = [4p + \max(0, \tfrac{1}{2}(p - 2)(p - 3))]2 \tag{9.1}$$

(the serendipity type of elements are used and hence, for $1 \leq p \leq 8$, the second term is not too large and the number of the shape functions is closer to $\approx p$ than p^2). The number of Gaussian points used is $2[p/2] + 2$ where $[a]$ denotes the integral part of a. Careful programming leads to $\approx p^4$ operations in the given range. In Figure 9.2 we show the time required for computations of 25 local stiffness matrices ($m = r = 5$) depending on p in

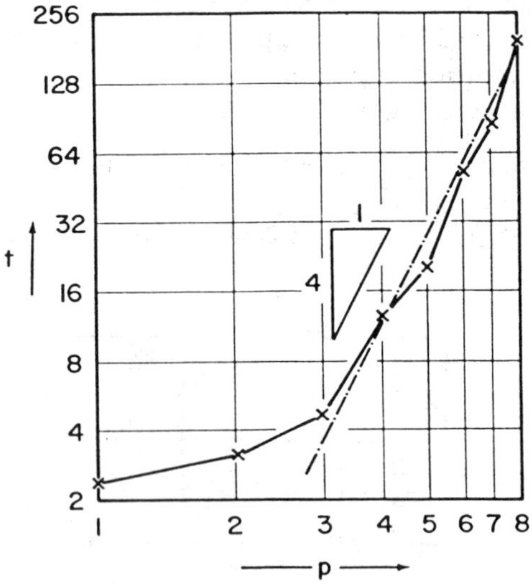

FIGURE 9.2. The computer time for computation of 25 local stiffness matrices.

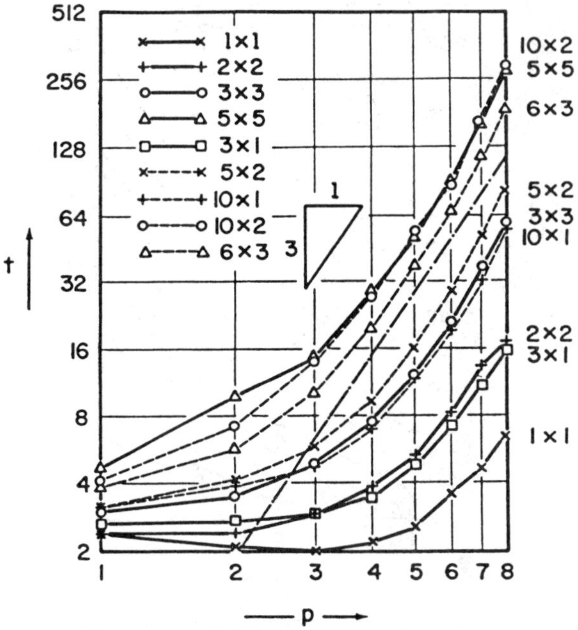

FIGURE 9.3. The time of assembly and elimination in dependence on p, m, r.

the double logarithmic scale (time units on VAX). The straight line with shape 4 confirms the rate p^4 we expected. The minor "wobbling" of the time curve is caused by using only even number Gaussian points.

(b) *The assembly and elimination form $m \geq r$.* The frontal solver that joins the elimination and the assembly has been used. The front is of order rp when properly counting (condensation approach) the dealing with internal shape functions. Hence, we can expect the time to be of the order $\approx (rp)^2 mrp = mr^3 p^3$. Figure 9.3 shows the time depending on p for various values m and r. Slope p^3, which agree with the expectation, is depicted in the figure.

Figure 9.4 shows the time depending on m with $m = r$ and different p. The slope of m^3 is shown in the figure. We see that in the given range, the rate is closer to m^3 than to m^4, as we would expect. For small p and m the tests are outside the asymptotic range that could be expected.

Analyzing Figures 9.2–9.4, we see that the number of operations for the h-p version with the number of layers roughly proportional to p (see, e.g., Figure 7.9), leads to a number of operation of order N^α where $1.5 \leq \alpha \leq 2.5$, with p increasing as α increases. The conventional h version with $p = 1, 2$ will practically never lead to an accuracy of order of 1 percent for any reasonable computational effort if the problem is not very simple. Nevertheless, for small accuracy, a small p is desirable (see Figure 7.9). Hence, the proper design of

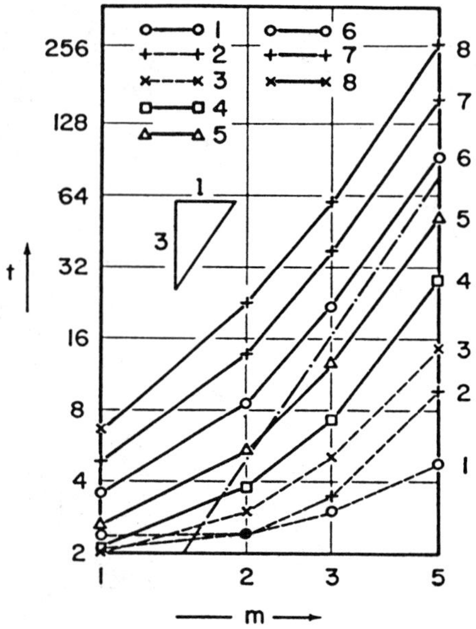

FIGURE 9.4. The time of assembly and elimination for $r = m$.

the mesh and assignment of the degree of elements are very important. This point will be discussed in the next two sections.

The method is very robust with respect to the mesh design when some basic rules are observed. The accuracy can be obtained by using higher degree without changing the mesh. Obviously the p version has quite strong element by element parallel character. For additional comparisons of the performance of elements of different degrees we refer to [15].

10.10. A Posteriori Assessment of the Quality of Computed Data

An essential part of computational analysis is the a posteriori assessment of the quality of the obtained results of interest. (For the survey of the main ideas for the quality assessment, we refer to [40]). Although the energy norm is the basic measure of accuracy, usually it is not the most important measure for engineering purposes.

The basic and practically most efficient method for assessment of the accuracy of the data of interest is comparing computed data obtained for various p and using the extrapolation method. The computed data can be of various types, including various equilibration checks. Although PROBE provides many reliability checks, theoretically a posteriori quality assessment is the least mathematically understood part of the computation.

TABLE 10.1

1	2	3	4	5	6	7	8
p	N	FE	E^- FE	$\frac{\|e\|_E}{\|u\|_E}\%$	β	CPRE	PRE
1	41	3.88608797	2.684 (−1)	25.41	—	—	25.41
2	119	4.12483255	2.971 (−2)	8.45	—	—	8.45
3	209	4.14811501	6.429 (−3)	3.93	1.93	5.34	3.91
4	335	4.15265042	1.893 (−3)	2.13	2.76	2.02	2.09
5	497	4.15362580	9.084 (−4)	1.47	3.05	1.01	1.41
6	695	4.15297464	5.695 (−4)	1.37	2.45	.80	1.10
7	929	4.15413900	4.052 (−4)	.98	1.83	.75	.89
8	1199	4.15423777	3.064 (−4)	.85	1.39	.74	.74
∞	∞	4.15454422					

The assessment of the error measure in the energy form is the basic one and relatively easiest to obtain because $\|e\|_E^2 = \mathscr{E}_E - \mathscr{E}_{FE}$ where $\mathscr{E}_E$ (resp. $\mathscr{E}_{FE}$) is the strain energy for the exact (resp. finite element) solution, and $\mathscr{E}_{FE}$ increases as p increases (for homogeneous essential boundary conditions). As we have seen, the rate of convergence of the p version with properly designed mesh is higher than the algebraic one. Hence, we can, for example (as implemented in PROBE), assume that

$$\|e\|_E^2 = (\mathscr{E}_E - \mathscr{E}_{FE}) = C(p)p^{-\beta(p)},$$

with C and β computed from three consecutive degrees $p-2, p-1, p$. Because of the reverse S property of the accuracy mentioned earlier, β increases with p and then decreases to the limiting value when an algebraic rate is achieved. The desired accuracy should be practically achieved in the range of p when β is about maximal. This gives an indication of the quality of the mesh for any given accuracy.

We will illustrate the approach on our L-shaped problem with the mesh of two layers ($n = 2$) discussed in Section 10.7. Table 10.1 gives the basic data. Column 5 is the value of the energy norm as a percentage of the true error. Column 7 shows the relative error as a percentage computationally predicted (CPRE) when computed from the current values $p-2, p-1, p$ of the degrees. Column 8 shows the error predicted (PRE) computed from the predicted strain energy for $p = 6, 7, 8$.

Returning to Figure 7.8 we see that the value $p = 4 \div 5$ that is indicated by β as the optimal one is close to the envelope of the curves. The predicted error CPRE is more reliable in the phase when β is increasing than when it is decreasing, as is obvious because of (10.1). Nevertheless, as a whole CPRE is very reliable.

Obviously, PRE is very accurate for lower values of p.

The reliability of the estimation we have shown is closely related to the monotone behavior of the error. Nevertheless, it is also necessary to estimate other values of interest. We shall show the computation of the stress intensity

TABLE 10.2

1 p	2 N	3 K
1	41	0.95268
2	119	1.02177
3	209	1.00250
4	335	1.00073
5	497	0.99991
6	695	0.99985
7	929	0.99987
8	1199	0.99990
∞	∞	1.00000

factor K for the case of the L-shaped domain analyzed in Section 10.7. The method of computation of this stress intensity has been developed in [12] and [56]. It has been shown that this method gives an accuracy roughly equivalent to the accuracy of the strain energy (i.e., the square of the energy norm error). Table 10.2 shows the values of the stress intensity factor (normalized to value 1). For more, see [56]. In contrast to the energy, we see that the error changes sign and is not monotonic. This, of course, complicates the extrapolation procedure.

Finally, let us show a more complicated example (where the exact solution is not known) of the cracked panel as shown in Figure 10.1 with the used mesh.

Table 10.3 shows the values of the first and second stress intensity factors K_{I} and K_{II}. See also [55].

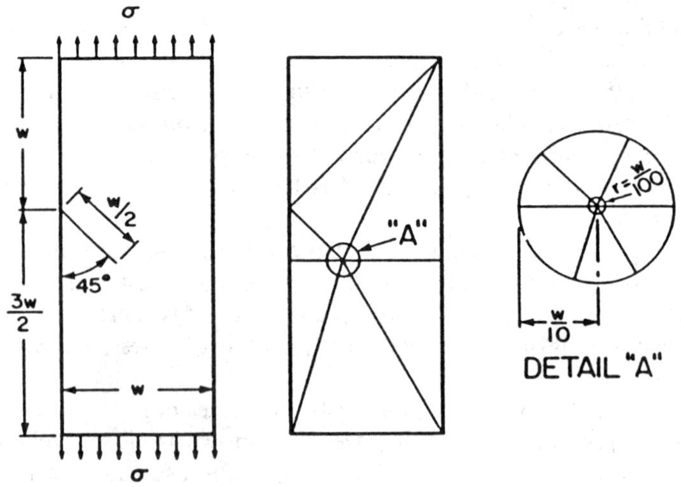

FIGURE 10.1. Scheme of the cracked panel.

TABLE 10.3

1 p	2 N	3 K_{I}	4 K_{II}
1	43	0.42259	−0.37480
2	125	0.55588	−0.25578
3	221	0.56161	−0.28951
4	355	0.59232	−0.28319
5	527	0.59825	−0.29398
6	737	0.60043	−0.28897
7	985	0.60119	−0.29196
8	1271	0.60132	−0.29042

The program PROBE has built in additional quality tests based on the equilibrium testing so that the user can place maximal confidence in the data of interest.

Let us underline that because of the hierarchic structure of the program the computation for different degrees can be arranged in a computationally very effective ways.

As we have seen, it is important to select the parameters of the h-p version, i.e., the mesh and the degree of the elements that lead to the desired accuracy of the computed data of interest, effectively. This can be done by a feedback (adaptive) approach or with the aid of an expert system or by a combination of both.

10.11. The Expert System and the Feedback Approach

Remark 11.1. In [31] we developed and theoretically studied a feedback system that leads to the simultaneous selection of the mesh and (nonuniform) degrees of elements. This study has shown that the adaptive approach for the h-p version, based on similar principles to those for the h version, is very complicated and that another approach probably has to be considered.

The feedback system for the p version with uniform p (i.e., for fixed mesh) can be relatively easily implemented. We have seen in Sections 10.8, and 10.9 that the hierarchic structure of the p version effectively allows us to change the degree. Hence, the feedback of the p version for the given mesh consists of successive computations for increasing p until the value of the interest has admissible accuracy. The main question is determining the ratio of the cost of the entire computation to the cost of the final one. Assuming that the most ineffective computation mode, i.e., independent computation for every p starting from $p = 1$, and that the cost $W(p)$ of the computation for degree p is $W(p) = Cp^\alpha$, the ratio for $(1 \le p \le 8)$ for $\alpha = 3$ (resp. $\alpha = 4$) ranges from 1 to 2.5 (resp. 2.14). Hence, a more effective computation arrangement exploiting the hierarchic structure leads to a ratio in the range 1.5–2 at the most. This ratio is highly comparable to the ratio in the adaptive approaches based on the h version for energy norm accuracy.

Previously we have assumed that all computations starting from $p = 1$ have been made and that the mesh is given. The aim is obviously to construct the mesh and $p = p_0$ so that the desired accuracy is achieved and to compute only the solutions for $p \leq p_0$ (which is very cheap because of the hierarchical structure) to get the error estimates based on the extrapolation procedure.

Recently, a lot of progress has been made in the general development of knowledge based systems (expert systems) in general (see, e.g., [34, 64]) and various attempts have been made to apply the system for computational mechanics problems (see, e.g., [27, 59]). In [14] and [46] an expert system that advises the user how to use the h-p version in an effective way is developed. One of the main problems is to develop the set of rules that is then utilized in the system.

Finite element analysis can be essentially separated into two parts, the decision phase and the execution phase that carries out the decision, although these two phases are often interwoven. The adaptive approach combines these two phases very closely when the user supplies only the basic data and the program takes care of the entire process. The approach in the expert system mentioned previously clearly separates the phases of decision and execution and adds instruction and communication. It uses the theoretical mathematical results mentioned in previous sections in combination with heuristics. The system was developed for treating polygonal domains and accuracy measured in the energy norm. We will now give an example. For more details, refer to [14].

Let the problem be the plane strain of the "cracked wrench" (symmetric) domain shown in Figure 11.1. The first task is the construction of the geometry mesh. Figure 11.2 shows it. From the theory of the h-p version, it is known

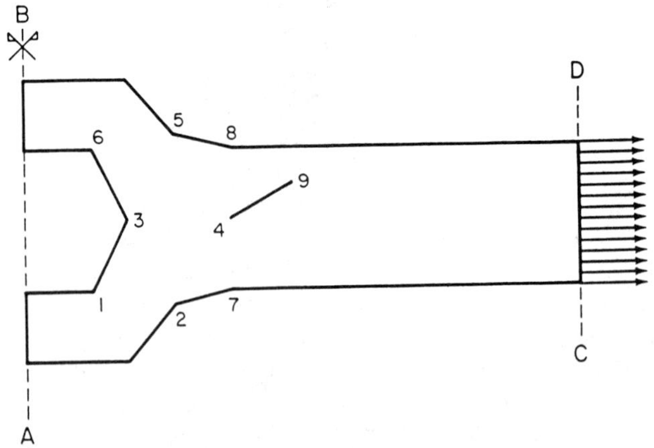

FIGURE 11.1. The cracked wrench problem.

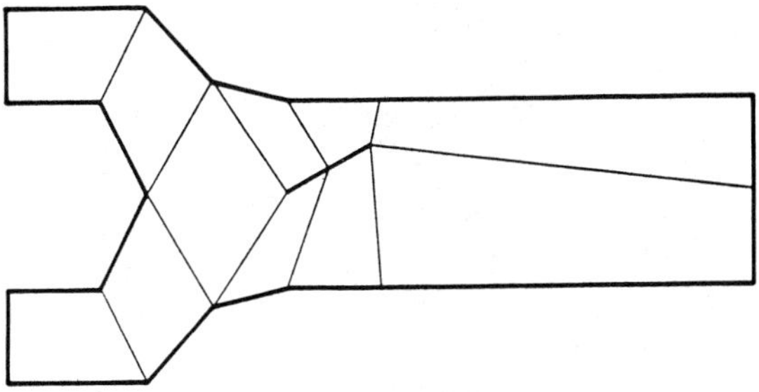

FIGURE 11.2. The geometry mesh for the cracked wrench problem.

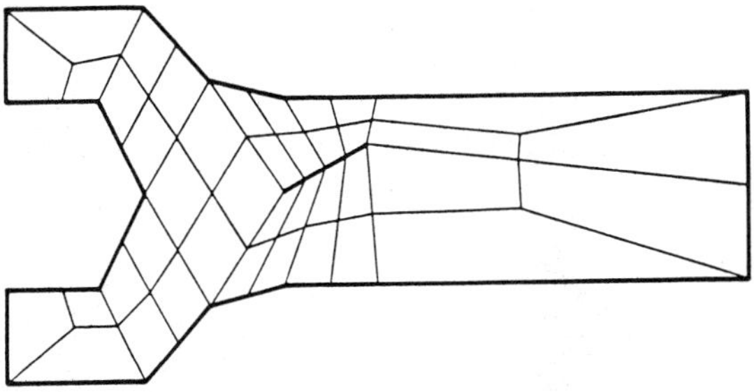

FIGURE 11.3. The basic mesh for the cracked wrench problems.

that areas in the neighborhood of the corners (entrant) are critical and have to be dealt with. The user is advised about it and a basic mesh that separates the critical areas is constructed. Figure 11.3 shows this basic mesh. The program then computes a rough solution on the basic mesh, computes the few stress intensity factors for every corner, and computes an optimization problem leading to the highest accuracy for the minimal cost, and gives for a given p the (optimal) distribution of layers. On the basis of this computation the expert presents the user with the relation of predicted error to cost (N) for the optimal mesh computation. Figure 11.4 shows this graph. The user selects the accuracy, and the program constructs the mesh for the final solution. Figure 11.5 shows this mesh (not all layers are shown). For example, the mesh for an error of 1.8 percent (resp. 1.3) leads to the predicted optimal degree $p = 4$, $N = 2101$ (resp. 2517) and the number of layers shown in Table 11.1. We have denoted by SN (Column 1) the number of the critical area as shown in Figure 11.1 and by L the suggested number of layers in this area. The computation on this mesh leads to errors of 3.1 and 2.6 percent, respectively.

We see that the predicted errors are of good quality although not perfect, as might be expected. Nevertheless, the optimal distribution of the layers was predicted correctly.

If an error of 3.1 percent is not acceptable, then p can be increased in the

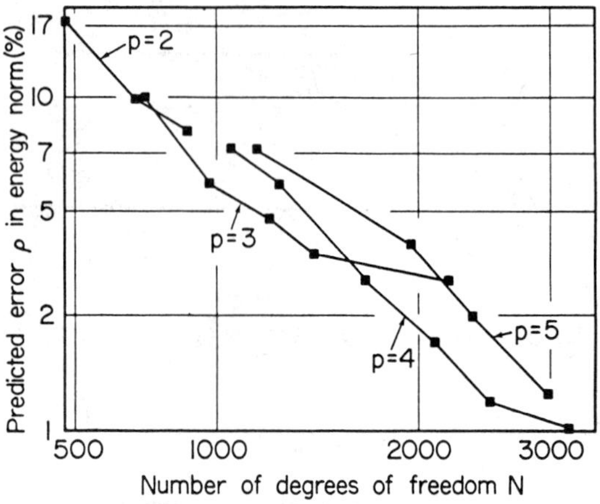

FIGURE 11.4. The predicted error for the cracked wrench problem.

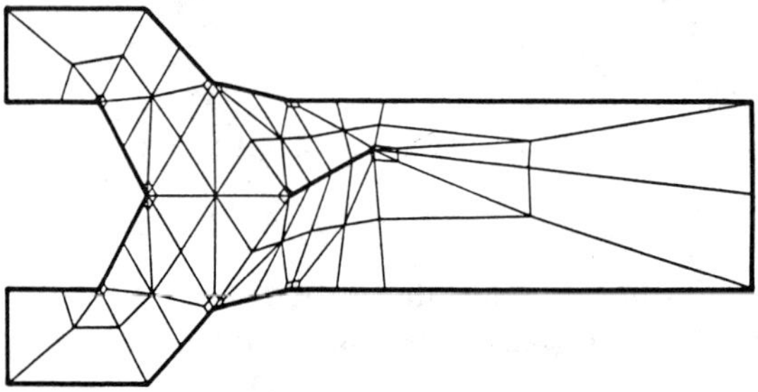

FIGURE 11.5. The final mesh for the cracked wrench problem.

TABLE 11.1

N = 2101				N = 2517			
1	2	1	2	1	2	1	2
SN	*L*	*SN*	*L*	*SN*	*L*	*SN*	*L*
1	3	6	3	1	3	6	3
2	2	7	2	2	3	7	2
3	2	8	2	3	4	8	2
4	3	9	3	4	4	9	3
5	3			5	3		

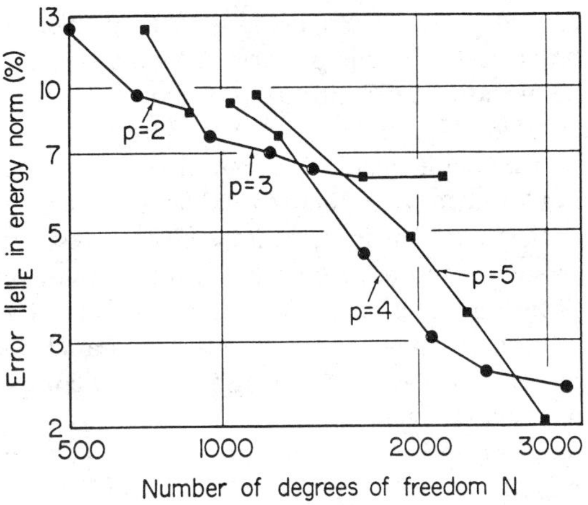

FIGURE 11.6. The actual error for the cracked wrench problem.

adaptive mode. In our case, this would mean increasing N to 3059, which will lead to an error of 1.9 percent. (The other possibility is to construct the optimal mesh for higher accuracy.)

Let us mention that the expert system we have outlined costs about 10–20 percent of the entire computer cost.

Figure 11.6 shows the actual error for the meshes and degrees shown in Figure 11.4.

Finally, let us underline that the expert system and the problem of the optimal mesh design are in a very early stage.

10.12. Industrial Experience with the *p* and *h-p* Versions of the Finite Element Method

As was stated in Section 10.1 the *p* version of the finite element method was first introduced in the U.S.A. when the FIESTA program was released [44]. The first release of FIESTA had elastostatic capabilities for solving three-dimensional problems. The *p* extension (i.e., increase of *p*) was possible through the specification of six groups of hierarchic basis functions corresponding to polynomial degrees ranging from 1 to 4. The program appealed to engineers mainly because it allowed the use of coarse meshes and therefore simplified data preparation tasks. Visual checking of the mesh in three dimensions is generally very difficult and often impossible when the *h* version is used.

Industrial users are usually far more concerned about human time than machine cycle requirements, but the greatest concern is the *elapsed real time* between problem definition and delivery of a *reliable* solution. This is because the costs associated with numerical analysis are usually negligible as compared with the consequences of not being able to deliver correct solutions on time within the project schedule prescribed by management.

The first release of the PROBE program was introduced to the aerospace industry in June 1985 by Noetic Technologies Corporation, a St. Louis firm. This release had linear plane elasticity capabilities only. The second release also has capabilities for linear two-dimensional heat transfer and axisymmetric and planar thermoelasticity. The first release was of interest mainly to airframe designers, whereas the second release is also of interest to engine designers.

The PROBE program was extensively tested and is used very successfully in the industry, e.g., in the analysis of orthotropic panels, various cut-cuts fracture mechanics, and problems involving laminated composites where the peeling stress is of primary interest.

Engineering users consider various features important and judge the programs and methods accordingly. The features are as follows:

(1) Increased level of confidence in the computation. The usual observation of the convergence of the data of interest (not only the energy norm) and various additional tests as equilibrium checks (having physical meaning), etc., give the user confidence in the computational analysis. (For some comments, see Section 10.8.)
(2) Lower human time requirement due to simplicity and flexibility of input. In an industrial test 30 to 40-fold savings in human time were reported in comparison with conventional finite element technology [22].
(3) Rapid convergence and flexibility, e.g., when large aspect ratios have to be used, as in the case of composite joints.
(4) Flexibility of mesh design, e.g., in fracture mechanics following the crack growth, series of short cracks, optimal design problem (e.g., [47]).
(5) Easy learning and robust performance

The *p* and *h-p* versions are responding very well to needs of this type. Nevertheless, additional new features relating to plate shells, fabricated plates, three-dimensional problems, and nonlinear analysis must be developed.

10.13. Relation to Some Other Methods

The *p* and *h-p* versions of the finite element method are related to today's form of the spectral method. The spectral method (see, e.g., [29]) expands the solution of the problem in high order Fourier or polynomial series, the coefficients of which are determined by weighted residual projections. The spectral method is used in fluid mechanics, for example, in the problem of transitional and low Reynold's number, turbulent incompressible fluid flow in simple domains. Recently, research on the spectral method is focused on extensions to more complex domains. The spectral method uses a Galerkin-type approach and assumes that condition (2.5) holds. The spectral method is, with the use of polynomial approximation, closely related to the *p* version with differences in applications and implementations. See [41] and the references given there. The pseudospectral method can be viewed as the Galerkin method with the numerical integration technique.

The spectral method traditionally has been used in the context of smooth solutions. Only recently the needs of coping with the singularities and geometries in the spirit of the *h-p* version is seen.

The *p* version is also related to the global element method in [23].

The *p* and *h-p* versions of the finite element method are relatively well developed and tailored to the needs of general problems in structural mechanics, as we have been seen in previous sections. The family of spectral methods applied in the field of fluid mechanics has various aspects that are related to the *p* and *h-p* versions discussed in this paper.

Acknowledgments. This work was partially supported by the Office of Naval Research under Grant N-00014-85-K-0169.

References

1. D. Arnold, I. Babuška, and J. Osborn, "Finite element methods: Principles for their selection," *Comput. Methods Appl. Mech. Engrg.*, **45** (1984), 57–96.
2. B. Anderson and U. Falk, "Finite element analysis of three-dimensional structures using adaptive *p*-extensions," (presented at Nato ASI (Lisbon 1986). The Aeronautical Research Institute of Sweden, FFA5-H-800, 1986.
3. I. Babuška and A.K. Aziz, "Survey lectures on the mathematical foundations of the finite element method," in *The Mathematical Foundations of the Finite Element Method with Applications to Partial Differential Equations* (A.K. Aziz, ed.), Academic Press, New York, 1972.
4. I. Babuška and M.R. Dorr, "Error estimates for the combined *h* and *p* versions of the finite element method," *Numer. Math.*, **37** (1981), 257–277.
5. I. Babuška, W. Gui, and B.A. Szabó, "Performance of the *h*, *p* and *h-p* versions of the finite element method," in *Research in Structures and Dynamics* (R.J. Hayduk and A.K. Noor, eds.), NASA Conference Publication 2325, 1984, pp. 73–94.
6. I. Babuška and B. Guo, "Regularity of the solution of elliptic problems with piecewise analytic data, I: Boundary value problem for linear elliptic equations of second order," *SIAM J. Math. Anal.* **19** (1988), 172–203.
7. I. Babuška and B. Guo, "The *h-p* version of the finite element method for domains with curved boundaries," Technical Note BN-1057, Institute for Physical Science and Technology, University of Maryland, College Park, MD, *SIAM J. Numer. Anal.* (1988), to appear.
8. I. Babuška and B. Guo, "The *h-p* version of the finite element method for problems with nonhomogeneous essential boundary conditions," Technical Note BN-1069, Institute for Physical Science and Technology, University of Maryland, College Park, MD, 1987.
9. I. Babuška and B. Guo, "The theory and practice of the *h-p* version of the finite element method in advances in computer methods for partial differential equations

VI," in *Proceedings of the Sixth IMACS International Symposium on Computer Methods for Partial Differential Equations* (*Lehigh University, Bethlehem, PA, June* 1987) (R. Vichnevetsky and R.S. Stepleman, eds.), pp. 241–247.

10. I. Babuška and B. Guo, "Regularity of the solution of elliptic problems with piecewise analytic data, II. Boundary value problem for linear elliptic equations of second order in three dimensions," in preparation.
11. I. Babuška, B. Guo, and J. Osborn, "The *h-p* version of the finite element method for solving eigenvalue problems," Technical Note BN-1073, Institute for Physical Science and Technology, University of Maryland, College Park, MD, 1987.
12. I. Babuška and A. Miller, "The post-processing approach in the finite element method, 2: The calculation of stress intensity factors," *Internat. J. Numer. Methods Engrg.*, **20** (1984), 1111–1129.
13. I. Babuška and Hae–Soo Oh, "Pollution problem for the *p* version and the *h-p* version of finite element method," *Comm. Appl. Numer. Meth.* **3** (1987), 553–561.
14. I. Babuška and E. Rank, "An expert-system-like feedback approach in the *h-p* version of the finite element method," *Finite Elements Anal. Design.*, **3** (1987), 127–147.
15. I. Babuška and T. Scapolla, "Computational aspects of the *h*, *p* and *h-p* versions of the finite element method in computer methods for partial differential equations VI," in *Proceedings of the Sixth IMACS International Symposium on Computer Methods for Partial Differential Equations* (*Lehigh University, Bethlehem, PA, June* 1987), (R. Vichnevetsky and R.S. Stepleman, eds.), pp. 233–240.
16. I. Babuška and M. Suri, "The optimal convergence rate of the *p* version of the finite element method," *SIAM J. Numer. Anal.*, **24** (1987), 199–238.
17. I. Babuška and M. Suri, "The *h-p* version of the finite element method with quasi-uniform meshes," *Math. Modelling Numer. Anal.* (*RAIRO*) **21** (1987) 199–238.
18. I. Babuška and M. Suri, "The treatment of nonhomogeneous Dirichlet boundary conditions by the *p* version of the finite element method," Technical Note BN-1063, Institute for Physical Science and Technology, University of Maryland, College Park, MD, 1987.
19. I. Babuška and M. Suri, "The *p* version of the finite element method for constraint boundary conditions," Technical Note BN-1064, Institute for Physical Science and Technology, University of Maryland, College Park, MD, 1987.
20. I. Babuška and B.A. Szabó, "Rates of convergence of the finite element method," *Internat. J. Numer. Methods Engrg.* **18** (1982), 323–341.
21. I. Babuška, B.A. Szabó, and I.N. Katz, "The *p* version of the finite element method. *SIAM J. Numer. Anal.*," **18** (1981), 512–545.
22. M.A. Barnhart and J.R. Eisenmann, "Analysis of a stiffened plate detail using *p* version and *h* version finite element techniques," paper presented at the *First World Congress on Computational Mechanics*, The University of Texas at Austin, 1986.
23. L.M. Delves and C. Phillips, "A fast implementation of the global element method," *J. Inst. Math. Applic.*, **25** (1980), 177–197.
24. M.R. Dorr, "The approximation theory for the *p* version of the finite element method," *SIAM J. Numer. Anal.*, **21** (1984), 1181–1207.
25. M.R. Dorr, "The approximation of solutions of elliptic boundary value problems via the *p* version of the finite element method," *SIAM J. Numer. Anal.*, **23** (1986), 58–77.

26. D.A. Dunavant and B.A. Szabó, "A-posteriori error indicators for the p version of the finite element method," *Internat. J. Numer. Methods Engrg.*, **10** (1983), 1851–1870.
27. J.S. Fenves, "A framework for cooperative development of a finite element modeling assistant," in *Reliability of Methods for Engineering Analysis* (K.J. Bathe and R.J. Owen, eds.), Pineridge Press, Swansea, Great Britain, 1986, pp. 475–486.
28. I.M. Gelfand and G.E. Shilov, *Generalized Functions*, Vol. 2, Academic Press, New York, 1986.
29. D.O. Gottlieb, M.Y. Hussaini, and S.A. Orsazag, "Theory and applications of spectral methods," in *Spectral Methods for Partial Differential Equations* (R.G. Voigt, D. Gottlieb, and M.Y. Hussaini, eds.), SIAM, Philadelphia, 1984, pp. 1–54.
30. P. Grisvard, *Elliptic Problems in Nonsmooth Domains*, Pitman, London, 1985.
31. W. Gui and I. Babuška, "The h, p and h-p versions of the finite element methods in 1 dimension, I: The error analysis of the p version, II: The error analysis of the h and h-p versions, III: The adaptive h-p versions," *Numer. Math.*, **49** (1986), 577–612, 613–657, 659–683.
32. B. Guo, "The h-p version of the finite element method for elliptic equations of order $2m$," *Numer. Math.* (1987), to appear.
33. B. Guo and I. Babuška, "The h-p version of the finite element method, I: The basic approximation results, II: General results and application," *Computational Mechanics*, **1** (1986), 1, 21–41, 203–220.
34. F. Hayes-Roth, D. Waterman, and D. Lenat (eds.), "*Rule-Based Expert Systems*," Addison-Wesley, Reading, MA., 1987.
35. K. Izadpanah, "Computation of the stress components of the finite element method, Ph.D. dissertation, Washington University St. Louis, 1984.
36. I.N. Katz, A.G. Peano, and M.P. Rossov, "Nodal variables for complete conforming finite elements of arbitrary polynomials order," *Comput. Math. Appl.*, **4** (1978), 85–112.
37. I.N. Katz, and E.W. Wang, "The p version of the finite element method for problems requiring C^1-continuity," *SIAM J. Numer. Anal.*, **22** (1985), 1082–1106.
38. V.A. Kondrat'ev, "Boundary value problem for elliptic equations in domain with conic or angular points," *Trans. Moscow Math. Soc.*, **16** (1967), 227–313.
39. V.A. Kondrat'ev and O.A. Olejnik, "Boundary value problems for partial differential equations in nonsmooth domains," *Russian Math. Surveys*, **38**, No. 2 (1983), 1–86.
40. A.K. Noor and I. Babuška, "Quality assessment and control of finite element solutions," *Finite Elements Anal. Design*, **3** (1987), 1–26.
41. A.T. Patera, "Advances and future directions of research on spectral methods," Paper presented to *ASME Winter Annual Meeting, Symposium on Future Directions in Computational Mechanics*, 1986.
42. A.G. Peano, "Hierarchies of conforming finite elements," Ph.D. dissertation, Washington University, St. Louis, 1975.
43. A.G. Peano, "Hierarchies of conforming finite elements for plane elasticity and plate bending," *Comput. Math. Appl.*, **2** (1976), 211–224.
44. A.G. Peano and J.W. Walker, "Modeling of solid continua by the p version of the finite element method," in *Proceedings of the 3rd World Congress on Finite Element Methods*, (*Beverly Hills*, CA), 1981.
45. A. Pinkus, *n-Widths in Approximation Theory*, Springer-Verlag, New York, 1985.
46. E. Rank and I. Babuška, "An expert system for the optimal mesh design in the

h-p version of the finite element method," *Internat. J. Numer. Anal.*, **24** (1987), 2087–2106.

47. J.E. Schiermeier and B.A. Szabó, "Optimal design of a torque arm: A case study," *Finite Elements Anal. Design*, **3** (1987), 93–107.
48. M. Suri, "The *p* version of the finite element method for elliptic equations of order 2*l*," Research Report N87-17, Department of Mathematics, University of Maryland, Baltimore County, 1987.
49. B.A. Szabó, "Some recent developments in finite element analysis," *Comput. Math. Appl.*, **5** (1979), 99–115.
50. B.A. Szabó, "PROBE: Theoretical manual," NOETIC Tech, St. Louis, 1985.
51. B.A. Szabó, "Mesh design for the *p* version of the finite element method," *Comput. Methods Appl. Math. Engrg.*, **55** (1986), 181–197.
52. B.A. Szabó, "Implementation of a finite element software system with *h*- and *p*-extension capabilities," *Finite Elements Anal. Design*," **2** (1986), 177–194.
53. B.A. Szabó, "Estimation and control of error based on *P*-convergence," in *Accuracy Estimates and Adaptive Refinements in Finite Element Computations* (I. Babuška, J. Gago, E.R. de A. Oliveira, and D.C. Zienkiwicz, eds.), Wiley, New York, 1986, pp. 61–70.
54. B.A. Szabó, "Computation of stress field parameters in areas of steep stress gradients," *Comm. Appl. Numer. Methods*, **2** (1986), 133–137.
55. B. Szabo, "Solution of electrostatic problems by the *p* version of the finite element method," Technical Note WU/CCU-87/1, Center for Computational Mechanics, Washington University, St. Louis, MO, 1987.
56. B.A. Szabó and I. Babuška, "Computation of the amplitude of stress singular terms for cracks and reentrant corners," Rep. WU/CCM-86/1, Center for Computational Mechanics, Washington University, St. Louis, 1986.
57. B.A. Szabó and A.K. Mehta, "*P*-convergent finite element approximations in fracture mechanics," *Internat. J. Numer. Methods Engrg.*, **12** (1978), 551–560.
58. B.A. Szabó and A.G. Peano, "Hierarchic finite elements," Rep WU/CCM-83/1, Center for Computational Mechanics, Washington University, St. Louis, 1983.
59. I.C. Taig, "Expert aids to reliable use of finite element analysis," in *Reliability of Methods for Engineering Analysis* (K.I. Bathe and R.J. Owen, eds.), Pineridge Press, Swansea, 1986, pp. 457–474.
60. D. Vasilopoulos, "Treatment of geometric singularities with the *p* version of the finite element method," Ph.D. dissertation, Washington University, St. Louis, 1984.
61. M. Vogelius, "An analysis of the *p* versions of the finite element method for nearly incompressible materials—uniformly valid, optimal estimates," *Numer. Math.* **41** (1983), 39–53.
62. D.W. Wang, I.N. Katz and G.A. Szabó, "Implementation of C^1 triangular elements based on the *p* version of the finite element method," *Comput. & Structures*, **10** (1984), 381–392.
63. D.W. Wang, I.N. Katz, and B.A. Szabó, "*H* and *p* versions finite element analysis of a rhombic plate," *Internat. J. Numer. Methods Engrg.*, **20** (1984), 1399–1405.
64. D.A. Waterman, *A Guide to Expert Systems*, Addison-Wesley, Reading, Mass., 1986.

65. O.C. Zienkiewicz, B.M. Irons, I.C. Scott, and J.S. Cambell, "Three-dimensional stress analysis," *Proceeding of the IUTAM Symposium on High Speed Computing of Elastic Structures* (*Liege, Belgium*, 1970).
66. O.C. Zienkiewicz, J.P. de S.R. Gago, and D.W. Kelly, "The hierarchical concept in finite element analysis," *Computers & Structures*, **16** (1983), 53–65.

CHAPTER 11

Transient Analysis Methods in Computational Dynamics

K.C. Park

11.1. Introduction

Transient response analysis of structural and mechanical systems is now routinely carried out via a direct time integration procedure. Typically, the size of discrete equations that must be solved can range from 10 to 10^5, depending upon modeling accuracy desired and computational resources available. As a result of increasing demand for performing the transient response analysis of structural and mechanical systems in recent years, most of the existing structural analysis computer programs have a direct time integration capability. This increased demand on computational dynamics has been due to its programming simplicity, its capability for handling nonlinearities, and its adaptability to automatic error control at each time integration step, thus obviating the difficult task of including sufficient modes in modal-based dynamics analysis.

Despite phenomenal progress made in computational structural dynamics during the last decade [1–4], typical time integration modules in many existing structural programs still lack robustness and user friendliness. To many structural analysts, quality of a time integration module may be assessed from user convenience, reliability, and efficiency rather than from theoretical aspects. The user convenience and reliability alone are adequate if either the problem size is relatively small or computational cost is insignificant compared to the analyst's effort in modeling, checking out input data, and interpreting the analysis results. Given the present rate of computer technology advances, it may be that in the near future this will be the case. However, computational efficiency concerns many analysts at present, and in fact it curtails many realistic dynamic simulation studies that are necessary for the design of large-scale structural systems. A case in point is the simulation of space structures dynamics and control.

User convenience is influenced by ease of input preparation and verification; flexibility for the user to pause, interrupt, and continue the time integration sequences; and, above all, modularity of the integration package for its easy adoption to special applications. The *reliability* of a time integration package refers to its capability to cope with logical abort or to deliver specified accuracy to the user throughout the entire integration time steps. It therefore calls for effective error and stepsize controls and for overcoming solution breakdowns without degrading the solution accuracy.

The *efficiency* depends upon the choice of integration formulas and strategies for error and stepsize controls, but primarily upon computer implementation, which is perhaps least discussed in the open literature on computational dynamics.

Although the three quality factors indicated are equally important in affecting the performance of direct time integration modules, user convenience is not discussed in this paper. We therefore concentrate on reliability and efficiency and use the term *time integration method* to refer to the combined properties of the integration formula, of error and stepsize control strategies, and of the details in computational sequences.

The present survey is largely a revision of an earlier article [5] for Sections 11.2 through 11.7. In the remainder of the paper the writer has made an attempt to include selectively new developments since 1980.

11.2. Behavior of Structural Dynamics Equations

The semidiscrete equations of motion for a general structural system can be written as

$$M\ddot{u} + D\dot{u} + S(u) = f(t), \tag{2.1}$$

where $M, D, S(\cdot)$ represent the mass, damping matrices, and stiffness operator, respectively; u is the displacement vector, a superscript dot denotes time (t) differentiation, and $f(t)$ is the external force vector. In the present paper, we do not wish to discuss the semidiscretization formulations and procedures that lead to the preceding ordinary differential equation (2.1).

The numerical solution of (2.1) begins with the initial conditions

$$u(0) = u^0 \qquad \text{and} \qquad \dot{u}(0) = \dot{u}^0 \tag{2.2}$$

to march in time to obtain the response history of a structural system modeled by (2.1). Obviously, this task requires a thorough understanding of the characteristics of the differential operator associated with (2.1). It is well known that the response components of (2.1) consist of widely different oscillatory components and the amplitudes of such oscillatory components also continually change as a result of loading conditions, attendant nonlinearities, and possible dynamic localizations. Therefore, some components of the solution vector u oscillate very rapidly and others may oscillate very slowly. From a physical viewpoint, this is due to differences in stiffness, although semidiscretization procedures do contribute to this phenomenon to some extent. When the response is primarily dominated by the high frequency components, equation (2.1) is called *a wave propagation problem*, whereas it is called a *structural dynamics problem* when the response is primarily dominated by the low frequency components. In many practical situations, the response is often interlaced with these two extreme responses. For example, the impulse response of a structural system is initially dominated by the high frequency components. As the response time progresses, however, the low frequency components gradually begin to dominate the response.

The foregoing observations suggest that, in order for direct time integration methods to be widely applicable to both wave propagation and low frequency response analysis, it is essential for us to understand the role the characteristic roots play on the stability and accuracy of integration formulas. This role in the formula synthesis can be examined from the eigenvalue analysis of the homogeneous linear (or linearized) equation

$$M\ddot{u} + D\dot{u} + Ku = 0, \tag{2.3}$$

which can be decoupled into

$$\ddot{x} + (\alpha + \beta\Lambda)\dot{x} + \Lambda x = 0, \tag{2.4}$$

where

$$u = Tx, \qquad T^t K T = \Lambda = \operatorname{diag}(\omega_1^2, \ldots, \omega_n^2), \qquad \text{and} \qquad T^t M T = I \tag{2.5}$$

in which ω_k is the kth undamped modal frequency and the damping matrix is assumed to have the expression of

$$D = \alpha M + \beta K. \tag{2.6}$$

We can obtain from (2.4) the characteristic equation for the kth decoupled component x_k from

$$s_k^2 + (\alpha + \beta\omega_k^2)s_k + \omega_k^2 = 0, \tag{2.7}$$

where s_k is the kth eigenvalue of (2.4). The solution for the kth component, x_k, can now be obtained as

$$x_k = a_k e^{st} + \bar{a}_k e^{\bar{s}t}, \tag{2.8}$$

where a_k and $\bar{a}_k$ are constants of conjugate pairs. The complete root locus of (2.4) was perhaps first analyzed by Jensen [6], whose lucid graphical representation is adopted in Figure 11.1, where the complex frequency s_k is normalized by the integration stepsize h. We now examine four physically important cases.

(a) *Undamped case* ($\alpha = \beta = 0$). In this case, the root locus of (sh) lies strictly on the imaginary axis, which incidentally is known to impose a severe restriction on the choice of integration formulas.

(b) *Mass-proportional damping* ($\alpha \neq 0$ and $\beta = 0$). The root locus of (sh) starts at the origin (0, 0) and $(-\alpha h, 0)$ for $\omega = 0$. It coalesces at $(\omega h) = \alpha h/2$. As (ωh) is further increased, the locus becomes a straight line.

(c) *Stiffness-proportional damping* ($\alpha = 0$, $\beta \neq 0$). The locus becomes a circle with radius $r = h/\beta$ centered at $(-h/\beta, 0)$. Thus, the locus passes the real axis at the origin (0, 0) and $(-2h/\beta, 0)$. If (ωh) is further increased one root approaches the origin and the other branches out toward the infinite negative real axis.

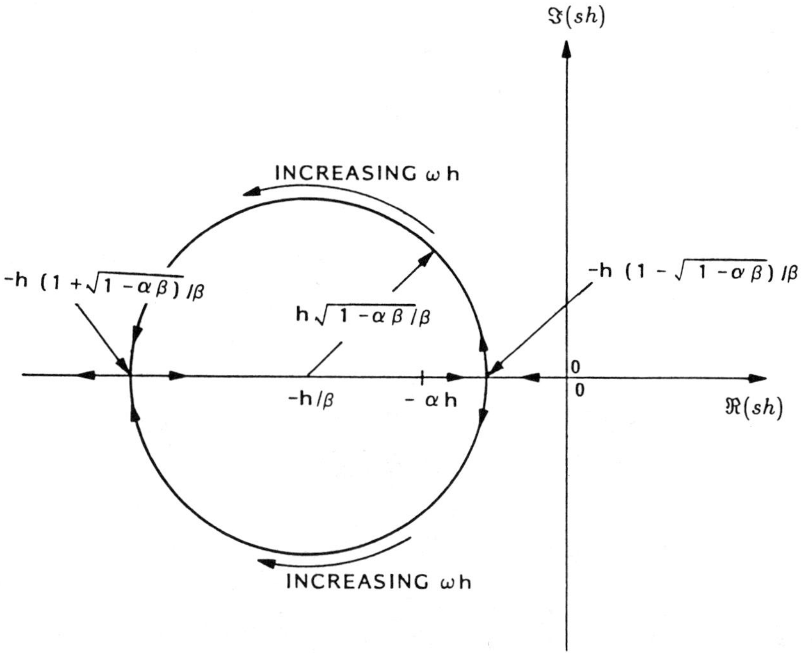

FIGURE 11.1. Locus of characteristic roots of structural dynamics equations.

(d) *Rayleigh damping* (*both* α *and* β *are nonzero*). The locus is again shown in Figure 11.1. At both low and high ωh values, the equations are heavily damped, whereas at an intermediate ωh value the equations are underdamped. This is in contrast to the stiffness-proportional damping case, which introduces very light damping at low ωh values. On the other hand, the mass-proportional damping case starts with heavy damping, and as ωh is increased the damping becomes lighter.

The preceding geometrical representation of the characteristic roots of structural dynamics equations provides the analyst with necessary stability margins in the selection (or synthesis) of integration formulas so as to capture the predominant physical behavior of (2.1). The relationship of the locus of the characteristic roots of the equations of motion to the intrinsic characteristics of integration formulas will be dealt with in the next section.

11.3. Characteristics of Integration Formulas

It is generally agreed that it is not difficult to construct a time integration formula for a specific problem, but rather to choose one from a staggering number of available integration formulas. They include a family of Runge–Kutta formulas, series expansion formulas, exponentially fitted formulas, linear multistep formulas, and other special nonlinear formulas. This is not to

TABLE 11.1. Comparison of three formulas for one degree of freedom problem

Problem: $\begin{cases} \ddot{u} + (2\pi)^2 u = 0. \\ u(0) = 1.0, \quad \dot{u}(0) = 0. \\ \text{Step size} = 0.125. \end{cases}$

Time (t)	Exact	Trapezoidal	Euler implicit	Euler explicit
	$u = \cos 2\pi t$	$u^{n+1} = u^n + h(\dot{u}^{n+1} + \dot{u}^n)/2$	$u^n = u^{n-1} + h\dot{u}^n$	$u^{n+1} = u^n + h\dot{u}^n$
0.0000	1.00000	1.00000	1.00000	1.00000
0.125	0.70711	0.73278	0.61849	1.00000
0.250	0.308E − 07	0.7394E − 01	0.14656	0.38315
...	...	...	...	...
...	...	...	...	...
3.000	1.00000	0.63088	−0.3019E − 02	−0.3.757E + 03
3.125	0.70711	0.99025	−0.146E − 02	−0.2405E + 03

say that there is no need to develop better integration methods, but to emphasize that a majority of structural dynamists are more concerned with choosing an appropriate formula than with developing a new one. For each class of formulas, two distinct categories exist: explicit and implicit integration formulas. If we set aside for the moment the computational overhead of each integration step, then numerical stability and accuracy characteristics become the two most important factors in the selection of integration formulas. For example, Table 11.1 presents numerical solutions of a single degree of freedom oscillator problem by three different integration formulas. The solution by the trapezoidal rule is the most accurate in the sense that the amplitude is preserved whereas the phase is shifted about 1/8 of the period after three cycles. The Euler backward formula dampens out the solution so rapidly that after one cycle hardly any amplitude, let alone significant phase distortion, is detected. Finally, the Euler forward formula gives an unstable solution, and at the end of the third cycle, the displacement becomes −307.0 at time $t = 3$.

A systematic means for studying the numerical stability and accuracy such as the ones observed in preceding Table 11.1 is to construct stability regions of various integration formulas on the complex (sh)-plane and to measure numerical amplitude decay and frequency distortion in terms of (ωh). This necessarily involves some algebraic manipulations, which are briefly described later.

11.3.1. Characteristic Equation for Integration Formulas

Let us consider a model system of linear differential equations

$$\dot{u} = Su \tag{3.1}$$

by a linear multistep formula in the form

$$u^n = \sum_{j=1}^{m} \alpha_j u^{n-j} + h \sum_{j=0}^{m} \beta_j \dot{u}^{n-j}, \tag{3.2}$$

where α_j and β_j are formula-dependent real coefficients. Assuming that (3.1) has a complete set of its eigenvalues ($s_1 \ldots s^n$) and its associated eigenvector matrix T, we have the following difference equations from (3.1) and (3.2).

$$\sum_{j=0}^{m} (\alpha_j - \beta_j sh)x_{n-i} = 0. \tag{3.3}$$

The characteristic equation for (3.3) can be obtained by seeking a nontrivial solution in the form

$$x^n = \lambda^k x^{n-k}, \tag{3.4}$$

where λ is the *solution amplification factor*. We can obtain from (3.3) and (3.4) its characteristic equation

$$\sum_{j=0}^{m} (\alpha_j - \beta_j sh)\lambda^{m-j} = 0. \tag{3.5}$$

The preceding equation will be called the *characteristic equation for integration formula* (3.2). We will now examine the computational stability of linear multistep formula (3.2) from its characteristic equation (3.5).

11.3.2. Stability Region

The numerical solution of (3.1) by formula (3.2) remains bounded only if

$$|\lambda_j| \leq 1.0, \qquad j = 1, \ldots, m, \tag{3.6}$$

as can be seen from (3.4). In particular, the case $|\lambda| = 1.0$, i.e., unit circle in the λ-plane, determines the boundary at which the solution neither grows nor decays. Therefore, the boundary on the complex (sh)-plane is the mapping of the unit circle $\{\lambda = e^{i\theta}; 0 < \theta < 2\pi, i = \sqrt{-1}\}$ on the (sh)-plane, viz.,

$$sh = \frac{\sum_{j=0}^{m} \alpha_j e^{i(m-j)\theta}}{\sum_{j=0}^{m} \beta_j e^{i(m-j)\theta}}. \tag{3.7}$$

Figure 11.2 shows stability regions of several integration formulas including the three formulas used for computing the example problem in Table 11.1. Note that in the example problem, the normalized roots are $sh = \pm 0.25\pi i$. For this case, the roots are located in the stable region of the Euler backward formula, in the unstable region of the Euler forward formula, and right on the stability boundary of the trapezoidal rule. Generally speaking, the amplitude of the solution is preserved (decays, grows) when the normalized roots of the differential equations lie on (inside, outside) the stability boundary of the integration formula.

Therefore, the implicit methods such as the Euler backward and the trapezoidal formulas are stable for the entire left-hand complex plane. On the other hand, explicit formulas such as the Euler forward formula are stable only inside stability boundaries. From the computational viewpoint, the price paid for the favorable stability regions of implicit formulas is to solve simultaneous matrix difference equations at each integration step, whereas the limited

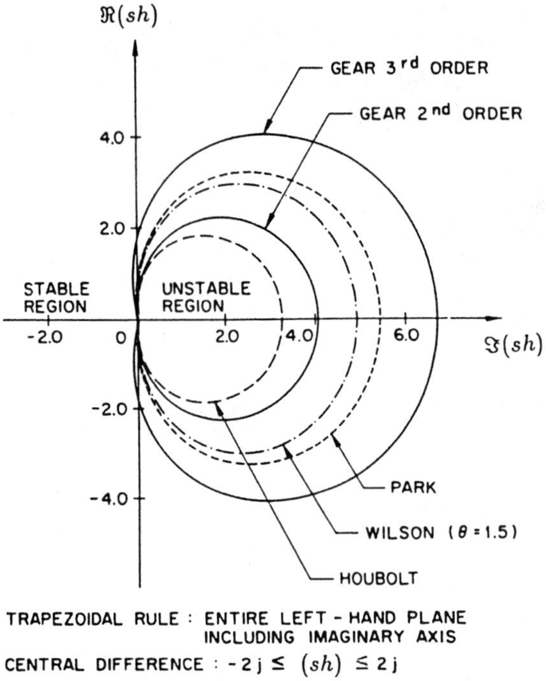

FIGURE 11.2. Stability regions of several implicit formulas.

stability regions of explicit formulas are to be compensated by their simplicity in computations, which require only vectorial calculations. In particular, implicit integration formulas whose stability regions contain the entire left-hand complex (sh)-plane, including the imaginary axis, are called *A-stable* in mathematical parlance or called *unconditionally stable* [7–15] in engineering literature. Furthermore, if the formulas are unconditionally stable and $|\lambda| \to 0$ as $|sh| \to \infty$, such formulas are called *L-stable* or *stiffly* unconditionally stable [16].

It is noted that the mapping of the normalized characteristic roots of the structural dynamics equations and of the stability regions of integration formulas on the same complex plane enables us to evaluate the computational stability of direct time integration process.

11.3.3. Explicit Formulas

In selecting an explicit formula among competing explicit integration formulas for structural dynamics equations, a proper selection criterion is the maximal imaginary stability limit per function evaluation. Note that the negative axis stability limit holds no significance for undamped structural systems, as the characteristic roots lie strictly on the imaginary axis. For example, the fourth-order Runge–Kutta formula has its imaginary stability limit $sh \leq \pm 2.82i$ and requires four function evaluations per each integration step. Hence, its imaginary stability limit per function evaluation is $\pm 0.707i$. Similarly, the imaginary stability limit per function evaluation of the second-order Adams–Moulton formula is $\pm 1.414i$. Finally, the widely used central difference formula gives its stability limit $\pm 2i$ per function evaluation. This is why the latter has been so popular in structural dynamics analysis. We note that the Euler explicit formula, although unconditionally unstable for undamped structural problems, has its stability limit along the negative real axis. Hence, it can be used for parabolic problems without instability, provided the stepsize does not exceed its stability limit, to be $\Re\{sh\} \leq -2$.

11.3.4. Implicit Formulas

As pointed out in Section 11.3.2, implicit formulas necessarily involve the simultaneous solution of matrix algebraic equations at each integration step. Therefore, unconditional stability is essential for all implicit formulas. This necessity in turn restricts the achievable order of accuracy at most to two [17], and consequently, all higher-order implicit formulas may engender instability unless some forms of stabilization are introduced [18, 19]. Therefore, high-order stiffly stable formulas whose unstable regions necessarily include some portions of the imaginary axis on the (sh)-plane should be used with caution for integrating the structural dynamics equations. This is because the success of the general ordinary differential equation (ODE) integration packages is primarily in parabolic equations, whose characteristic roots lie strictly on the negative real axis. Finally, we mention that stiff-unconditional stability often becomes a desirable property when a strong nonlinear coupling from the high-frequency modes to the low-frequency modes is prevalent throughout the response period. For example, the trapezoidal formula does not possess such a property and is known to exhibit difficulty for treating the equations of motion whose response is dominated by that class of coupling [19, 20].

11.3.5. Accuracy of Integration Formulas

Unlike parabolic problems for which integration formulas with maximal real-axis stability limits are suitable, a desirable stability characteristic of integrators for structural dynamic problems is a maximal imaginary stability limit. In addition, in practical terms, we would like to know how big a time stepsize should be if the response is to be traced accurately with respect to a dominant frequency ω_d and both the computed amplitude and phase errors are not to exceed, say, 1 percent.

Figures 11.3 and 11.4 show numerical damping and phase distortion errors, respectively, for various integration formulas. Note that both the central difference and the trapezoidal formulas exhibit no numerical damping. Note also that all implicit (explicit) formulas lengthen (shorten) the period. The

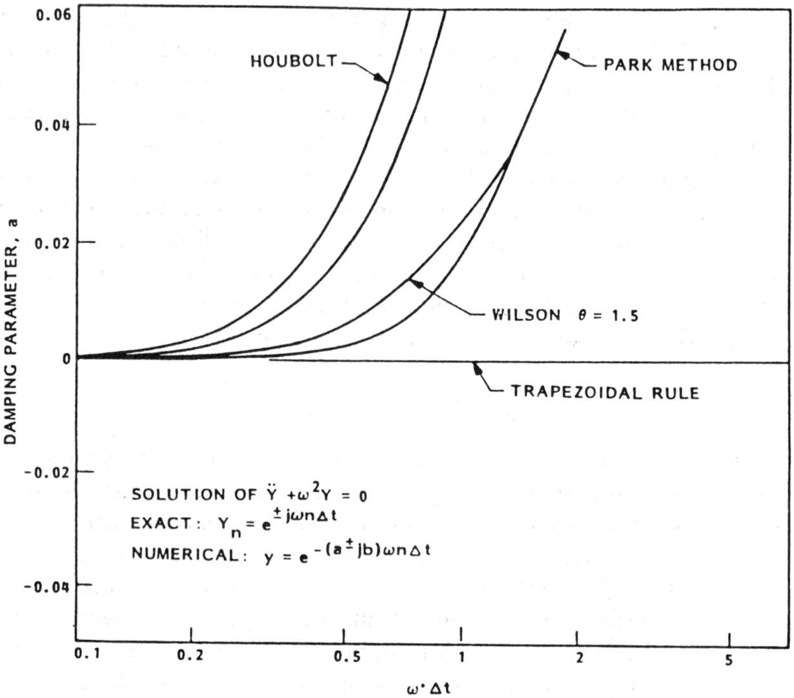

FIGURE 11.3. Numerical damping characteristics of integration formulas.

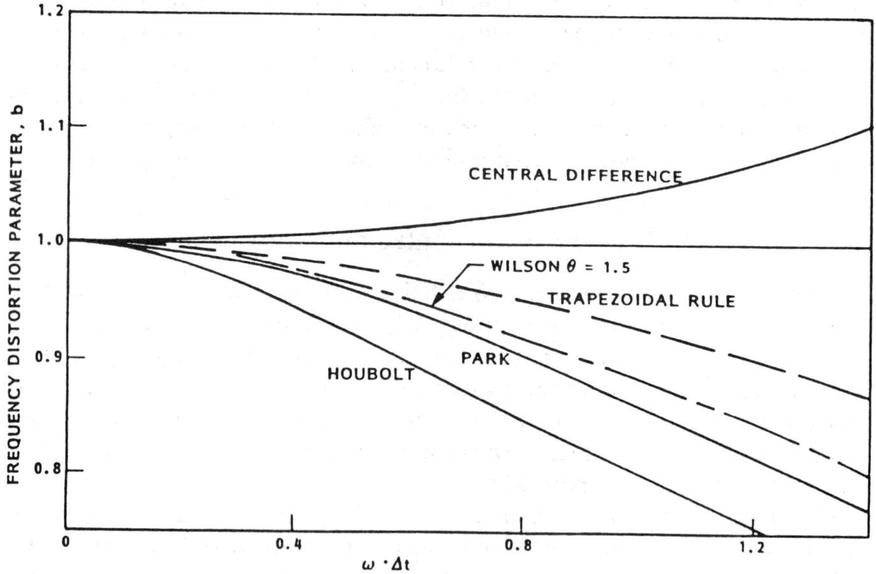

FIGURE 11.4. Numerical frequency distortion of integration formulas.

absolute phase errors of both the central difference and the trapezoidal formulas are, however, favorably compared. This is why, in addition to their simplicity in computer implementation, the two formulas have been widely used in structural dynamics analysis. However, when there is a strong non-linear coupling from the high-frequency modes to the low-frequency modes, the trapezoidal rule does not yield a satisfactory solution unless the stepsize is sufficiently small [19–20]; hence, formulas that possess stiffly unconditional stability have been preferred by several investigators [21–26].

11.4. Control of Error and Stepsize

The task of control of error and stepsize still remains as the least matured aspect in computational dynamics, and to date no satisfactory general approach is available. Consequently, a host of ad hoc or empirical strategies are implemented in various structural analysis codes. Here again the purely oscillatory response characteristic imposes a severe restriction on desirable strategies. For example, most available ODE solution packages developed by numerical analysts adopt truncation error-based stepsize control strategies in conjunction with variable-order formulas [27]. It should be pointed out that implicit integration formulas for structural dynamics analysis cannot exceed their order of accuracy beyond two because of unconditional stability requirements [17]. Likewise, developers of explicit formulas have yet to come up with one whose imaginary stability limit in the (sh)-plane per function evaluation exceeds that of the central difference formula, $\pm 2i$. The exception occurs when a high accuracy (say, 10^{-6} in the solution error) becomes an overriding concern for some specific requirements. However, given the errors already introduced in the modeling and semidiscretization procedures, such a high accuracy requirement is seldom justified in most structural analysis cases.

11.4.1. Stepsize Control for Explicit Formulas

The selection of stepsize for explicit formulas should be dictated by satisfying simultaneously both accuracy and stability requirements. In most cases, the initial response is dominated by the high-frequency components. This is because the high-frequency components can participate in the response more quickly than the low-frequency components. Hence, the stepsize during the initial response period is dictated by accuracy requirements, for which explicit formulas are better suited. As the response time progresses further, the low-frequency components gradually become dominant, and often the stepsize can be increased beyond the stability limit of explicit formulas from the accuracy viewpoint. But the stepsize of explicit formulas should be restricted within their stability limits in order to keep the solution bounded.

Experience suggests that stepsize control strategies based on the truncation error are not effective under this condition except when an extremely high accuracy is demanded. The reason for this reservation is as follows. The solution of a typical set of undamped structural dynamics equations is characterized by oscillatory components. Therefore, an error introduced in a step propagates throughout the subsequent steps because of the energy-conserving nature of the problem. To say that a stepsize exceeded the stability limit of the problem is to imply that a numerical creation of energy whose magnitude is exponentially proportional to the stepsize has occurred in the system. Hence, if the truncation-based error criterion detects numerical instability only after several steps of unstable integration have been performed (this often is the case since accuracy measure does not necessarily detect instability), spurious energy, which often contributes more errors to the solution than the errors tolerated by the truncation error measure, has been added to the system. Under such circumstances, it is desirable to carry out integration with a step size that is close to or right on the stability limit.

In what follows, we present a stepsize control strategy that is based on the system response frequency characteristics. This strategy assumes that the dominant error component has the same frequency as the dominant response component. In other words, the error behaves in the same way as the dominant solution responds. Hence, this strategy makes use of two computed solution characteristics; a measure of accuracy is monitored by the system dominant frequency, and the stability is controlled by the so-called maximum perturbed apparent frequency [28, 29]. For a single degree of freedom case, the local frequency (or apparent frequency), ω_a, can be approximated by

$$\omega_a^2 = \frac{|\ddot{u}^n - \ddot{u}^{n-1}|}{|u^n - u^{n-1}|}. \tag{4.1}$$

For multidegrees of freedom problems the maximum perturbed apparent frequency, $(\omega_a)_{\max}$, is obtained by

$$(\omega_a)_{\max} = \text{Max}\{(\omega_a)_1, \ldots, (\omega_a)_N\}. \tag{4.2}$$

The dominant frequency, ω_d, can be computed by

$$\omega_d^2 = \frac{\dot{u}^t K \dot{u}}{\dot{u}^t M \dot{u}}. \tag{4.3}$$

Once the dominant frequency is determined, the required accuracy can be maintained by specifying the stepsize from Figures 11.2 and 11.3. Finally, the stepsize to be used is obtained from

$$h \leq \text{Min}\left\{\frac{c}{\omega_d}, \frac{2}{(\omega_a)_{\max}}\right\} \tag{4.4}$$

where c is a constant (0.01 ~ 0.5, typically) to be determined from Figures 11.2 and 11.3.

11.4.2. Error and Stepsize Control Strategies for Implicit Formulas

Error control criteria proposed for implicit formulas to date include the truncation-based error, the difference between the two-half-step and the one-full-step solutions, the difference between the interpolated and the evaluated residual errors at the midpoint [30], an energy balance check, and global error prediction based on the number of samples per dominant period. Among these, the most reliable (hence, the most expensive) is the one based on the difference between the two-half-step solution and the one-full-step solution [31]. Strategies based on the truncation error or the acceleration difference are often sensitive to computer implementation and coding details and lead to a conservative selection of stepsizes [28–29]. The residual error criterion appears to be reliable for small-scale problems [26] but requires one additional calculation of stiffness and damping terms per step. The criterion based on the energy balance at each integration step [32] appears to satisfy the necessary condition for a bounded solution and work well in some numerical problems.

The global error criterion is an old technique [33], and its theoretical justification in an ODE parlance is just beginning to emerge [34]. Simply stated, the theory states that the stepsize selection based on the dominant response frequency becomes reliable if the dominant error source also has approximately the same frequency as the dominant response mode. Hence, the global error control criterion is limited to continuous, smoothly varying systems; it is not, for example, generally applicable to impact-contact problems.

The following strategy is mainly based on this writer's experience, and no claim of its superiority over the other strategies is implied. The strategy is aimed at the hybrid integration sequences, viz., using either an explicit or implicit formula sequentially in time; at each integration step the following quantities are computed [35]:

Apparent highest frequency of the system:

$$\omega_a^2 = \mathrm{Max}\left\{\left|\frac{\Delta \ddot{u}_i}{\Delta u_i}\right|, i = 1, \dots, N\right\}. \tag{4.5}$$

Dominant frequency:

$$\omega_d^2 = \frac{|\Delta u^t M \Delta \ddot{u}|}{\Delta u^t M \Delta u}. \tag{4.6}$$

In order to maintain a given accuracy with respect to the dominant frequency, it is necessary to restrict the stepsize, h_d

$$h_d \le \frac{c}{\omega_d}, \tag{4.7}$$

where $c = (0.05 \sim 0.3)$ typically. For example, the choice $c = 0.3$ corresponds approximately to twenty samples per the dominant response period. The maximum stepsize of the central difference formula, h_a, is given by

$$h_a \le \frac{2}{\omega_a}. \tag{4.8}$$

If the overhead (ovhd) is defined as the cost associated with the calculation and data transfer tasks per integration step, and if the condition

$$\text{ovhd}_{\text{exp.}} > \frac{h_a}{h_d} \cdot \text{ovhd}_{\text{imp}} \tag{4.9}$$

holds consistently, then an implicit formula can be invoked. Otherwise, an explicit formula (preferably the central difference formula) is used. In addition, if the ovhd margin is small (say less than 2), an explicit formula is to be preferred.

11.5. Time-Discretization of Structural Dynamics Equations

In this section, the difference equations that need be solved at each integration step are derived. To this end, we first cast (2.1) into a set of first-order differential equations by introducing the following auxiliary vector v [6]:

$$v = AM\dot{u} + Bu, \tag{5.1}$$

where A and B are arbitrary matrices. Equations (2.1) and (5.1) can be presented as

$$\begin{bmatrix} AM & 0 \\ AD - B & I \end{bmatrix} \begin{Bmatrix} \dot{u} \\ \dot{v} \end{Bmatrix} + \begin{bmatrix} B & -I \\ AS(\cdot) & 0 \end{bmatrix} \begin{Bmatrix} u \\ v \end{Bmatrix} = \begin{Bmatrix} 0 \\ Af \end{Bmatrix}. \tag{5.2}$$

Multistep integration formulas (3.2) for vector $\lfloor u, v \rfloor^t$ can be written as

$$\begin{Bmatrix} u^n \\ v^n \end{Bmatrix} = \begin{Bmatrix} \delta_u \dot{u}^n \\ \delta_v \dot{v}^n \end{Bmatrix} + \begin{Bmatrix} h_u^n \\ h_v^n \end{Bmatrix}, \tag{5.3}$$

where $\delta = \beta_0 h$ and the historical vector h_x^n has the form of

$$h_x^n = \sum_{j=1}^{m} (-\alpha_j x_{n-j} + h\beta_j \dot{x}_{n-j}) \tag{5.4}$$

in which $\alpha_0 = 1.0$ is assumed for simplicity. Substitution of (5.3) into (5.2) for $\lfloor \dot{u}, \dot{v} \rfloor^t$ gives

$$\begin{bmatrix} AM + \delta_u B & -\delta_u I \\ AD - B + \delta_u AS(\cdot) & \delta_u/\delta_v I \end{bmatrix} \begin{Bmatrix} u^n \\ v^n \end{Bmatrix} = \begin{Bmatrix} AMh_u^n \\ (AD - B)h_u^n + \delta_u/\delta_v h_v^n + \delta_u Af^n \end{Bmatrix}. \tag{5.5}$$

Note that the size of (5.5) is twice the governing equation (2.1), and this is the usual setting of many ODE solution packages. As v^n is an auxiliary vector (physically it is a generalized momentum vector), we can eliminate it to reduce the size of (5.5) to obtain

$$Cu^n + \delta_v \delta_u S(u^n) = g^n, \tag{5.6}$$

where

$$C = M + \delta_v D + (\delta_u - \delta_v)A^{-1}B, \tag{5.7}$$

$$g^n = (M + \delta_v(D - A^{-1}B))h_u^n + \delta_u A^{-1} h_v^n + \delta_u \delta_v f^n. \tag{5.8}$$

Remark 1. Observe that different choices of A and B lead to different difference equations, the two most commonly used choices being $A = I$ and $B = D$, and $A = M^{-1}$ and $B = 0$; see [6, 36].

Remark 2. The foregoing time-discretization has been achieved by applying one-derivative linear multistep integration formulas, i.e., formulas involving only displacements and velocities, in (3.2). It should be mentioned that the preceding procedure is also applicable to formulas involving higher derivatives, provided they are expressed in summed forms [36]. For example, the Houbolt formula [37]

$$u^n = 2.5u^{n-1} - 2u^{n-2} + 0.5u^{n-3} + 0.5h^2\ddot{u}^n \tag{5.9}$$

can be expressed in a summed form as a pair of

$$\begin{aligned} u^n &= 1.5u^{n-1} - 0.5u^{n-2} + 0.5h\dot{u}^n, \\ v^n &= v^{n-1} + h\dot{v}^n. \end{aligned} \tag{5.10}$$

Hence, the preceding discretization is equally applicable to two-derivative formulas as well.

Equation (5.5) also embodies the special case of the summed-form central difference formulas, viz.,

$$\begin{aligned} \dot{u}^{n-1/2} &= \dot{u}^{n-3/2} + h\ddot{u}^{n-1}, \\ u^n &= u^{n-1} + h\dot{u}^{n-1/2}, \end{aligned} \tag{5.11}$$

if we choose $\delta_u = h$, $\delta_v = 0$ and

$$\begin{aligned} h_u^n &= u^{n-1}, \\ h_v^n &= (\dot{u}^{n-3/2} + h\ddot{u}^{n-1}). \end{aligned} \tag{5.12}$$

11.6. Treatment of Nonlinear Terms

For explicit integration formulas, the presence of the nonlinear term $S(u)$ in (2.1) poses no difficulty as the integration process requires an extrapolation involving only the previous solution vectors. For implicit formulas, however, nonlinear terms present both computational and implementational difficulties. For illustrative purposes, we deal with only $S(u)$, which may be approximated by [38, 39]

$$S_{(k+1)}^n = S^{n-1} + J_{(m)}\Delta u_{(k+1)}^n + r_{(k)}^n, \tag{6.1}$$

where $J_{(m)}$ is the Jacobian of $S(u)$ evaluated at $u^n_{(m)}$ and $r^n_{(k)}$ is the residual term of $S(u^n_{(k)})$. Note that the choices

$$J_{(m)} = J_{(k)}, \tag{6.2}$$

$$r^n_{(k)} = S^n_{(k)} - S^{n-1}, \tag{6.3}$$

result in the full Newton procedure. If we adopt

$$J_{(m)} = J(u^{n-1}) \tag{6.4}$$

together with (6.3), then the tangent modulus approximation [26, 38, 39] with iterative correction results. In practice, the evaluation of the Jacobian and the subsequent refactoring of the solution matrix $E^N = E + \delta_v \delta_v J$ can become intolerable for large-scale problems. For these cases, one alternative is to use the Jacobian evaluated at a previous step [say, $J(u^n_m)$] until the convergence deteriorates. At this point, a new Jacobian is evaluated and the refactoring of E^N is performed.

An alternative is to generate the Jacobian based on a mean response state and use it throughout the dynamics analysis. The convergence is controlled by the integration stepsizes, and if the stepsize becomes too small to the extent that the explicit integration mode is deemed more economical, then the explicit integration mode is invoked. This is known as the *modified pseudoforce method* [35, 38], which is apparently an improvement over the pseudoforce method [24–26] in which the nonlinear stiffness matrix $J = 0$ is assumed. This approach is particularly attractive in an adaptive integration package, where the explicit integration mode is invoked initially until such a time later in the computation when the integration mode is switched to the implicit one. In this mode, the use of the Jacobian evaluated at the time of the switch often provides an adequate convergence rate, thus avoiding reevaluations of the Jacobian and the subsequent refactoring of the solution matrix E^N.

Implementation of the nonlinearities in the direct time integration procedure can now be made by substituting (6.1) into (5.6), which yields

$$E^N u^n_{(k+1)} = g^n - \delta_v \delta_u (S^{n-1} - J_{(m)} u^n_{(k)} + r^n_{(k)}), \tag{6.5}$$

where

$$E^N = M + \delta_v D + (\delta_u - \delta_v) A^{-1} B + \delta_v \delta_u J_{(m)}. \tag{6.6}$$

Note that the nonlinear term $S(u)$ consists of both geometric and material contributions, which are computed in two separate computational modules; i.e.,

$$S = S_G + S_P, \tag{6.7}$$

$$J = J_G + J_P, \tag{6.8}$$

$$r = r_G + r_P, \tag{6.9}$$

where the subscripts G and P denote geometrical and material nonlinearities, respectively. Experience indicates that if iterations are necessary at each time-advancing cycle, it is better to keep the material nonlinear term out of the iteration loop. Once the iterations on a geometric nonlinear term yield a converged solution, then the material nonlinear term may be computed by the subincremental procedure [40, 41]. Finally, the Jacobian of the material nonlinear term J^P may be generated by the standard procedure; see, for example, [42, 43].

To date, the criteria for selecting a problem-dependent treatment of nonlinear terms are not available, and most existing nonlinear structural analysis codes adopt either the tangent modulus or the pseudoforce procedure; very rarely existing structural analysis programs adopt concurrently the procedures outlined previously.

11.7. Partitioned Transient Analysis Procedures

So far we have discussed the behavior of structural dynamics equations, the characteristics of time integration formulas, including numerical stability and accuracy, error and stepsize control strategies, time-discretization of the ordinary differential equations of motion to the difference equations, and different treatments of nonlinear terms. The remaining two major aspects of direct time integration methods are the solution procedures and computer implementation: the former governs the calculation sequences to be carried out, and the latter deals with efficient ways of handling and transferring data in carrying out integration steps.

Solution procedures for explicit integration formulas need not be discussed any further as they involve only vectorial calculations; see Remark 2. It is the solution procedures for implicit integration formulas, viz., the case $\delta_u \neq 0$ and $\delta_v \neq 0$ in (5.3), that require simultaneous algebraic solutions of the coupled-difference equation (5.6).

Traditionally, solution procedures implemented in a typical analysis program adopt either a totally implicit or a totally explicit integration mode. Recently, the concurrent use of both implicit and explicit integration formulas for different spatial domains or in alternating temporal sequences has been introduced to improve computational efficiency. In addition, for the dynamic analysis of coupled-field problems such as fluid-structure interaction problems, software modularity requirements dictate that the interaction systems be treated by two separate analysis submodules (fluid and structural) rather than by one combined computational module.

The selection of spatially or temporally different solution formulas becomes, therefore, desirable in order to exploit phenomena such as weak coupling, widely different response characteristics, isolation of discontinuous and nonlinear effects, availability of modular analysis software, and combinations of these. Solution procedures that exploit the preceding attributes are called *partitioned transient analysis procedures* [44]. Therefore, the solution of the entire equations of motion for complex dynamic systems can be performed by either parallel or sequential execution of different computational methods and/or different single-field analysis modules. The essential features and some applications of partitioned procedures are presented later.

The essence of partitioned procedures consists of the following sequences of formulation and implementation. First, the governing equations of motion for the entire system are assembled as one entity. For example, in the case of fluid-structure interaction problems, the governing equations of motion for fluid and structures and the interaction terms are combined in such a way that the total equations are written as one entity. Second, the assembled equations of motion are time discretized by an (or a set of) implicit time integration formula(s), thus yielding coupled difference equations. Third, the solution matrix E in (6.6) is partitioned according to the need for employing either different integration formulas or different analysis modules for some parts of the solution vector. Fourth, the partitioned terms that are transferred to the right-hand side of the difference equations are approximated via an extrapolator. Finally, appropriate computer implementation is carried out to accommodate the required calculations.

The first two steps are already covered in Sections 11.2 through 11.5. Hence, the remaining three aspects, viz., partition, extrapolation, and implementation, will be dealt with in this section. We recall from (6.5) that the difference equations for the entire system can be expressed as

$$E^N u^n = b^n, \tag{7.1}$$

where

$$\begin{aligned} E^N &= M + \delta D + \delta^2 K^N, \qquad K^N = J^n_{(k)}, \\ b^n &= g^n - \delta^2(S^{n-1} - Ku^{n-1} - r^n_p), \end{aligned} \tag{7.2}$$

in which $\delta = \delta_u = \delta_v$; i.e., the same integration formula for both $\dot{u}$ and $\dot{v}$ is adopted for simplicity.

A general partition of the solution matrix E^N is now introduced in the form of

$$E^N = E_1^N + E_2^N, \tag{7.3}$$

where

$$\begin{aligned} D &= D_1 + D_2, \\ K^N &= K_1^N + K_2^N, \\ E_1^N &= M + \delta D_1 + \delta^2 K_1^N, \\ E_2^N &= \delta D_2 + \delta^2 K_2^N. \end{aligned} \tag{7.4}$$

Substitution of (7.3) into (7.1) and transfer of $E_2^N u^n$ to the right-hand side yields

$$E_1^N u^n = b^n - E_2^N u_p^n, \tag{7.5}$$

where u_p^n is an appropriate extrapolator for u^n.

Given the stiffness operator in the form

$$K^N = \begin{bmatrix} K_{xx} & K_{xb} & \\ K_{bx} & K_{bb}^x + K_{bb}^y & K_{by} \\ 0 & K_{yb} & K_{yy} \end{bmatrix} \tag{7.6}$$

for two fields x and y with the boundary b, the general partitioned procedure as given by (7.3)–(7.5) can embody several special cases. It is called an *implicit–implicit procedure* if K_2^N contains only off-diagonal entries, an *implicit–explicit procedure* if K_2^N contains diagonal entries, and, in particular, a *semiexplicit procedure* if K_2^N consists of strictly upper (or lower) triangular matrices. A classification of partitioned procedures [44–52] is given later. Note that combinations of those partitions are also possible and may be desirable for complex problems.

11.7.1. Node-by-Node Implicit–Explicit Procedure

The node-by-node partition is effected if K_2 is of the form

$$K_2 = \begin{bmatrix} 0 & 0 & 0 \\ 0 & 0 & 0 \\ 0 & K_{yb} & K_{yy} \end{bmatrix} \tag{7.7}$$

that pertains to the y-field nodal vector. This procedure can be made equivalent to the implicit–explicit algorithm of Belytschko and Mullen [45, 46] if the trapezoidal rule in conjunction with the extrapolator

$$u_p^n \approx u^{n-1} \tag{7.8}$$

is adopted, and v^n and $\dot{u}^n$ are updated from integration operations rather than differentiation operations; see [36].

11.7.2. Element-by-Element Implicit–Explicit Procedure

If we choose K_2 in the form

$$K_2 = \begin{bmatrix} 0 & 0 & 0 \\ 0 & K_{bb}^y & K_{by} \\ 0 & K_{yb} & K_{yy} \end{bmatrix} \tag{7.9}$$

an element-by-element partition results. For this case, the stiffness force associated with K_2 corresponds to the y-field nodal vector plus the elemental boundary vector belonging only to the y-field. This procedure specializes to the implicit–explicit algorithm in finite elements of Hughes and Liu [47] if the trapezoidal rule, the extrapolator (7.8), and the integration-operation updates of v^n and $\dot{u}^n$ are adopted.

11.7.3. DOF-by-DOF Implicit–Explicit Procedure

The DOF-by-DOF implicit–explicit procedure results if we envision x-, y-field vectors as collections of certain degrees of freedom (DOF), e.g., separation of translational degrees of freedom from rotational ones. Hence, it enables us to treat explicitly translational degrees of freedom that are usually associated with low frequencies while treating implicitly high frequency infested rotational degrees of freedom. Specifically, we have for K_2 the following partition [44]:

$$K_2 = \begin{bmatrix} 0 & 0 & 0 \\ 0 & 0 & 0 \\ 0 & 0 & K_{yy} \end{bmatrix} \tag{7.10}$$

11.7.4. Staggered Implicit–Implicit Procedure

This partition has K_2 in the form

$$K_2 = \begin{bmatrix} 0 & 0 & 0 \\ 0 & 0 & 0 \\ 0 & K_{yb} & 0 \end{bmatrix}. \tag{7.11}$$

Physically, this corresponds to extrapolating the boundary force vector on the y-field equations $K_{yb}u^b$ and to solving the y-field solution vector u_y. Once u_y is available, the boundary vector u_b and the x-field vector u_x can be solved implicitly. This procedure finds its applications in two- (or more than two) field interaction problems that are to be analyzed by two implicit integration modules.

11.7.5. Element-by-Element Implicit–Implicit Procedure [44]

One can construct an implicit–implicit procedure that makes use of the element-by-element partition (see Section 11.7.2). It involves the following sequence of computations

$$(M + \delta D_1 + \delta^2 K_1^N)\hat{u}^n = b^n - (\delta D_2 + \delta^2 K_2^N)u_p^n, \tag{7.12a}$$

$$(M + \delta D_2 + \delta^2 K_2^N)u^n = b^n - (\delta D_1 + \delta^2 K_1^N)\hat{u}^n. \tag{7.12b}$$

From the computational viewpoint, this procedure solves the boundary solution vector u_b twice. Nevertheless, this procedure makes the use of finite element-based subsystem analyzers easier than the staggered implicit–implicit procedure.

11.7.6. Semiexplicit Procedure [49–52]

The solution matrix E_1 of either implicit–explicit or implicit–implicit partitioned procedures must be factored and the solution is obtained by back and forward substitution operations. For nonlinear problems, this leads to a Newton-like method. For certain systems whose accuracy-governing parts can be treated by explicit integration formulas, the solution matrix E_1 can be further partitioned into

$$E_1 = E_1^L + E_1^D + E_1^U, \tag{7.13}$$

where E_1^L and E_1^U are strictly lower and upper triangular matrices, respectively, and E_1^D consists of diagonal entries of E_1. For example, (7.12a) can be further partitioned as

$$\begin{aligned}(M + \delta D^L + \delta D_1^D + \delta^2 K_1^L + \delta^2 K_1^D)\hat{u}^n \\ = b^n - [\delta(D_2 + D_1^U) + \delta^2(K_2 + K_1^U)]u_p^n\end{aligned} \tag{7.14}$$

in which

$$K_1^L = \begin{bmatrix} 0 & 0 & 0 \\ K_{bx} & 0 & 0 \\ 0 & K_{yb} & 0 \end{bmatrix}. \tag{7.15}$$

For nonlinear problems, iterations on u^n may be desirable and can be performed by involving only vectorial operations. Equation (7.12b) can be similarly partitioned further. The justification of the preceding semiexplicit procedure comes from the closeness of u_p^n to $\hat{u}^n$, and no difficulty in step size changes occurs as it involves only vectorial operations, hence the name *semiexplicit procedure*. That is, it works as an explicit procedure does, but it is stable for linear problems. A somewhat different algorithm based on the alternating direction method for linear structural dynamics equations was proposed by Trujillo [52].

11.7.7. Software Considerations

The choice of a particular partitioned procedure inevitably impacts computer implementation. For example, the element-by-element implicit–explicit procedure requires an elemental matrix representation for K_1, whereas it prefers an elemental vector form for $K_2 u^n$. Here the term *elemental* is meant to form either K_1 or $K_2 u^n$ on an element contribution basis at partition boundaries. Similar considerations apply for other partitions. Therefore, partitioned procedures, in general, not only are aimed at improving computational efficiency, but also can make use of existing subsystem analysis modules that adopt different solution procedures. The latter attribute leads to software modularity in programming practice and at the same time increases interface flexibility. In other words, special analysis modules, which were built for analyzing special subsystems, can be used as subsystem analyzers in the analysis of complex systems via partitioned procedures.

11.8. Equation Solvers

The partitioned procedures presented in the preceding section are aimed at improving either computational efficiency or software modularity. Thus, computational efficiency is achieved by exploiting different physical response characteristics and/or semidiscretization techniques rather than by adopting more efficient numerical analysis methods. The modularity is maintained by a proper modeling of the problems and formulations of the equations of motion in such a manner that either subsystems can be loosely coupled or different analysis methods become mandatory as a result of the distinct nature of some of the subsystems.

Computational efficiency can be further improved by adopting better numerical techniques for solving algebraic equations per each integration step. One class of such an effort is centered around some modifications of Newton or Newton-like methods that are known in the literature as *quasi-Newton, variable metric, or secant update methods* [38, 39, 53]. The other is a class of alternating direction methods (also known as *fractional step methods, factorization methods*, etc.) [54].

Quasi-Newton methods have been extensively used in unconstrained minimization problems. The application of quasi-Newton methods in structural dynamics problems is at present in experimental stages [55, 56], and more experiments are needed before a consensus as regards the efficiency of the methods compared with the traditional Newton–Raphson method can be reached.

The alternating direction methods and their allied methods have been successfully used in fluid flow problems in conjunction with the finite difference spatial discretization techniques. Their use in structural dynamics problems has been attempted only along the line of the finite difference method [57], and chances are that research activities directed toward their application in finite element analysis may prove fruitful.

Finally, iterative solution techniques are beginning to emerge as a viable alternative to direct elimination techniques (such as Newton-like and alternating direction methods) for very large problems, e.g., problems with 10^4 or more. This is because in spite of the computer hardware progress there is a limit beyond which the cost associated with the high-speed memory and the auxiliary storage devices becomes dominant in the total computations. As almost all iterative solution techniques can be shown to be equivalent to some forms of dynamic relaxation techniques [58, 59, 60], iterative solution techniques applied to direct time integration process can be viewed as casting the difference equations at each time step to a nonphysical set of dynamic equations.

The preceding observations may open up a possibility for the program developers to implement both static and dynamics analysis capabilities within a single dynamics analysis module.

11.9. Recent Research Directions in Computational Dynamics

Since 1980 there have been three major research thrusts in transient analysis methods. They are parallel/concurrent algorithms [61–64], methods for constrained dynamical systems that involve contact–impact problems [65–73], adoption of domain decomposition and fractional/subcycling time marching techniques [74–76]. Of these, parallel algorithms appear to impact computational structural dynamics most significantly. Nevertheless, as dynamical systems to be solved are becoming more complex in such a way as to include time-varying boundary conditions—i.e., contact–impact conditions, internal constraints, and other coupling effects such as control, thermal field, and fluid interactions—development of suitable algorithms for such complex dynamical systems will continue to play an important role in computational dynamics. Research on domain decomposition and fractional stepping algorithms in itself may be considered an old profession; however, their intelligent union with parallel computing techniques may advance state-of-the-art transient analysis methods significantly.

11.10. Computer Implementation

A prerequisite for wide acceptance of a time integration module (for that matter, other analysis modules such as plasticity modeling modules) is a careful modular design and adherence to the modular software architecture by the methods engineer. The fruit of such a disciplined software development is the module's flexibility for accommodating different software interface conditions, i.e., its apparent transportability from one computing (or analysis) environment to another facility, and ease of software maintenance. Above all, from the method developer's viewpoint, the modular software architecture makes it relatively easier for introducing new methods (ideally they are more efficient and enhance modularity further) into an existing module. The widening time lag from the development of new analysis methods to their implementation into existing production-level computer programs is, the writer believes, partially due to current embedded-type implementation practices. A new software environment now advocated by NASA/Langley scientists under *Computational Mechanics Initiative* [77] appears to provide much of the necessary software modularity to methods developers [78]. It is hoped that an engineering software environment such as NICE system [79–81] will eventually be made available to most methods developers in computational dynamics.

11.11. Concluding Remarks

In this paper, several theoretical and practical aspects of the design and implementation of direct-time integration methods as a modular capability have been reviewed; for other types of reviews, consult [82–84]. So far, predominant research activities in computational structural dynamics have been directed toward improving algorithmic behavior of formulas and of solution procedures.

There is lack of both theory and systematic numerical experiments as to how to control error and stepsizes; hence, existing error and stepsize control strategies are mostly empirical. This is where systematic comparative studies of various strategies should provide useful information for enhancing solution reliability.

As is often the case for most maturing fields, little consideration was initially paid to implementation details and modularity requirements in general-purpose structural analysis programs. This is one area where a substantial pay-off can be achieved by a selective adoption of advances in computer software architecture.

To date, there virtually exists no black-box-type time integration package that is modular, reliable, transportable, and easy to use. The inhibiting factors for this include the embedded programming practice, the subservient role of transient structural analysis to static analysis in the past, the lack of adequate theories for the black-box integration module, and the delayed recognition by both the funding agencies and individual investigators of the importance of the software aspects and numerical experiments on par with the development of computational methods. Numerical experiments were often conducted only to demonstrate the validity of some methods and very rarely to simulate the thought processes of researchers. This is because the term *research* in computational mechanics has by-and-large been coined with methods developments, and a good portion of money expended on the software design and implementation efforts must still remain under the rug for a typical computational mechanics research program.

Finally, an interchangeability era of software and hardware processor boards is on the horizon. Given the present projection that dynamics response analyses will be increasingly important and be performed as often as static analyses, there is even a possibility of building dedicated transient analysis machines for structural dynamics. This does not, however, eliminate the need for the development of modular transient analysis software systems, as the design of hardware transient analysis processors will have to be based on the corresponding simulators (or software modules).

Acknowledgments. The present paper was prepared under support from NASA/Langley Research Center, Contract NAS1-17660. The author thanks Jeff Stroud and Jerry Housner of NASA/Langley for their interest in and encouragement of the present survey article.

References

1. T. Belytschko, J.R. Osias, and P.V. Marcal (eds.), *Finite Element Analysis of Transient Nonlinear Structural Behavior*, AMD Series, Vol. 14, ASME, New York, 1975.
2. J. Donea (ed.), *Advanced Structural Dynamics*, Applied Science, Essex, England, 1980.
3. T. Belytschko and T.J.R. Hughes (eds.), *Computational Methods for Transient Analysis*, North-Holland, Amsterdam, 1983.
4. W.K. Liu, T. Belytschko, and K.C. Park (eds.), *Innovative Methods for Nonlinear Problems*, Pineridge Press, Swansea, Great Britain, 1985.
5. K.C. Park, "Time integration of structural dynamics: A survey," in *Pressure Vessels and Piping Design Technology—A Decade of Progress*, ASME, New York, 1982, bk. 4.2.
6. P.S. Jensen, "Transient analysis of structures by stiffly stable methods," *Comput. & Structures*, **4** (1974), 67–94.
7. N.M. Newmark, "A method of computation for structural dynamics," *Proc. ASCE*, **85**, EM3 (1959), 00.67-94.
8. D.E. Johnson, "A proof of the stability of the Houbolt method," *AIAA J.*, **4** (1966), 1450–1451.
9. R.E. Nickell, "On the stability of approximation operators in problems of structural dynamics," *Internat. J. Solids and Structures*, **7** (1971), 301–319.
10. K.J. Bathe, and E.L. Wilson, "Stability and accuracy analysis of direct integration methods," *Internat. J. Earthquake Engrg. Struct. Dynamics*, **1** (1973), 283–291.
11. R.D. Krieg and S.W. Key, "Transient shell response by numerical time integration," *Internat. J. Numer. Methods Engrg.* **7** (1973), 273–286.
12. G.L. Goudreau and R.L. Taylor, "Evaluation of numerical methods in elastodynamics," *J. Comput. Methods Appl. Mech. Engrg.*, **2** (1973), 69–97.
13. J.H. Argyris, P.C. Dunne, and T. Angelopoulos, "Dynamic response by large step integration," *Internat. J. Earthquake Engrg. Struct. Dynamics*, **2** (1973), 185–203.
14. K.C. Park, "An improved stiffly stable method for direct integration of nonlinear structural dynamic equations," *J. Appl. Mech.*, **42** (1975), 464–470.
15. H.M. Hilber and T.J.R. Hughes, "Collocation, dissipation and overshoot for time integration scheme in structural dynamics," *Internat. J. Earthquake Engrg. Struct. Dynamics*, **6** (1978), 99–117.
16. G. Hall and J.M. Watt (eds.), *Modern Numerical Methods for Ordinary Differential Equations*, Clarendon Press, Oxford, 1976, p. 128.
17. G. Dahlquist, "A special stability problem for linear multi-step methods," *BIT*, **3** (1963), 27–43.
18. P.S. Jensen, "Stiffly stable methods for undamped second-order equations of motion," *SIAM J. Numer. Anal.*, **13** (1976), 549–563.
19. K.C. Park, "Practical aspects of numerical time integration," *Comput. & Structures*, **7** (1977), 343–353.
20. K.C. Park, "Evaluating time integration methods for nonlinear dynamics analysis," in *Finite Element Analysis of Transient Nonlinear Structural Behavior* (T. Belytschko *et al.*, ed.), ASME Applied Mechanics Symposia AMD, Vol. 14, 1975, pp. 35–58.
21. W.H. Enright, T.E. Hull, and B. Lindberg, "Comparing numerical methods for stiff systems of O.D.E.'s," *BIT*, **15** (1975), 10–48.
22. R.K. Brayton, F.G. Gustavson, and G.D. Hachtel, "A new efficient algorithm for

solving differential-algebraic systems using implicit backward differentiation formulas," *Proc. IEEE*, **60** (1972), 98–108.

23. C.W. Gear, "The automatic integration of stiff ordinary differential equations," *Proceedings of the IFIP Congress*, 1969, pp. 187–193.
24. J.A. Stricklin, J.E. Martinez, J.R. Tillerson, J.R. Hong, and W.E. Haisler, "Nonlinear dynamic analysis of shells of revolution by the matrix displacement method," *AIAA J.*, **9** (1971), 629–636.
25. R.W.H. Wu and E.A. Witmer, "Nonlinear transient responses of structures by the spatial finite-element method," *AIAA J.* **11** (1973), 1087–1104.
26. J.F. McNamara, "Solution schemes for problems of nonlinear structural dynamics," *J. Pressure Vessels Tech. ASME*, **96** (1974), 96–102.
27. W.H. Enright and T.E. Hull, "Test results on initial value methods for nonstiff ordinary differential equations," *SIAM J. Numer. Anal.*, **13** (1976), 944–961.
28. K.C. Park and P.G. Underwood, "A variable-step central difference method for structural dynamics analysis: Theoretical aspects," *Comput. Methods Appl. Mech. Engrg.*, **22** (1980), 241–258.
29. P.G. Underwood and K.C. Park, "A variable-step central difference method for structural dynamics analysis: Implementation and performance evaluation," *Comput. Methods Appl. Mech. Engrg.*, **23** (1980), 259–279.
30. B. Lindberg, "Error estimates and stepsize strategy for the implicit midpoint rule with smoothing and extrapolation," Report NA 7259, Department of Information Processing, Royal Institute of Technology, Stockholm, 1972.
31. H.D. Hibbit and B.I. Karlsson, "Analysis of pipe whip," Paper 79-PVP-122, ASME, New York, 1979.
32. T. Belytschko and D.F. Schoeberle, "On the unconditional stability of an implicit algorithm for nonlinear structural dynamics," *J. Appl. Mech.*, **97** (1975), 865–869.
33. V. Blæs, "Zur angenaherten Lösung gewöhnlicher Differentialgleichungen," *VDI-Z*, **81** (1937), 587–596.
34. B. Lindberg, "Characterization of optimal stepsize sequences for methods for stiff differential equations," *SIAM J. Numer. Anal.*, **14** (1977), 859–887.
35. K.C. Park and P.G. Underwood, "An adaptive direct time integration package for structural dynamics analysis: Design considerations," in preparation.
36. C.A. Felippa and K.C. Park, "Computational aspects of time integration procedures: Implementation," *J. Appl. Math. Mech.*, **45** (1978), 595–602.
37. J.C. Houbolt, "A recurrence matrix solution for the dynamic response of an elastic aircraft," *J. Aeronaut. Sci.*, **17** (1950), 540–550.
38. C.A. Felippa, "Procedures for computer analysis of large nonlinear structural systems, in *Large Engineering Systems* (A. Wexler, ed.), Pergamon, Oxford, 1977, pp. 60–101.
39. C.A. Felippa, "Solution of nonlinear static equations," in *Large Deflection and Stability of Structures* (K.J. Bathe, ed.), North-Holland, Amsterdam, 1986.
40. D. Bushnell, "A strategy for the solution of problems involving large deflections, plasticity and creeps," *Internat. J. Numer. Methods Engrg.*, **11** (1977), 683–708.
41. H.L. Schreyer, R.F. Kulak, and J.M. Kramer, "Accurate numerical solutions for elastic-plastic models," Preprint ASME Paper No. 79-PVP-107.
42. P.V. Marcal, "A stiffness method for elastic-plastic problems," *Internat. J. Mech. Sci.*, (1965), 229–238.
43. J.C. Nagtegaal, D.M. Parks, and J.R. Rice, "On numerically accurate finite

element solutions in the fully plastic range," *Comput. Methods Appl. Mech. Engrg.*, **4** (1974), 153–177.

44. K.C. Park, "Partitioned transient analysis procedures for coupled-field problems: Stability analysis," *J. Appl. Math. Mech.*, **47**, No. 2 (1980), 370–376.
45. T. Belytschko and R. Mullen, "Mesh partitions of explicit–implicit time integration," in *Formulations and Computational Algorithms in Finite Element Analysis*, (J. Bathe *et al.* ed.), MIT Press, Cambridge, MA, 1977.
46. T. Belytschko and R. Mullen, "Stability of explicit–implicit mesh partitions in time integration," *Internat. J. Numer. Methods Engrg.*, **12** (1978), 1575–1586.
47. T.J.R. Hughes and W.K. Liu, "Implicit–explicit finite elements in transient analysis: Stability theory," *J. Appl. Math. Mech.*, **45** (1978), 375–378.
48. K.C. Park, C.A. Felippa, and J.A. DeRuntz, "Stabilization of staggered solution procedures for fluid-structure interaction analysis," in *Computational Methods for Fluid-Structure Interaction Problems* (T. Belytschko and T.L. Geers, eds.), ASME Applied Mechanics Symposia, AMD Vol. 26, ASME, New York, 1977, pp. 94–124.
49. T. Belytschko, H.-J. Yen, and R. Mullen, "Mixed methods for time integration," *Comput. Methods Appl. Mech. Engrg.*, **17** (1979), 259–275.
50. O.C. Zienkiewicz, E. Hinton, K.H. Leung, and R.E. Taylor, "Staggered time marching schemes in dynamic soil analysis and a selective explicit extrapolation algorithm," in *Innovative Numerical Analysis for the Applied Engineering Science* (R. Shaw *et al.* eds.), The University of Virginia Press, Charlottesville, 1980, pp. 525–530.
51. K.C. Park and J.M. Housner, "Semi-explicit transient analysis procedures for structural dynamics problems," *Internat. J. Numer. Methods Engrg.*, **18** (1982), 609–622.
52. D.M. Trujillo, "An unconditionally stable explicit algorithm for structural dynamics," *Internat. J. Numer. Methods Engrg.*, **11** (1977), 1579–1592.
53. J.E. Dennis Jr. and J.J. More, "Quasi-Newton methods, motivation and theory," *SIAM Rev.*, **19** (1977), 46–89.
54. N.N. Yanenko, *The Method of Fractional Steps*, Springer-Verlag, New York, 1971.
55. H. Mathies and G. Strang, "The solution of nonlinear finite element equations," *Internat. J. Numer. Methods Engrg.*, **14** (1979), 1613–1626.
56. M. Geradin and M. Hogge, "Quasi-Newton iteration in nonlinear structural dynamics," *Proc. SMiRT-5*, Paper No. M7/1, Berlin, August 1979.
57. A.N. Konovalov, "Application of the method of splitting to the dynamic solution of problems of the theory of elasticity," *U.S.S.R. Comput. Math.*, **4** (1964), 192–198.
58. S.P. Frankel, "Convergence rates of iterative treatments of partial differential equations," *Math. Tables, Aids to Comput.*, **4** (1950), 65–70.
59. J.S. Brew and D.M. Brotton, "Nonlinear structural analysis by dynamic relaxation," *Internat. J. Numer. Methods Engrg.*, **3** (1971), 463–483.
60. P.G. Underwood, "Dynamic relaxation," in *Computational Methods for Transient Dynamic Analysis* (T. Belytschko and T.J.R. Hughes, eds.), North-Holland, Amsterdam, 1983, Chapter 5.
61. G.S. Patterson, "Large scale scientific computing: future directions," *Comput. Phys. Comm.*, **26** (1982), 217–225.
62. H. Allik, J. Crawther, J. Goodhue, S. Moor, and R. Thomas, "Implementation of

the finite element methods on the butterfly parallel processor," *BBN Lab. Inc.*, Cambridge, MA, July 1985.

63. G.A. Lyzenga, A. Raefsky, and B.H. Hager, "Finite elements and the method of conjugate gradient on a concurrent processor," a CalTech Preprint, 1985.
64. B. Nour-Omid and K.C. Park, "Solving structural mechanics problems on the CalTech hypercube machine," *Comput. Methods Appl. Mech. Engrg.*, **61**, (1981), 161–176.
65. P. Lötstedt and L.R. Petzold, "Numerical solution of nonlinear differential Equations with Algebraic Constraints," Sandia National Laboratories, Report SAND 83-8877, Livermore, CA, 1983.
66. R.A. Wehage and E.J. Haug, "Generalized coordinate partitioning for dimension reduction in analysis of constrained dynamic systems," *ASME J. Mech. Design*, **104** (1982), 247–255.
67. P. Lötstedt, "On a penalty function method for the simulation of mechanical systems subject to constraints," TRITA-NA-7919, 1979, Royal Institute of Technology, Stockholm.
68. C. Fuehrer and O. Wallrapp, "A computer-oriented method for reducing linearized multibody system equations by incorporating constraints," *Comput. Methods Appl. Mech. Engrg.*, **46** (1984), 169–175.
69. P.E. Nikravesh, "Some methods for dynamic analysis of constrained mechanical systems: a survey," in *Computer Aided Analysis and Optimization of Mechanical Systems Dynamics* (E.J. Haug, ed.), NATO ASI series, F9, Springer-Verlag, Berlin, 1984, pp. 351–367.
70. K.C. Park, "Stabilization of computational procedures for constrained dynamical system: Formulation," AIAA Paper No. AIAA-86-0926-CP, presented at the 27th SDM Conference, San Antonio, Texas, May 1986.
71. N. Kikuchi and J.T. Oden, "Contact problems in elastostatics," in *Finite Elements: Special Problems in Solid Mechanics* (J.T. Oden and G. Carey, eds.), Prentice-Hall, Englewood Cliffs, NJ, 1984.
72. B. Nour-Omid and P. Wriggers, "Solution methods for contact problems," *Comput. Methods Appl. Mech. Engrg.*, **52** (1983), 31–44.
73. J.T. Oden, "Exterior penalty methods for contact problems in elasticity," in *Nonlinear Finite Element Analysis in Structural Mechanics* (W. Wunderlich, P. Stein, and K.J. Bathe, eds.), Springer-Verlag, Berlin, 1981.
74. T.J.R. Hughes, I. Levit, and J.M. Winget, "An element-by-element solution algorithm for problems of structural and solid mechanics," *Comput. Methods Appl. Mech. Engrg.*, **36** (1983), 223–239.
75. T. Belytschko, P. Smolinski, and W.-K. Liu, "Stability of multi-step partitioned integrators for first-order finite element systems," *Comput. Methods Appl. Mech. Engrg.*, **52** (1985), 281–298.
76. M. Ortiz and B. Nour-Omid, "Unconditionally stable concurrent procedures for transient finite element analysis," *Comput. Methods Appl. Mech. Engrg.* (1986), to appear.
77. N.F. Knight and W.J. Stroud, "Computational structural mechanics: A new activity at the NASA Langley Research Center," NASA Tech. Mem. 87612, Hampton, VA, 1985.
78. C.G. Lotts (*compiler*), "Introduction to the CSM Testbed: NICE/SPAR," NASA/Langley Research Center, June 1986.
79. C.A. Felippa, "Architecture of a distributed analysis network for computational

mechanics," *Comput. & Structures*, **13** (1981), 405–413.

80. C.A. Felippa, "A command language for applied mechanics processors," 1–3, LMSC-D78511, Lockheed Palo Research Laboratory, Palo Alto, CA, 1983.
81. C.A. Felippa and G.M. Stanley, "NICE: A utility architecture for computational mechanics," in *Finite Elements for Nonlinear Problems* (P.G. Bergan, K.J. Bathe, and W. Wunderlich, eds.), Springer-Verlag, New York, 1986, pp. 447–463.
82. T. Belytschko, "Transient analysis," in *Structural Mechanics Computer Programs* (W. Pilkey *et al.* eds.), The University Press of Virginia, Charlottesville, 1974, pp. 255–276.
83. C.A. Felippa and K.C. Park, "Direct time integration methods in nonlinear structural dynamics," *Comput. Methods Appl. Mech. Engrg.*, **17/18** (1979), 277–313.
84. T.J.R. Hughes and T. Belytschko, "A précis of developments in computational methods for transient analysis," *J. Appl. Math. Mech.*, **50** (1985), 1033–1041.

CHAPTER 12

Adaptive Finite Element Methods for Problems in Solid and Fluid Mechanics

J. Tinsley Oden

12.1. Introduction

In this paper, a general class of h methods for adaptive finite element approximations is described that has been applied successfully to three classes of problems in solid and fluid mechanics:

(1) supersonic flow of compressible gases in two-dimensional domains;
(2) flow interaction in supersonic rotor–stator simulations;
(3) quasi-static problems in viscoplasticity of metals under elevated temperatures.

After this section, we describe features of the adaptive method for two-dimensional meshes of quadrilaterals and triangles. This is followed by discussions of algorithms for each class of problems and by results of numerical experiments.

12.2. h Methods for Two-Dimensional Problems

As is well known, the purpose of an adaptive finite element method (FEM) is to change the structure of a finite element approximation so as to improve the quality of the solution. This concept naturally entails two basic issues: the dynamic assessment of the quality of the solution, by which we generally mean the a posteriori estimation of local errors in appropriate norms, and the change of the structure of the approximation itself, by which we generally mean the refinement of the mesh (an h method), the enrichment of the local shape functions to produce higher accuracy (p methods or spectral methods), a redistribution of nodal points (a moving mesh or r method), or a combination of these strategies. We focus on the latter issue (adaptive improvement) here. For a survey of the literature on these methods, see [7]. Although we have developed and tested methods in each of these categories (see [8]), the present discussion focuses on h methods for classes of two-dimensional problems in nonlinear solid and fluid mechanics.

12.2.1. Steady Problems—A Refinement/Unrefinement Scheme

Consider a finite element mesh consisting of four-node quadrilateral elements or triangular elements over which the local fields are approximated by (isoparametric images of) bilinear or linear shape functions. Suppose that an initial coarse mesh that has a sufficient number of elements to adequately represent the geometry of the domain on which the problem is defined is generated. Suppose that a trial solution of the problem is obtained on the initial mesh and that it is used to compute local error estimators θ_e, $1 \leq e \leq E_0$ over each element.

The question that then arises is one of data management and algorithm structure: how can we refine the mesh so as to reduce errors below preassigned limits, while coping with changes in element numbering, node numbering, element connectivity, band width size, and storage requirements? Also, if the error is "small," can the mesh be "unrefined"; i.e., can fewer elements be used in the approximation? The following algorithm resolves these questions.

12.3. Refinement/Unrefinement Algorithm

The scheme employs successive bisection to refine a mesh of quadrilaterals and a collapse of groups of four bisected quads to unrefine. All refinement/unrefinement steps are followed up to various hierarchical "levels," the number being specified a priori by the analyst.

(1) A "premesh" is generated, not necessarily the "initial" mesh, which consists of element *groups* which contain four subelements. These groups can be further bisected several times to obtain the so-called initial -0-level mesh. The meshes of sequentially larger elements define levels 1, 2, 3, etc., until the lowest level is attained, and this is the original premesh level. This is then the coarsest mesh size the calculation will admit, no matter how small the error. It could conceivably consist of only one element.

(2) An initial numerical solution is obtained on the 0-level initial mesh, and error indicators θ_e are computed over each element, $1 \leq e \leq E_0$. Then the maximum error indicator is calculated:

$$\theta^0_{\text{MAX}} = \max_{1 \leq e \leq E_0} \theta_e.$$

(3) The -1-level of element groups is scanned and the group error indicators are computed:

$$\theta^k_{\text{GROUP}} = \sum_{j=1}^{4} \theta_{ej}, \qquad k = 1, 2, \ldots P.$$

Here, P is the number of 1-level groups.

(4) Error tolerances are defined by real numbers, α and β; $0 \leq \alpha, \beta \leq 1$. If

$$\theta_e \geq \beta \theta^0_{\text{MAX}}$$

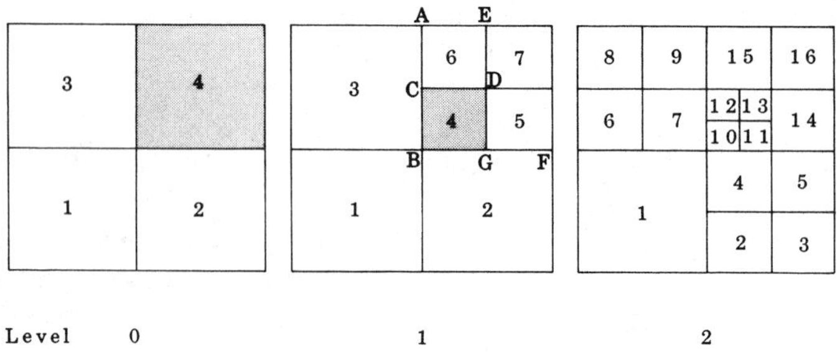

FIGURE 12.1. Two typical refinement steps for subdivision of bilinear elements.

element e is refined (bisected). If

$$\theta^k_{\mathrm{GROUP}} \geq \alpha\theta_{\mathrm{MAX}},$$

we *unrefine* group k by replacing this group with a single element at the 1-level. After refinement and unrefinement, a new mesh containing E_1-elements is obtained.

(5) The problem is resolved on the new mesh, and a new set of error indicators θ_e, $1 \leq e \leq E_1$, is computed.

(6) Go to 2). (The process is continued until errors are below the assigned tolerance throughout the mesh.)

The data management scheme is crucial in the implementation of this algorithm. Two steps in our refinement procedure are illustrated in Figure 12.1. The element that is shaded in the figure is that in a group of four for which $\theta_4 \geq \alpha\theta_{\mathrm{MAX}}$. Element four is thus bisected, as shown, but only one "active" node is actually added to the mesh (node D). The value of the solution at node C is contrained to be the average of the values at the endpoints of that sides; e.g.,

$$u_C = \tfrac{1}{2}(u_A + u_B).$$

We do not allow more than one node on any side of an element. Thus, if element four in level one must still be refined, it is necessary also to bisect elements two and three. All midside nodes are still "constrained" nodes; they support values that are averages of the values at the ends of each element side. *The degrees of freedom of an element are thus not necessarily the values of the solution at the element nodes, but are only the active nodes associated with the element.* For example, for element four in level one in Figure 12.1, the degrees of freedom are the nodal values at nodes A, B, D, and F (not B, C, D, G).

The dynamic numbering scheme is handled by introducing several connectivity arrays that handle the node and labeling issue locally over each element. After a mesh refinement/unrefinement loop has been completed, each element in the current mesh structure is assigned an element number NEL = 1, 2, ..., NELM and to each global node a label NODE. A Boolean array

NODES(J, NEL) ($J = 1, 2, 3, 4$ for a quadrilateral) relates the local node number J to the labels NODE. The global coordinates X_J, Y_J of J are also filed in two arrays, NODES(J, NEL) and $X(K, L) =$ the array of K coordinates of global node L, $K = 1$ or 2.

The arrays NODES($\cdot$, $\cdot$), $X(\cdot, \cdot)$ and another pair of arrays NELCON($\cdot$, $\cdot$) and LEVEL($\cdot$) (where LEVEL monitors the refinement level attained) provide for a dynamic renumbering of elements and nodes. For example, if element four, level 2, of Figure 12.1 is to be refined, we run through the following steps:

(1) Loop over the neighbors of element four (which is made possible with the NELCON array); check the level of the neighboring elements in relation to the level of element four.
(2) If any neighboring element has a level lower than 4, then the element cannot be refined at this state.
(3) If 4 can be refined (as is the case in Figure 12.1), we generate new element numbers (thus changing NELEM and new node numbers for unconstrained nodes).
(4) Compute the connectivity matrix NELCON for the new elements.
(5) Adapt the connectivity matrices for the neighboring elements (since the refinement of 4 has now changed this connectivity).
(6) Interpolate the solution between the unconstrained nodes.

It is clear that some strategy is needed to test whether a designated element is appropriately connected for a refinement to take place.

The following subroutine determines refinements prerequisite to referring an element A:

SUBROUTINE DIVIDE(NEL1, NEL2)
NEL1 = the input element that needs to be refined

$$\text{NEL2} = \text{output element} = \begin{cases} \text{NEL1 if NE11 has been divided; or} \\ \text{NELD} = \text{element that needs to be divided prior to NEL1.} \end{cases}$$

Then, symbolically, we have

```
Repeat
NEL1 = A
CALL DIVIDE(NEL1, NEL2)
WRITE(NEL2.NE.NEL1)
NEL1 = NEL2
CALL DIVIDE(NEL1, NEL2)
END WRITE
UNTIL(NEL2.EQ.A)
```

For triangular elements, we employ one of two subdivision techniques. In the first method, elements with errors $\theta_e \geq \alpha\theta_{\text{MAX}}$ are subdivided by inserting

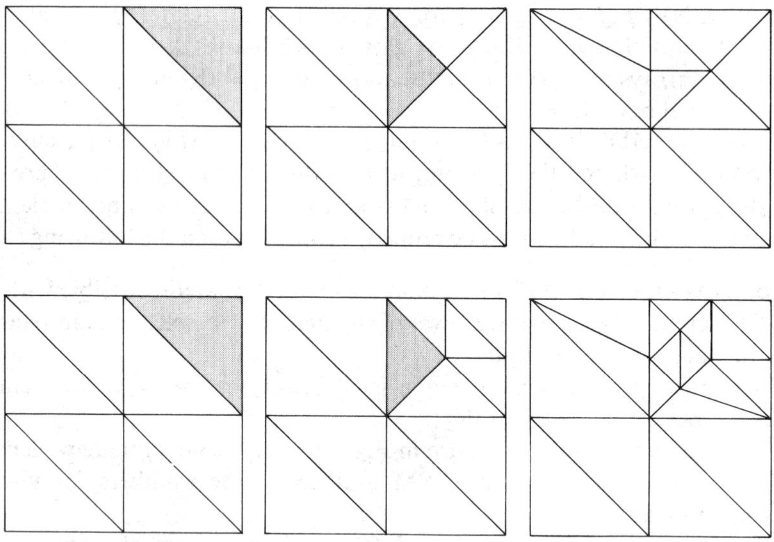

FIGURE 12.2. Two methods for successive refinement on triangular meshes.

a node at the midpoint of the longest side and connecting this point to the opposite vertex. Since no nodes are constrained in this case, a search is made for all elements that contain a midside node. These elements are also refined by connecting midside nodes to an opposite vertex. The second strategy consists of inscribing a triangle inside elements where $\theta_e \geq \alpha\theta_{\text{MAX}}$ by the addition of three unconstrained nodes that are then attached to the opposite vertices of bounding elements. A series of refinements by each method is illustrated in Figure 12.2.

12.4. Refinement/Unrefinement—Transient Problems

For time-dependent problems, a refinement/unrefinement strategy similar to those described must be implemented over a partition of the time domain as well. In calculations to be described later, the following algorithms were employed:

(a) Advance the solution N time steps.
(b) Perform the following steps until no more elements can be refined:
  (1) Compute the element error indicators θ_e.
  (2) Refine all elements with $\theta_e \geq \alpha\theta_{\text{MAX}}$
  (3) Integrate the last N time steps with the updated (refined) mesh.
  (4) Go to (1).
(c) Compute the element error indicators θ_e and unrefine all groups with

$$\theta^{\text{GROUP}} \leq \beta\theta_{\text{MAX}}.$$

(d) Go to step (a).

We note that the "do loop" in step (b) converges when no more elements can be refined (the maximum level of refinement is fixed). Although the iteration in step (b) guarantees a fully updated mesh it may lead to an expensive scheme if more than a few passes are required for convergence of the "do loop." A cheaper alternative, which we used in the calculations of the rotor–stator problem, is the following:

(a) Advance the solution N time steps.
(b) Compute the element error indicators θ_e.
(c) Refine all elements with $\theta_e \geq \alpha\theta_{\mathrm{MAX}}$.
(d) Integrate the last N time steps with the refined mesh obtained in (c).
(e) Compute the element error indicator θ_e and
  (1) unrefine all element groups with

$$\theta_m^{\mathrm{GROUP}} \leq \beta\theta_{\mathrm{MAX}}.$$

  (2) refine all elements with

$$\theta_e \geq \alpha\theta_{\mathrm{MAX}}.$$

(f) Go to step (a).

12.5. Algorithms for Problems in Solid and Fluid Mechanics

As noted in the introduction, we are interested here in applications of the h methods to three types of problems. We shall now outline briefly the finite element algorithms that are used to advance the solutions in time.

12.5.1. Supersonic Flow in Domains with Moving Bodies

We are interested in supersonic flow of ideal inviscid compressible gases in domains in which two or more bodies are in motion in relation to one another. This class of problems can be characterized by a space–time variational principle in which the dependent variable is the four-vector of conservation variables.

$$\mathbf{U}^T = (\rho, m_1, m_2, e) = \mathbf{U}^T(\mathbf{x}, t),$$

and the flux is the Euler flux

$$\mathbf{Q}(\mathbf{U}) = \begin{bmatrix} \mathbf{m} \\ \rho^{-1}\mathbf{m} \times \mathbf{m} + p(\mathbf{u})\mathbf{1} \\ \rho^{-1}\mathbf{m}(e + p(\mathbf{u})) \end{bmatrix}$$

appears. Here ρ is the mass density, $\mathbf{m}$ is the momentum vector, e the energy, p the thermodynamic pressure,

$$p(\mathbf{U}) = (\gamma - 1)\left(e - \rho^{-1}\mathbf{m} \otimes \frac{\mathbf{m}}{2}\right)$$

with γ the ratio of specific heats.

The Euler equations, Rankine–Hugoniot jump conditions, initial conditions, and boundary conditions for the dynamic behavior of a gas over a time interval $[\tau_1, \tau_2]$ are governed by the following weak formulation:

Find $\mathbf{U} = \mathbf{U}(\mathbf{x}, t) \in \mathbf{V}$ such that

$$-\int_{t_1}^{t_2}\int_{\Omega(t)} \mathbf{U}^T\phi_t\, d\Omega\, dt + \int_{\Omega(\tau_2)} \mathbf{U}^T(\mathbf{x}, \tau_2)\phi(\mathbf{x}, \tau_2)\, d\Omega$$
$$- \int_{\Omega(\tau_1)} \mathbf{U}^T(\mathbf{x}, \tau_1)\phi(\mathbf{x}_1, \tau_1)\, d\Omega$$
$$= -\int_{\tau_1}^{\tau_2}\int_{\Omega(t)} \mathbf{Q}(\mathbf{U}): \nabla\phi\, d\Omega\, dt - \int_{\tau_1}^{\tau_2}\int_{\Omega(t)} \phi^T(\mathbf{Q}(\mathbf{U}) - \mathbf{U}\otimes\mathbf{u}^G)\mathbf{n}\, ds\, dt$$
$$\text{for all} \quad \phi \in \mathbf{W}. \qquad (5.1)$$

Here, $\phi = \{\phi_1, \phi_2, \phi_3, \phi_4\}^T$, $d\Omega = dx$, $\phi_\alpha = \phi_\alpha(\mathbf{x}, t)\mathbf{x} \in \Omega(t)$, $a = 1, 2, 3, 4$ and $\phi_\tau = \partial\phi/\partial t$.

Here, $\mathbf{u}^G$ is the *grid velocity*: the velocity of a coordinate system moving in relation to a fixed spatial coordinate frame $\mathbf{x}$. If $\mathbf{u}^G = 0$, we obtain the standard Eulerian description of the motion of the gas. If $\mathbf{u}^G = \mathbf{U} = \mathbf{m}/\rho$, we have a Lagrangian description of the motion.

Consider, in particular, the situation illustrated in Figure 12.3, where two rigid bodies are moving in relation to one another with velocities $\mathbf{v}_1$, $\mathbf{v}_2$, respectively. A mesh of quadrilateral elements is generated around each body, and those meshes meet along a mesh interface Γ. To provide for simultaneous continuous flow across the interface and the smooth sliding of one element

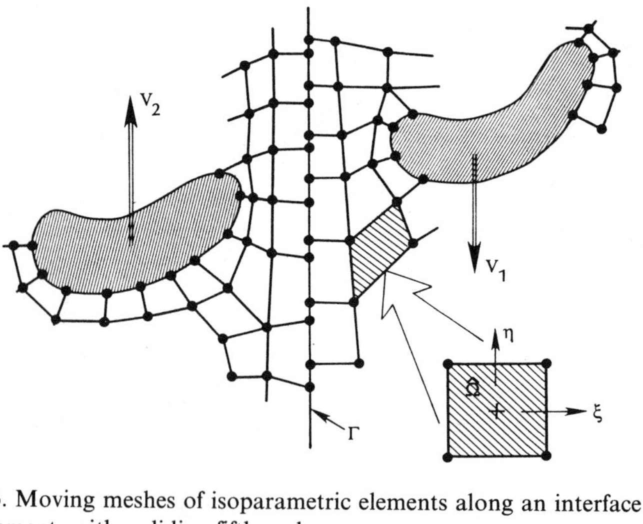

FIGURE 12.3. Moving meshes of isoparametric elements along an interface of special five-node elements with a sliding fifth node.

with respect to another, a special five-node interface element is used. This element is introduced in [6], where a more detailed discussion of its properties can be found.

The shape functions Ψ_j associated with a moving element Ω_e are images of bilinear functions $\Psi_j(\zeta, \eta)$ defined on a master element Ω. The moving local coordinates on a typical element are then given by the isoparametric map, $\mathbf{x} = \sum_j^4 x^j \Psi_j(\zeta, \eta)$.

An important property of the shape functions can be noted at this point. Since the Ψ_j assume values of one or zero at nodes, their value at any point in the moving distorted element must remain constant. The material time derivative of the shape functions is, thus, zero:

$$\frac{D\Psi_j}{Dt} = 0 = \frac{\partial}{\partial t}\bigg|_{\mathbf{x}\text{ fixed}} \Psi_j(\mathbf{x}) + \nabla_x \Psi_j \frac{\partial x}{\partial t}.$$

Here, $\nabla_x \Psi_j = [\partial \Psi_j / \partial x_k]$ and $\partial \mathbf{x}/\partial t$ is the grid velocity, $\mathbf{u}^G = \partial \mathbf{x}/\partial t$ (with $\mathbf{u}^G = \mathbf{v}_1$ for mesh 1 and $\mathbf{u}^G = \mathbf{v}_2$ for mesh 2). Thus,

$$\frac{\partial \Psi_j(\mathbf{x})}{\partial t} = -\mathbf{u}^G \cdot \nabla_x \Psi_j(\mathbf{x}).$$

Let U_α, $\alpha = 1, 2, 3, 4$, denote the αth component of the vector of conservation variables $\mathbf{U}$. The local finite element approximation of U_α over element Ω is a linear combination of the local shape functions of the form

$$U_\alpha^h(\mathbf{x}, t) = \sum_{j=1}^{N} U_\alpha^j(t) \Psi_j(\mathbf{x}),$$

where $U_\alpha^j(t) = U_\alpha^h(b_j(t))$ are the nodal values of U_α^h (at node b_j) at time t. Likewise, the discrete test functions are of a similar form. In strictly a formal way, the finite element approximation of the flow through either moving mesh is obtained from the weak statement (5.1) of the conservation laws, by interpreting Ω as a moving quadrilateral element, replacing $\mathbf{U}$ by the discrete approximation $\mathbf{U}^h$, and replacing the test functions ϕ by the discrete test functions ϕ_h. Introducing the local approximation of $\mathbf{U}^h$ and choosing an appropriate numerical integration rule to evaluate the time integral yield a local finite element model of the conservation laws.

A Taylor–Galerkin/Lax–Wendroff scheme is derived by using the midpoint integration rule:

$$\int_{\tau_1}^{\tau_2} f(t)\, dt = (\tau_2 - \tau_1) f\left[\frac{(\tau_1 + \tau_2)}{2}\right].$$

We partition the time interval of interest into a finite number of time steps,

$$0 = t_0 < t_1 < t_2 < \cdots < t_N = T,$$

and recognize that

$$\phi_t^h = -\mathbf{u}^G \cdot \nabla_x \phi^h.$$

Applying (5.1) (for the discrete fields) to a typical time interval $[t_n, t_{n+1}]$ and to all elements Ω_e, $e = 1, 2, \ldots, E$, and adding the resulting equations, we obtain the following global system of equations that determines $\mathbf{U}_h^{n+1}$:

$$\int_{\Omega^{n+1}} \phi^{hT}\mathbf{U}_h^{n+1}\, d\Omega = \int_{\Omega^n} \phi^{hT}\mathbf{U}_h^n\, d\Omega + \int_{\Omega^{n+1/2}} \mathbf{Q}(\mathbf{U}_h^{n+1/2}) : \nabla_x\phi^h\, d\Omega$$

$$- \Delta t \int_{\Omega^{n+1/2}} \mathbf{u}^G \otimes \mathbf{U}_h^T : \nabla_x\phi^h\, d\Omega - \Sigma_h$$

for all test functions ϕ^h.

Here,

$$\mathbf{U}_h^n = \mathbf{U}^h(\mathbf{x}, t_n); \qquad \Omega^n = \Omega(t_n); \qquad \mathbf{u}^G \cdot \mathbf{U}_h^T : \nabla_x\phi^h = \sum\sum u_i^G U_\alpha^h \frac{\partial\phi_\alpha^h}{\partial x_i},$$

where Σ_h denotes the boundary term,

$$\Sigma_h = \int_{\partial\Omega^{n+1/2}} [(\mathbf{Q}(\mathbf{U}_h^{n+1/2}) - \mathbf{u}^G \otimes \mathbf{U}_h^{n+1/2})\mathbf{n}]^T \phi^n\, dS.$$

The flux at $t_n + \Delta t/2$ is constant over each element and is computed from the average element quantities $\mathbf{U}_e^{n+1/2}$, which are calculated from,

$$A_e^{n+1}\mathbf{U}_e^{n+1/2} = \int_{\Omega_e} U^n\, d\Omega - \frac{\Delta t}{2}\left(\frac{A_e^{n+1}}{A_e^{n+1/2}}\right)\int_{\Omega_e} \operatorname{div} \mathbf{Q}(\mathbf{U}^n)\, d\Omega$$

$$+ \Delta t \int_{\partial\Omega_e} \mathbf{u}^G \otimes \mathbf{U}_h^n \mathbf{n}\, ds,$$

where $\mathbf{U}_e^{n+1/2}$ is an average of $\mathbf{U}^h$ over Ω_e and A_e^n is the area of the element at time t_n. The fluxes $\mathbf{Q}^n$ are computed by using the "group" approximation (i.e., $\mathbf{Q}^h$ is represented as a bilinear finite element interpolant using discrete values at the element nodes). In addition, artificial viscosity terms can be added to dampen possible spurious oscillations in the solution. The resulting moving mesh algorithm is, thus, a two-step scheme:

First step. For each element Ω_e, calculate a *constant* element vector $\mathbf{U}_{\alpha e}^{n+1/2}$ from

$$A_e^{n+1}U_{\alpha e}^{n+1/2} = \sum_{i=1}^{4}\left\{\left(\int_{\Omega^n} \phi_i\, d\Omega\right)U_\alpha^{i,n} - \frac{\Delta t}{2}\frac{A_e^{n+1}}{A_e^{n+1/2}}\right.$$

$$\left.\times\left(\int_{\Omega_e^{n+1/2}} \frac{\partial\phi_i}{\partial x_\beta}\, d\Omega\right)Q_{\alpha,\beta}^{i,n}\right\}$$

$$+ \sum_{i=1}\sum_{j=1} \Delta t \int_{\Omega_e^{n+1/2}} \frac{\partial(\phi_i\phi_j)}{\partial x_\beta}\, d\Omega\, u_\beta^{i,G}U_\alpha^{j,n}.$$

Second step. For each node calculate $U_\alpha^{j,n}$ by solving the following system of equations:

$$\sum_{j=1}^{N}\left(\int_{\Omega^{n+1}}\phi_i\phi_j\,d\Omega\right)U_\alpha^{j,n}=\sum_{j=1}^{N}\left(\int_{\Omega^n}\phi_i\phi_j\,d\Omega\right)U_\alpha^{j,n}$$
$$+\Delta t\int_{\Omega^{n+1}}Q_{\alpha\beta}^{n+1/2}\frac{\partial\phi_i}{\partial x_\beta}\,d\Omega$$
$$-\Delta t\int_{\partial\Omega^{n+1}}n_\beta(\bar{Q}_{ab}^{n+1/2}-\bar{Q}_{\alpha\beta}^{n})\phi_i\,dS$$
$$-\Delta t\int_{\partial\Omega^{n+1}}n_\beta\bar{Q}_{\alpha\beta}\phi_i\,ds.$$

Here $\bar{Q}_{\alpha\beta}^n$ denotes an elementwise averages value of the flux, and $\bar{Q}_{\alpha\beta}$ is defined by

$$\bar{Q}_{\alpha\beta}=\bar{Q}_{\alpha\beta}-u_\beta^G U_\alpha.$$

This scheme is used in conjunction with the adaptive strategy described earlier to produce a dynamic adaptive method for gas dynamics problems with moving boundaries.

12.5.2. Rate-Dependent Plasticity

Next we consider a very complex class of problems in nonlinear solid mechanics: the quasi-static behavior of a viscoplastic material under cyclic loading. These types of problems are especially difficult to handle with adaptive methods since small changes in stress can create dramatic changes in the response of the material.

We shall consider so-called internal state variable models of viscoplasticity, governed by constitutive equations of the form

$$\begin{aligned}\dot{\sigma}_{ij}&=E_{ijkl}\dot{\varepsilon}_{kl}^e,\\ \dot{\varepsilon}_{ij}^p&=f_{ij}(\sigma_{kl},z_m),\\ \dot{z}_i&=g_i(\sigma_{kl},z_m),\end{aligned}\tag{5.2}$$

where $\dot{\sigma}_{ij}$ is the Cauchy stress rate, E_{ijkl} the elastic constants, $\dot{\varepsilon}_{kl}^e$ the elastic part of the strain rate tensor, $\dot{\varepsilon}_{ij}^p$ the plastic strain rate, z_i the internal state variable (here a vector), and $f_{ij}(\cdot)$, $g_i(\cdot)$ given constitutive functionals. Each field is a function $\mathbf{x}$ and time t, $\mathbf{x}\in\Omega\subset\mathbb{R}^2$, $t\in[0,T]$. The total strain rate is then

$$\dot{\varepsilon}_{ij}=\tfrac{1}{2}(\dot{u}_{i,j}+\dot{u}_{j,i}),$$

where u_{ij} are the local velocity gradients, $(u_{i,j}=\partial u_i/\partial x_j)$.

The motion of the body is governed by the linear momentum equation,

$$\frac{\partial}{\partial x_j}[E_{ijkl}(\dot{u}_{k,l} - \dot{\varepsilon}^p_{ij})] = f_i,$$

where f_i are components of body force rate. Here $\dot{\varepsilon}^p_{ij} = f_{kl}(\sigma, \mathbf{z})$ and $\mathbf{z}$ is obtained from the evolution equation for the internal variable $\mathbf{z}$. On a portion $\partial\Omega_2$ we have a prescribed traction rate $T_i = \dot{\sigma}_{ij}n_i$.

The variational problem is then:

Find a displacement rate $t \to \mathbf{u}(\mathbf{x}, t)$ and a hidden variable field $t \to \mathbf{z}(\mathbf{x}, t)$ such that

$$\begin{aligned} &\int_\Omega E_{ijkl}(\dot{u}_{k,l} - f_{k,l}(\sigma(\mathbf{u}, \mathbf{z}), \mathbf{z})v_{i,j}\, dx = \int_\Omega f_i v_i\, dx + \int_{\partial\Omega_2} T_i v_i\, ds, \qquad \forall v_i \in V, \\ &\int_\Omega (z_i - g_i(\sigma, \mathbf{z})w_i\, dx = 0, \qquad \forall w_i \in W, \end{aligned} \tag{5.3}$$

where V and W are appropriate spaces of test functions (see [3]).

For the calculations discussed in the next section, we use the Bodner constitutive equations for z:

$$\dot{\varepsilon}^p_{ij} = D_0 \frac{S_{ij}}{\sqrt{J_2}} \exp\left(-0.5\left(\frac{h_{ij}}{\zeta J_2}\right)^n \left(n + \frac{1}{n}\right)\right),$$

$$\dot{z}^P = S_{ij}\dot{\varepsilon}^p_{ij},$$

where S_{ij} = the deviator components of stress, J_2 the second invariant of S_{ij}, h_{ij} a hardness tensor, D_0 a scale factor, and n a material parameter.

The discrete form of (5.3) is obtained by using standard finite element techniques. Using standard notations, the discrete momentum equation assumes the form

$$\int_\Omega \mathbf{B}^T\mathbf{E}\mathbf{B}\Delta\mathbf{u}\, dx = \int_\Omega \mathbf{B}^T\mathbf{E}\Delta\varepsilon^p\, dx + \Delta\mathbf{F},$$

where $\Delta\mathbf{u}$ and $\Delta\varepsilon^p$ are increments in $\mathbf{u}$ and ε^p over a time interval Δt.

The integration of this system in time is accomplished by using the method of Bass [2]:

1. Initialize σ, z_i; set $t_n = 0$; and select Δt.
2. Estimate $\Delta\varepsilon^n$ over Δt (using an adequate predictor method).
3. Solve the preceding equilibrium equation for $\Delta\mathbf{u}_h$.
4. Calculate $\Delta\varepsilon = \mathbf{B}\Delta\mathbf{u}_h$.
5. Assume $\dot{e} = \Delta\varepsilon/\Delta T$ is constant over ΔT.
6. Subincrement and integrate σ, z_i accurately over Δt, neglecting the previous estimate of $\Delta\mathbf{e}^n$.
7. Calculate a new guess of $\Delta\varepsilon^n = \Delta\varepsilon - \mathbf{E}^{-1}\Delta\sigma$.
8. Check for convergence by comparison of $\Delta\varepsilon^n$ with $\Delta\varepsilon^n$; if a preassigned

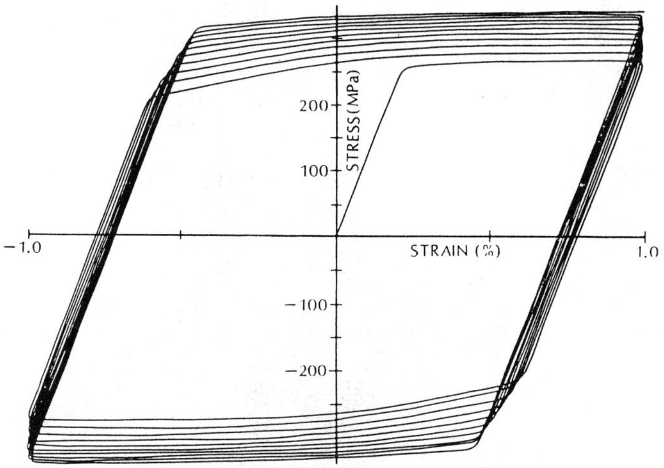

FIGURE 12.4. Cyclic stress–strain curves for titanium for ± 1 percent strain (Bodner's equations.)

tolerance is not met, set $\Delta\varepsilon^n = \Delta\varepsilon^n$ and go to (5.3); otherwise set $\Delta\sigma = \mathbf{E}(\Delta\varepsilon - \Delta\varepsilon^n)$.

9. If $t_n + \Delta T = T$, stop.

This computational method possesses the following desirable characteristics:

(i) A constant elastic stiffness matrix is used throughout the solution process;
(ii) different time steps at different integration points are allowed, and the overall ΔT is not restricted to be the smallest time step;
(iii) integration using subincrementation requires the coordinates and geometry of each element to be loaded much less frequently in the finite element simulation; and
(iv) by selecting a larger overall time step ΔT, fewer total time steps may be taken and thus fewer right-hand sides need be considered.

This algorithm is known to produce excellent results in cases of cyclic loading wherein virtually all other schemes fail. Figure 12.4 shows a stress history computed using this algorithm.

12.6. Numerical Experiments

We shall now present representative numerical results obtained by using the adaptive schemes and algorithms described earlier.

12.6.1. Supersonic Flow Interaction

The algorithms described previously were applied to a problem of supersonic rotor–stator interaction with $\alpha = 0.19$ and $\beta = 0.06$.

In order to capture shocks of variable strength, the "normalized" error

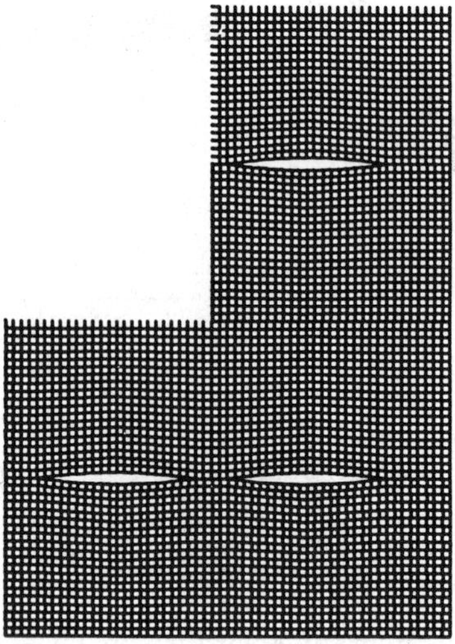

FIGURE 12.5. Initial mesh: uniform steady flow with initial data: $\rho = 1.4$ $m_1 = 4.2$, $m_2 = 0$, $\rho e = 8.8$.

indicators

$$\theta_e = h_e^3 \frac{1}{\rho} \sup_{\substack{i=1,2 \\ \mathbf{x} \in \Omega_e}} \left| \frac{\partial \rho}{\partial x_i}(\mathbf{x}) \right|$$

were used. It becomes clear from the numerical results that variable shocks are captured well and the mesh evolves dynamically to adapt to the solution of the rotor–stator problem.

Typical results are shown in Figures 12.5–12.11. The initial mesh is that shown in Figure 12.5. The first adaptive calculation for a steady-state condition, the adapted mesh, is shown in Figure 12.6, and the corresponding computed pressure contours are shown in Figure 12.7. Note the symmetry of the shock lines, the continuous pressure fields across the mesh interface, and the fact that both unrefinement and refinement of the mesh were required to achieve the accuracy limits specified. The rotor and stator blades are then allowed to move in relation to one another, and the mesh is dynamically refined. Plots are shown in Figures 12.8–12.11 of calculated adaptive meshes and pressure profiles for 1/8 and 1 cycle (period) of the motion, during which a rotor blade makes a complete revolution from its initial position in Figure 12.6 back to the same position.

Several features of the computed meshes and solutions are noteworthy. In the initial steady-state case, only 16 unrefined elements appear. The size of these large elements, indicating small local error, is limited in the present calculations by the distance from the tips of the rotor and stator blades: 2 elements in the present case since there must exist a sliding interface between

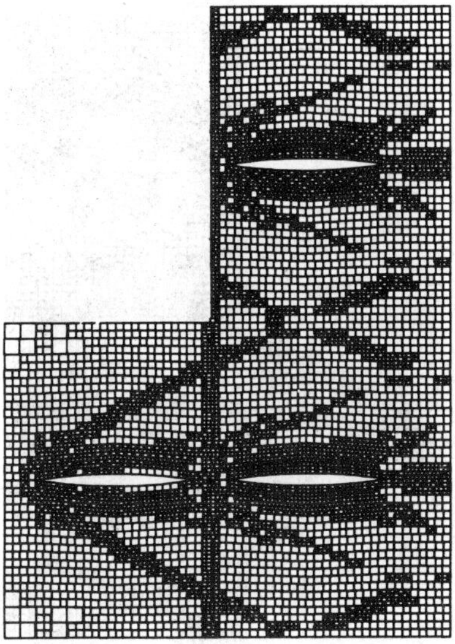

FIGURE 12.6. Initial adaptively refined mesh for steady flow through a rotor–stator configuration and the computed initial pressure contours.

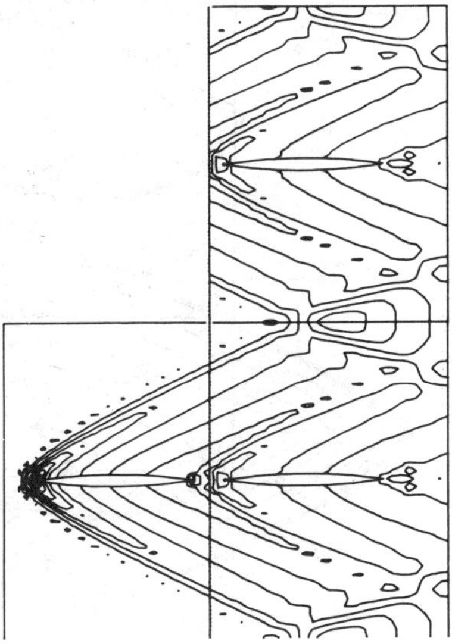

FIGURE 12.7. Computed pressure contours for steady-state case.

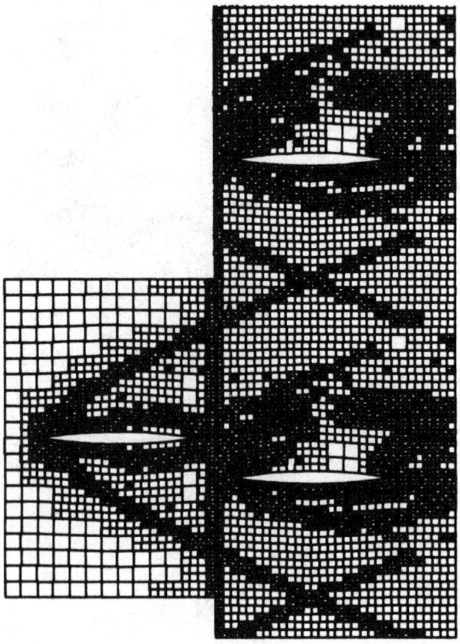

FIGURE 12.8. Computed mesh for preassigned local accuracy levels at 1/8 of a rotor cycle and corresponding pressure contours.

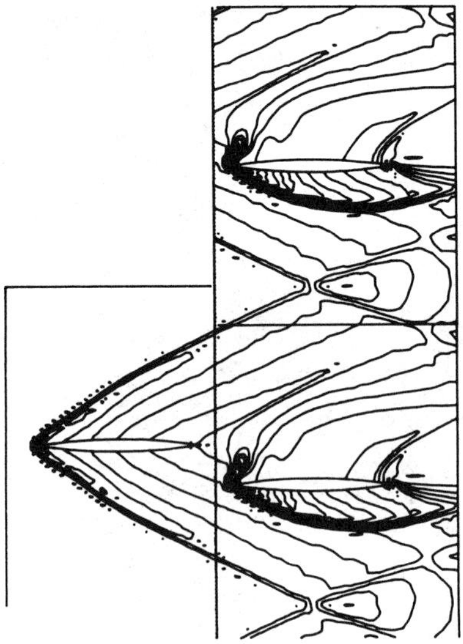

FIGURE 12.9. Pressure contours for 1/8 cycle.

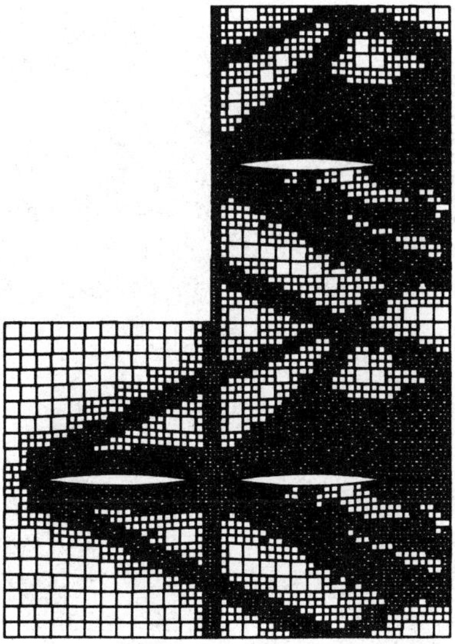

FIGURE 12.10. Adapted mesh at a full cycle.

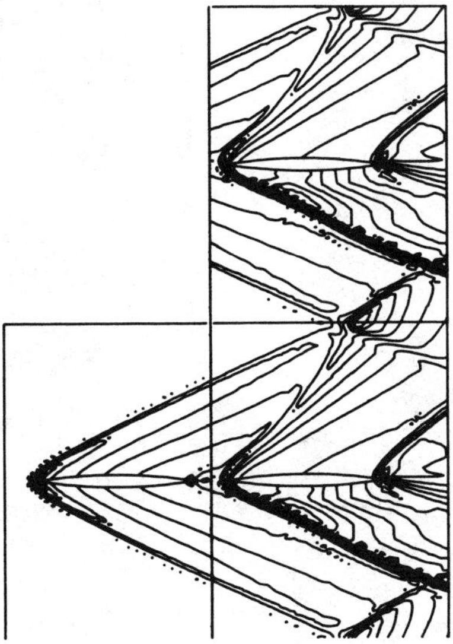

FIGURE 12.11. Computed pressure profiles at a full cycle.

them. A minor program modification could allow much larger elements in regions of small error. For the transient case, the number of larger elements (indicating substantial unrefinement) increases, and these regions of low error migrate over the mesh as the solution evolves in time. Conversely, substantial refinement of the mesh is indicated at the interface and along shock lines. The method successfully captures shock interactions and the increasing density of pressure profiles downstream from the moving blades. The ratio of the number of elements in the adaptive mesh to that in the uniform fine mesh varies in time but is typically 4,000/12,500, a reduction of 68 percent! The initial coarse mesh contains around 4000 cells and is incapable of delivering the required accuracy, a fact not easily recognized until after a lengthy calculation is made.

12.6.2. Viscoplasticity

The adaptive scheme outlined earlier has been implemented in a computer code that uses the time-marching scheme for rate-dependent plasticity outlined previously (Bass's method). The parameters α and β used in the refinement strategy were selected to be 0.75 and 0.50, respectively. Thus, only those elements with estimated error magnitudes greater than or equal to 75 percent of the maximum allowable error were refined, and no refinement occurred when more than 50 percent of the total number of elements had on error greater than 75 percent of the maximum error.

In these calculations, five different error estimations were used.

Indicator 1. Integral of the jump in displacement gradients

$$\theta_e^2 = CA_e \int_{\partial\Omega_e} \left[\left[\left[\frac{\partial \Delta_m u_1}{\partial n} \right]\right]^2 + \left[\left[\frac{\partial \Delta_m u_2}{\partial n} \right]\right]^2 \right] dS.$$

Indicator 2. Interpolated values

$$\theta_e^2 = CA_e \int_{\partial\Omega_e} (\Delta_m u_{1,11}^2 + \Delta_m u_{1,22}^2 + \Delta_m u_{2,11}^2 + \Delta_m u_{2,22}^2)\, d\Omega.$$

Indicator 3. Integral of stress jumps over the boundary

$$\theta_e^2 = CA_e \int_{\partial\Omega_e} ([[\sigma_1]]^2 + [[\sigma_2]]^2 + [[\sigma_{12}]]^2)\, dS.$$

Indicator 4. Maximum difference in the effective strain

$$\theta_e^2 = CA_e \Delta\varepsilon_{\max}^2,$$

where $\Delta\varepsilon_{\max}$ is the maximum difference of the increment in effective strain between Ω_e and surrounding elements over the mth step.

Indicator 5. Integral of the effective strain

$$\theta_e^2 = CA_e \int_{\Omega_e} \Delta\varepsilon^2\, d\Omega_e,$$

where $\Delta\varepsilon$ is the increment of the effective strain over the mth step. In these equations, A_e is the area of element Ω_e.

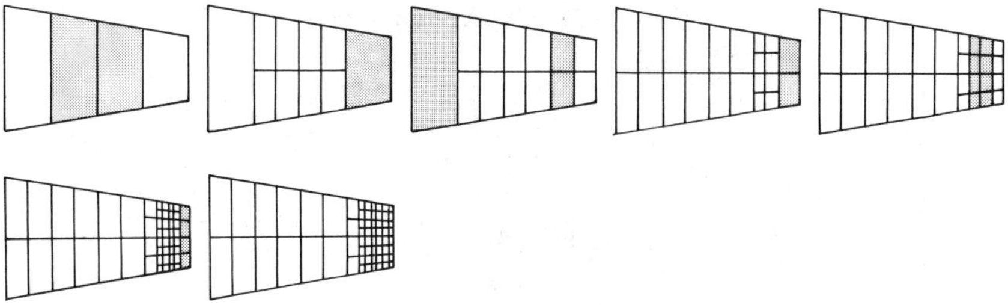

FIGURE 12.12. Adaptive refinement sequence, for a tapered bar, with local errors estimated by estimator 1.

We first cite an example discussed in [3]: a titanium bar 10 centimeters (cm) long and 2 cm wide at the thick end is linearly tapered at a rate of 5 percent per centimeter. The bar is simply supported at the end with the largest cross section at the opposite and is uniformly displaced at a rate of 0.004 cm/s for a period of 10 seconds (s). This loading produces a stress concentration in the slender region so that part of the bar behaves in a viscoplastic manner and the remainder behaves almost elastically.

The refinement times selected include 0.1, 1.0, 2.5, 5.0, 7.5, and 10 s. These times were chosen so that the first refinement occurred while the material still behaved "elastically," and the remaining times were selected to divide the loading period evenly.

Successive mesh refinements calculated by using Estimator 1 are shown in Figure 12.12. As expected, a fine mesh is generated at the end of the bar as the stress levels there hit peak values. Similar mesh sequences were obtained by using Indicators 2, 3, and 4, with only minor differences in refinement sequences occurring over this time interval. The use of Estimator 5, the integral of the effective strain, gave a quite different sequence of meshes, as indicated in Figure 12.13. This error indicator proved to be very sensitive to local changes in the inelastic deformation and, therefore, may be very useful in monitoring delicate changes in solutions common in viscoplasticity problems. Computed stress contours at $t = 0$ s obtained with the various estimators are

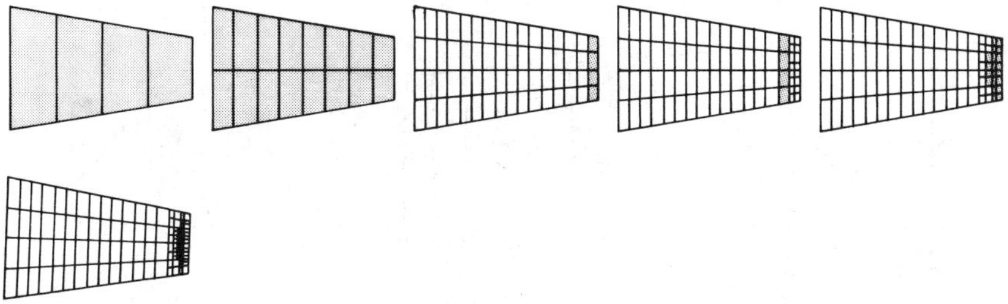

FIGURE 12.13. Adaptive refinement sequence, for a tapered bar, obtained by using Estimator 6.

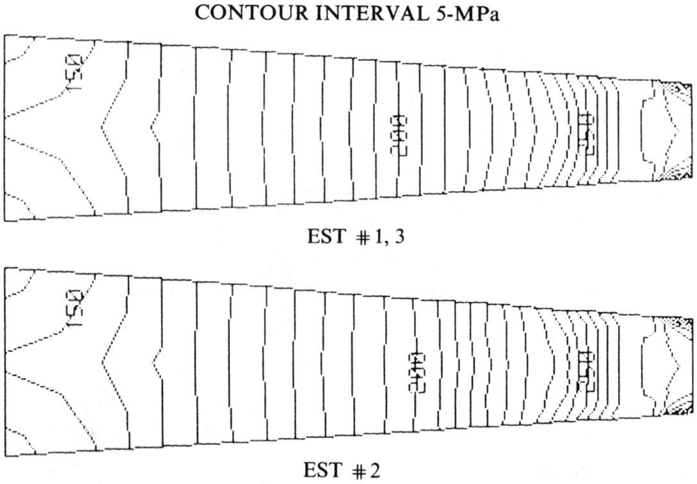

FIGURE 12.14. Sigma-x contours at 10 s for a tapered bar, obtained with Estimators 1–3.

shown in Figures 12.14 and 12.15. Estimators 4 and 5 were sensitive enough to capture the intricate stress distribution near the tip of the bar, even though slightly different results were obtained for each. The computed distribution of equivalent nonelastic strain obtained with Estimator 5 is shown in Figure 12.16. There are the very complex distribution of inelastic strains can be seen.

Results of other numerical experiments are shown in Figures 12.17–12.21.

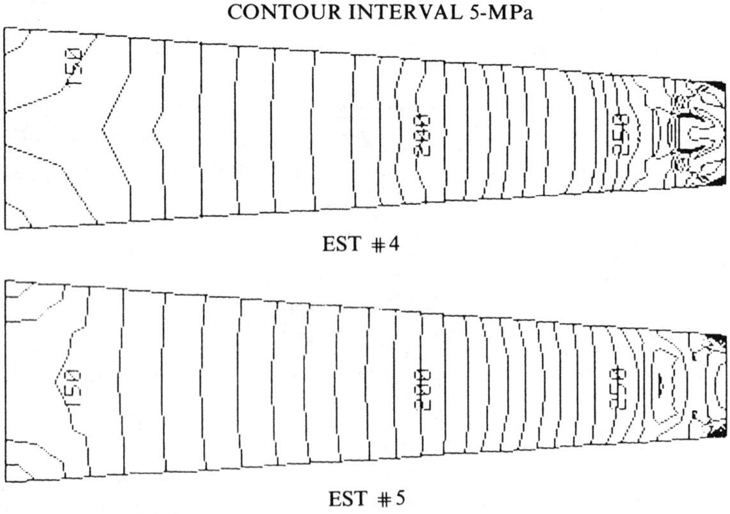

FIGURE 12.15. Sigma-x stress contours at 10 s, obtained by using Estimator 5.

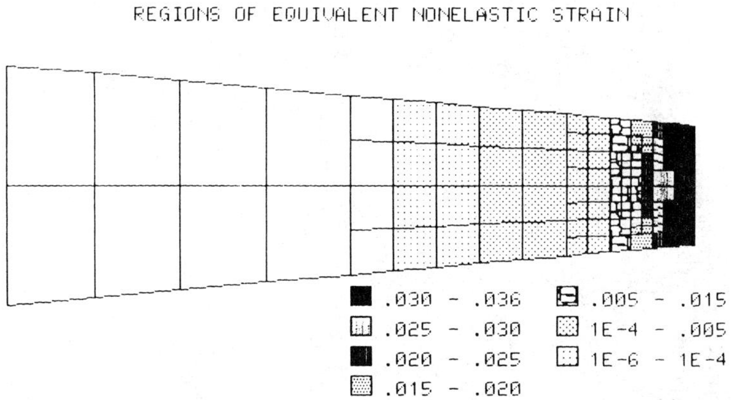

FIGURE 12.16. Regions of equivalent nonelastic strain at $t = 10$ s, obtained by using Estimator 6.

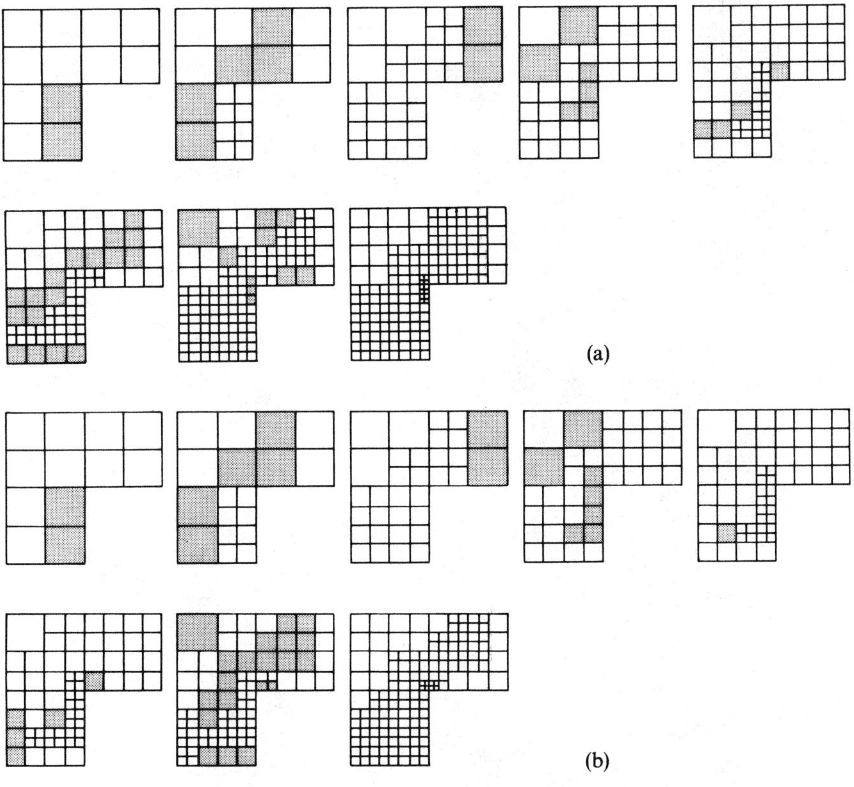

FIGURE 12.17. Adaptive refinement sequence for an L-shaped domain with local errors approximated by using error Estimator 1; (a) linear elastic case (b) viscoelastic case.

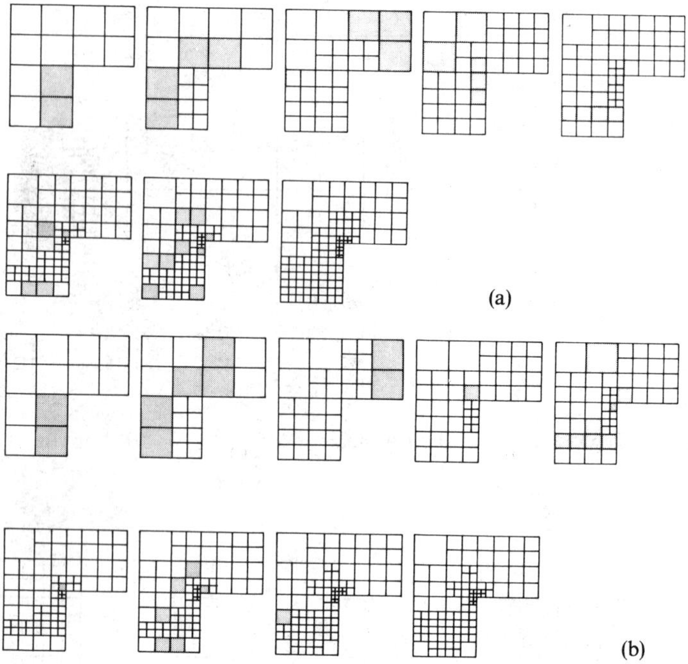

(a)

(b)

FIGURE 12.18. Adaptive refinement sequence for an L-shaped domain with local errors approximated by using error Estimator 2; (a) linear elastic case (b) viscoelastic case.

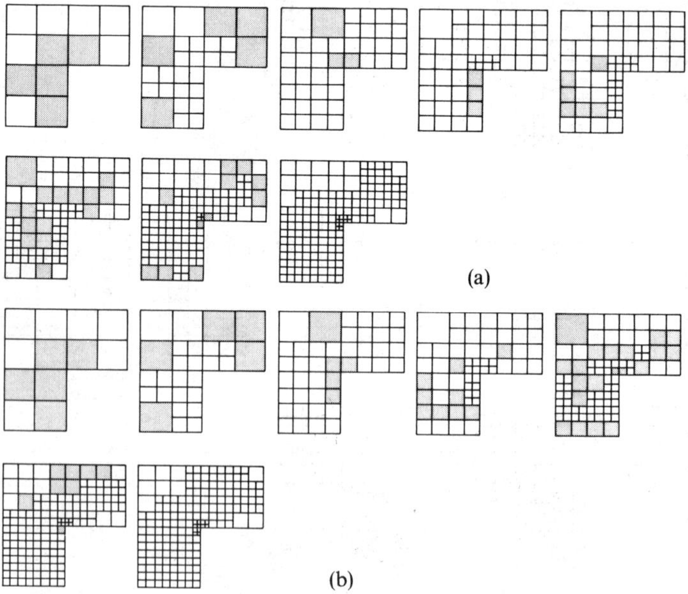

(a)

(b)

FIGURE 12.19. Adaptive refinement sequence for an L-shaped domain with local errors approximated by using error Estimator 3; (a) linear elastic case (b) viscoelastic case.

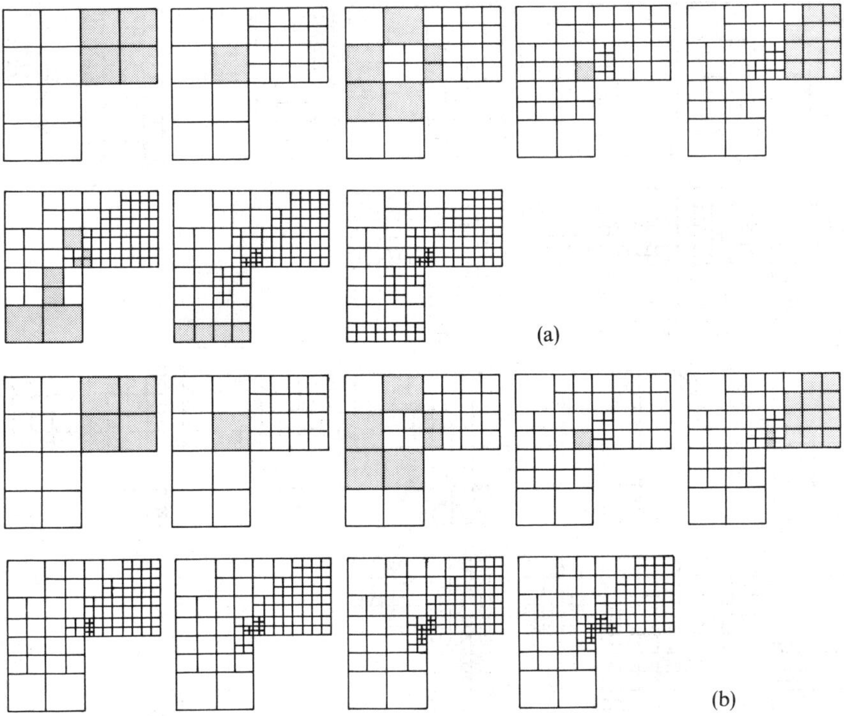

FIGURE 12.20. Adaptive refinement sequence for an L-shaped domain with local errors approximated by using error Estimator 1; (a) linear elastic case (b) viscoelastic case.

These figures depict mesh sequences obtained by stretching a titanium L-shaped plate subjected to a uniform traction along one side that jumps from 0 to 10 megapascals (MPa) over a small time step and is then linearly increased to 50 MPa over a 10-s time interval. The plate is supported and loaded as indicated in Figure 12.16 and suffers several inelastic deformations near the reentrant corner. Refinement times are chosen at 0.1, 1.0, 2.5, 5.0, 7.5, and 10.0 s.

Computed mesh sequences are shown in Figures 12.17–12.21 with results of the perfectly elastic case compared with the viscoplasticity case (for the same estimator). Note the similarity between "elastic and inelastic" mesh sequences. This, of course, is due to the local character of the inelastic response in this problem, the high stress rates being confined to regions near the reentrant corner. Substantially different results are obtained if cycle loading is considered, and such calculations are currently under study. Note again that Estimator 5 is apparently sensitive to changes in stress and inelastic strain and yields the finest mesh in the neighborhood of the singularity.

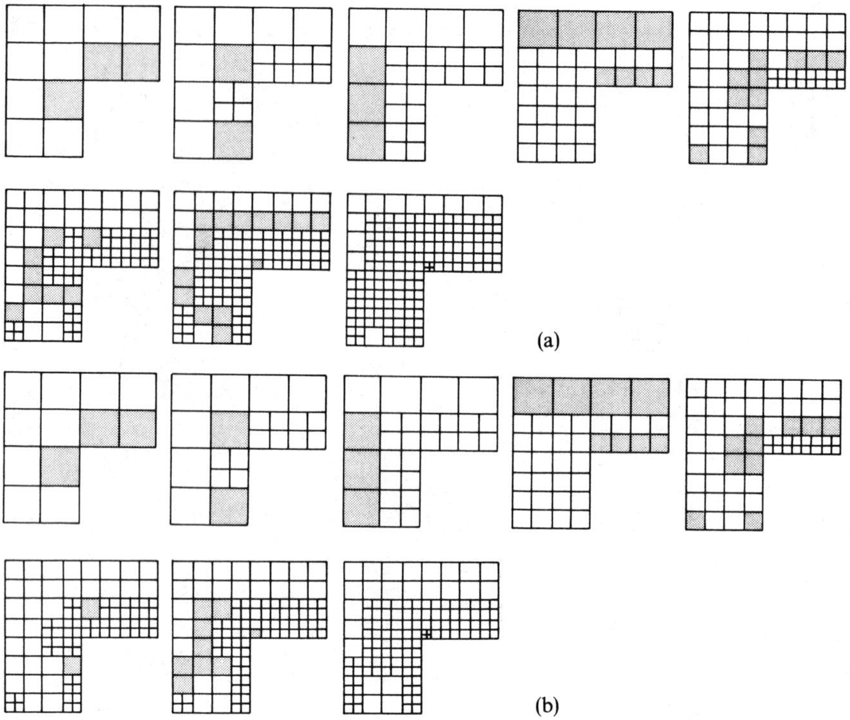

FIGURE 12.21. Adaptive refinement sequence for an L-shaped domain with local errors approximated by using error Estimator 5; (a) linear elastic case (b) viscoelastic case.

Acknowledgment. This work has been done in collaboration with Jon M. Bass, T. Strouboulis, and P. Devloo, colleagues and former students; also, early collaborative work with L. Demkowicz was helpful in developing various approaches to adaptive finite element methods. Some of the results cited here are extracted from our reports and papers, where more detailed accounts can be found. See, in particular, [1–8].

The research reported here was supported by the U.S. Office of Naval Research (particularly, the work on stress estimation and adaptive methods for viscoplasticity problems) under Contract N00014-84-K-0409 and by NASA (particularly, the adaptive methods for supersonic flow and flow interaction), under Contract NAS3-24849 with the Lewis Research Center. Joint work on adaptive methods for subsonic and supersonic flow problems has also been done with Lawrence W. Spradley and his associates at Lockheed Missile and Space Co., Huntsville, Alabama, under Contract NAS8-36647 with the NASA Langley Research Center.

References

1. J.T. Oden, T. Strouboulis, and P. Devloo, "Adaptive finite element methods for the analysis of inviscid compressible flow, I: Fast refinement/unrefinement and moving mesh methods for unstructured meshes," *Comput. Methods Appl. Mech. Engrg.*, **59**, No. 3 (1986), 327–362.
2. J.M. Bass, "Numerical implementation of constitutive models in rate-dependent plasticity," Ph.D. dissertation, Univ. of Texas, Austin, 1985.
3. J.M. Bass and J.T. Oden, "Adaptive finite element methods for a class of evolution problems in viscoplasticity," *Internat. J. Engrg. Sci.*, **25**, No. 6 (1987), 623–653.
4. J.T. Oden, T. Strouboulis, and P. Devloo, "Adaptive finite element methods for compressible flow problems," in *Numerical Methods for Compressible Flows—Finite Difference, Element and Volume Techniques* (T.E. Tezduyor and T.J.R. Hughes, eds.), AMD Vol. 78, ASME, New York, 1987, pp. 115–126.
5. L. Demkowicz and J.T. Oden, "An adaptive characteristic Petrov–Galerkin finite element method for linear and nonlinear convection-dominated diffusion problems in one space variable," *J. Comput. Physics*, **68**, No. 1, (1986), 188–273.
6. T. Strouboulis, P. Devloo, and J.T. Oden, "A moving-grid finite element algorithm for supersonic flow interaction between moving bodies," *Comput. Methods Appl. Mech. Engrg.*, **59**, No. 2, (1986), 235–255.
7. J.T. Oden and L. Demkowicz, "Advances in adaptive improvements," in *State of the Art Surveys in Computational Mechanics*, Special Publications, American Society of Mechanical Engineers, New York, to appear.
8. J.T. Oden, L. Demkowicz, T. Strouboulis, and P. Devloo, "Adaptive methods for problems in solid and fluid mechanics," in *Accuracy Estimates and Adaptive Refinements in Finite Element Computations*, (I. Babuska, O.C. Zienkiewicz, J. Gago, and E.R. Arantes Oliveira, eds.), Wiley, Chichester, 1986.

CHAPTER 13

Finite Element Methodology for Integrated Flow-Thermal-Structural Analysis

Earl A. Thornton and Pramote Dechaumphai

13.1. Introduction

Thermal deformations and stresses induced by aerodynamic heating are important considerations in the design of hypersonic flight vehicles. Aerodynamic heating has a significant effect on the performance of the structure, and effective techniques for predicting the heating rates and thermal-structural response are required. In the past, the heating and thermal-structural analyses have typically been uncoupled. Structural heating rates have been computed for viscous flows over undeformable aerodynamics surfaces with prescribed thermal boundary conditions. In realistic high speed flows, aerodynamic surfaces deform as a result of the thermal loads, and thermal conditions are not constant at the fluid-structure interface. During the aerodynamic heating of a structure the fluid-structure interface deforms, and the structure continuously absorbs thermal energy from the flow. These interaction effects may in fact alter the heating rates to the structure. In [1–2], experiments and analyses conducted at the NASA Langley Research Center show that actual heating rates experienced by metallic thermal protection systems can be significantly higher than rates based on undeformed flat plate computations. For example, for deformed heights less than the boundary layer thickness, heating rates can be increased by as much as 40 percent; moreover, heating rates increase rapidly when deformed heights exceed the boundary layer thickness.

The purpose of this paper is to describe initial efforts aimed toward the development of finite element methodology that will couple the flow-thermal-structural analyses for high speed flows over flight structures to include fluid-structure interaction effects. The computational approach is highlighted, and initial results from a coupled flow, thermal, and structural analysis for a Mach 3 flow over a thin panel are presented.

13.2. Problem Formulation

Consider an unsteady viscous flow over the thin panel shown in Figure 13.1. The fluid flow is described by the conservation of mass, momentum, and energy equations for a laminar flow of a perfect gas, assuming temperature-dependent viscosity and thermal conductivity. The thermal behavior of the panel is described by the conservation of energy equation, and the structural behavior is described by the quasi-static equations of motion for a solid. For simplicity in these initial analyses, radiation heat transfer from the panel is neglected, and the panel is assumed to respond elastically. Thermal and elastic properties of the panel are assumed to be constant, i.e., independent of temperature. The equations solved are written in conservation form.

Fluid

$$\frac{\partial\{U\}}{\partial t} + \frac{\partial\{E_\mathrm{I} - E_\mathrm{V}\}}{\partial x} + \frac{\partial\{F_\mathrm{I} - F_\mathrm{V}\}}{\partial y} = 0, \tag{2.1}$$

where $\{U\}$ is a vector of the conservation variables for the fluid, $\{E_\mathrm{I}\}$ and $\{F_\mathrm{I}\}$ are inviscid flux components, and $\{E_\mathrm{V}\}$ and $\{F_\mathrm{V}\}$ are viscous flux components. These vectors are given by

$$\begin{aligned}
\{U\}^\mathrm{T} &= [\rho \quad \rho u \quad \rho v \quad \rho\varepsilon],\\
\{E_\mathrm{I}\}^\mathrm{T} &= [\rho u \quad \rho u^2 + p \quad \rho uv \quad (\rho\varepsilon + p)u],\\
\{F_\mathrm{I}\}^\mathrm{T} &= [\rho v \quad \rho vu \quad \rho v^2 + p \quad (\rho\varepsilon + p)v],\\
\{E_\mathrm{V}\}^\mathrm{T} &= [0 \quad \sigma_{xx} \quad \tau_{xy} \quad (u\sigma_{xx} + v\tau_{xy} - q_x)],\\
\{F_\mathrm{V}\}^\mathrm{T} &= [0 \quad \tau_{xy} \quad \sigma_{yy} \quad (u\tau_{xy} + v\sigma_{yy} - q_y)],
\end{aligned} \tag{2.2}$$

where ρ is the fluid density, u and v are velocity components; ε is the total

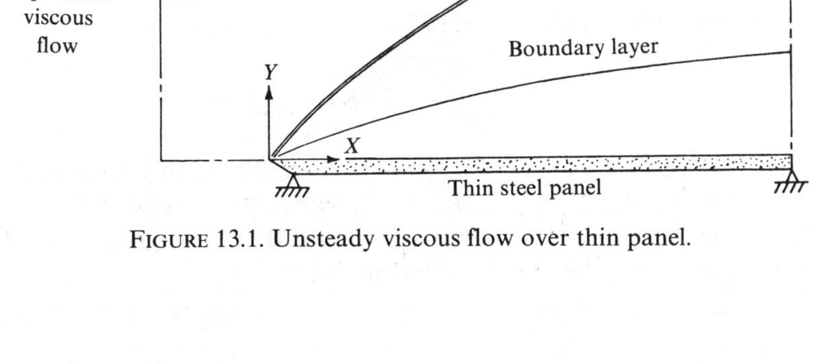

FIGURE 13.1. Unsteady viscous flow over thin panel.

energy; σ_{xx}, σ_{yy}, and τ_{xy} are viscous stress components; and q_x, q_y are heat fluxes. The superscript T denotes matrix transpose. In the inviscid flux components, the pressure p is related to the total energy for a perfect gas, assuming a constant ratio of specific heats. The stress components are related to the velocity gradients assuming the Stokes hypothesis, and the heat fluxes are related to the temperature gradients by Fourier's law. The temperature-dependent viscosity is computed from Sutherland's law, and the thermal conductivity is computed by assuming a Prandtl number equal to 0.72.

Solid

$$\frac{\partial\{U\}}{\partial t} + \frac{\partial\{E\}}{\partial x} + \frac{\partial\{F\}}{\partial y} = 0, \tag{2.3}$$

where $\{U\}$ is a vector of the conservation variables for the solid, and $\{E\}$ and $\{F\}$ are "flux" components for the solid. These vectors are given by

$$\begin{aligned} \{U\}^{\mathrm{T}} &= [cu \quad cv \quad \rho c_v T], \\ \{E\}^{\mathrm{T}} &= [-\sigma_x \quad -\tau_{xy} \quad q_x], \\ \{F\}^{\mathrm{T}} &= [-\tau_{xy} \quad -\sigma_y \quad q_y], \end{aligned} \tag{2.4}$$

where u and v are displacement component, ρ is the solid density, c_v is the specific heat, and T is the temperature. The coefficient c of the displacement components is a fictitious damping constant that is used to facilitate time marching to a steady-state quasi-static solution. The solid's stress components σ_x, τ_{xy}, and σ_y are related to the displacement gradients and temperature using generalized Hooke's law, and the solid's heat fluxes q_x and q_y are related to the temperature gradients by Fourier's law.

The sets of equations (2.1) and (2.3) are solved subject to appropriate boundary, interface, and initial conditions for the fluid and solid. Flow boundary conditions correspond to supersonic flow; hence all conservation variables are specified on in-flow surfaces. On supersonic outflow surfaces, the finite element formulation provides appropriate natural boundary conditions. Solid boundary conditions include either specified displacements or surface tractions. At the fluid–solid interface, flow velocity components are zero, fluid–solid temperatures are continuous, and heat fluxes normal to the interface are continuous. The fluid–solid temperature continuity condition is imposed explicitly in the solution algorithm, but continuity of the heat fluxes is assured implicitly by the finite element formulation. Initially, the flow is assumed to be uniform at free stream values, and the solid is assumed to be undeformed and unstressed at room temperature.

13.3. Solution Approach

The solution approach solves equations (2.1) and (2.3) using an explicit time marching scheme, the *Taylor–Galerkin algorithm*, that has been described in [3–6]. Herein, the essential features of the algorithm are highlighted, and the solution sequence for the coupled problem is presented.

13.3.1. Taylor–Galerkin Algorithm

The basic concept of the Taylor–Galerkin algorithm is to use:

(1) Taylor series expansions in time to establish recurrence relations for time marching; and
(2) the method of weighted residuals with Galerkin's criteria to develop the finite element matrix equations describing the spatial distribution of the dependent variables.

For simplicity, the finite element formulation will be given for a single scalar equation,

$$\frac{\partial u}{\partial t} + \frac{\partial E}{\partial x} + \frac{\partial F}{\partial y} = 0, \tag{3.1}$$

where the variables u, E, and F are analogous to the corresponding vector quantities in equation (2.1). Let $\{u\}^n$ denote the element nodal values of the flow variables $u(x, y, t)$ at time t_n. The time step Δt spans two typical time t_n and t_{n+1} in the transient response. The computation proceeds through two time levels, $t_{n+1/2}$ and t_{n+1}. At time level $t_{n+1/2}$, values for u that are constant within each element are computed explicitly. At time level t_{n+1}, the constant element values computed at the first time level are used to compute nodal values for u. In the time level t_{n+1} computations, element contributions are assembled to yield the global equations for nodal unknowns. The "mass" matrix for the resulting equations is approximately diagonalized to yield an explicit algorithm.

13.3.2. Time Level $t_{n+1/2}$

The constant element value $U_D^{n+1/2}$ is computed from

$$AU_D^{n+1/2} = \int_A [N]\, dA\{u\}^n - \frac{\Delta t}{2}\int_A \left[\frac{\partial N}{\partial X}\right] dA\{E\}^n - \frac{\Delta t}{2}\int_A \left[\frac{\partial N}{\partial y}\right] dA\{F\}^n, \tag{3.2}$$

where A denotes an element's area, Δt is the time step, and $[N]$ is a matrix of interpolation functions. Quantities in braces, such as $\{E\}^n$, represent nodal values at time t_n.

13.3.3. Time Level t_{n+1}

The nodal values for a single element are given by

$$[M]\{u\}^{n+1} = [M]\{u\}^n + \Delta t \int_A \left\{\frac{\partial N}{\partial x}\right\} dAE(t_{n+1/2}) + \Delta t \int_A \left\{\frac{\partial N}{\partial y}\right\} dAF(t_{n+1/2}) + \{R\}^{n+1/2}, \tag{3.3}$$

where $\{R\}^{n+1/2}$ represents known boundary terms. The matrix $[M]$ is the consistent mass matrix; to produce an explicit algorithm, $[M]$ is diagonalized. Following usual finite element procedures, the element matrices given in equation (3.3) are then assembled to form system equations.

The two time level Taylor–Galerkin algorithm is conditionally stable, and the time step must satisfy stability requirements based on the CFL condition and the Reynolds number. Artificial viscosity, the form due to Lapidus, is used to reduce oscillations. Further details of the algorithm are discussed in [3–6].

13.3.4. Solution Sequence

Experimental and computational results for flows over panels show that the flow initially approaches steady state in a much shorter time than required for the panel to respond thermally and structurally. Typically, heating rates approach steady state in a few milliseconds. A panel at this time remains virtually undeformed at temperatures only slightly higher than the initial temperature. After a few seconds, panel temperatures begin to rise appreciably, and significant deformations occur. During this time, thermal and deformation coupling effects alter the flow field. The coupling effects continue to alter the flow and panel behavior until the panel reaches thermal equilibrium.

On the basis of this sequence of events, the solution sequence shown in Figure 13.2 has been adopted. The solution sequence alternates between coupled flow and thermal analyses of the fluid/panel and thermal analysis of the panel. At selected times, temperatures of the panel are used to compute structural deformations. If significant, the structural deformations may be used to define a new flow boundary.

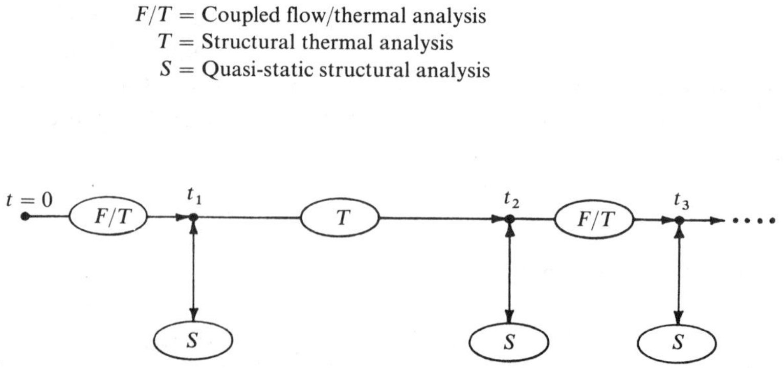

FIGURE 13.2. Solution sequence for coupled flow-thermal-structural analysis.

13.4. Numerical Example

A coupled flow-thermal-structural analysis was performed for a Mach 3 flow over a 0.1-inch-thick by 4-inch-long steel panel. The Mach 3 flow was selected because it had been previously analyzed in [5], and the finite element solution was shown to be in good agreement with published finite difference results by Carter [7]. The finite element mesh for the fluid and panel are shown in Figure 13.3. A relatively coarse 35 × 31 mesh is used for the flow, and only one element is used through the thickness of the panel. The analysis sequence is shown in Table 13.1. The table shows that initially a coupled flow/thermal analysis was performed for 0.010 s until the flow approached steady state. Then the heating rate distribution at the flow-panel interface was applied to the panel, and the panel was analyzed thermally for 600 s. Subsequently, a coupled flow/thermal analysis was performed for 0.005 s until the flow approached a new steady-state condition. Then the resulting panel temperatures were used as input to a structural analysis of the panel. The results of the analyses are shown in Figures 13.4–13.10.

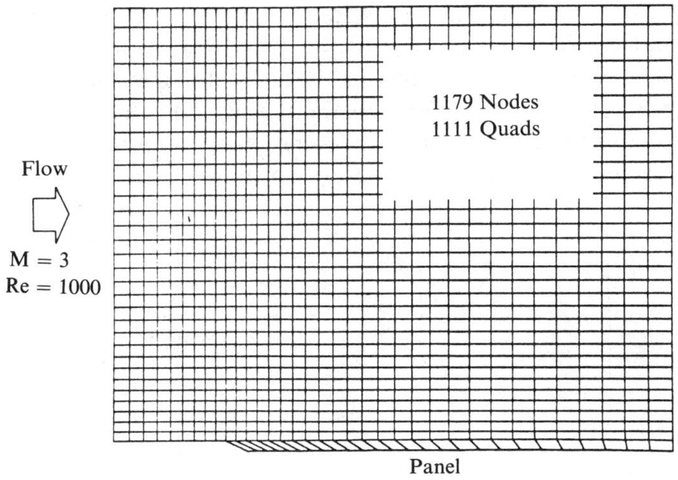

FIGURE 13.3. Finite element model for analysis of thin panel.

TABLE 13.1. Analysis sequence for Mach 3 flow over panel.

Analysis	Time duration, s	Time step, s
Coupled flow/thermal	$0.000 < t < 0.010$	10^{-6}
Panel thermal	$0.010 < t < 600.010$	10^{-1}
Coupled flow/thermal	$600.010 < t < 600.015$	10^{-6}

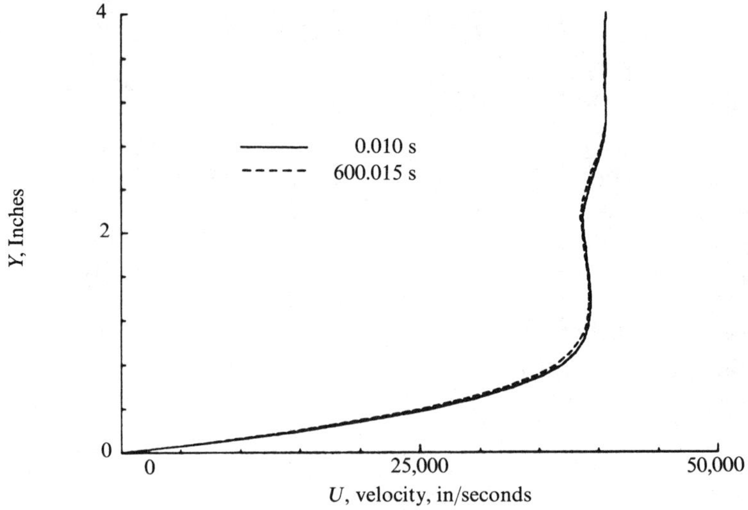

FIGURE 13.4. Flow exist u velocity profiles at two times.

Figures 13.4–13.6 show thermal interaction effects for the exist velocity, density, and temperature profiles of the flow. Results are shown for $t =$ 0.010 s and $t =$ 600.015 s. The thermal interaction alters the velocity profile only slightly but has more effect on the density and temperature profiles. The temperature profiles (Figure 13.6) are changed most because the surface temperature of the panel rises from an initial value of 530 °R to over

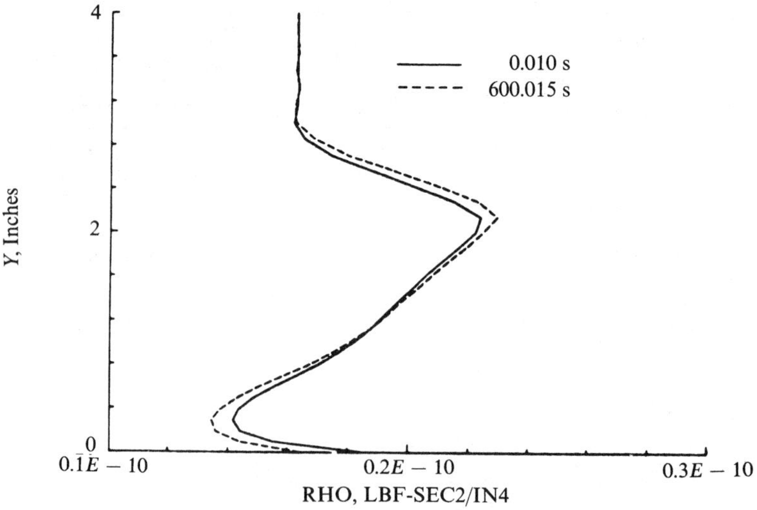

FIGURE 13.5. Flow exist density profiles at two times.

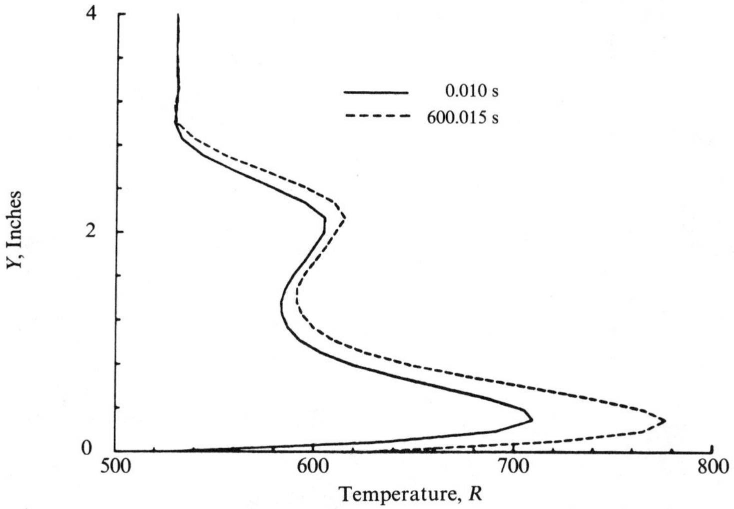

FIGURE 13.6. Flow exist temperature profiles at two times.

600 °R during the analysis. Figure 13.7 shows the rise in the flow/panel interface temperature distribution at four times during the response. Figure 13.8 shows the thermal interaction effects on the panel heating rates, and Figure 13.9 shows the interaction effects on the panel pressure distribution. The heating rates are altered only slightly, but the pressure distribution is increased significantly near the panel leading edge. However, the pressure

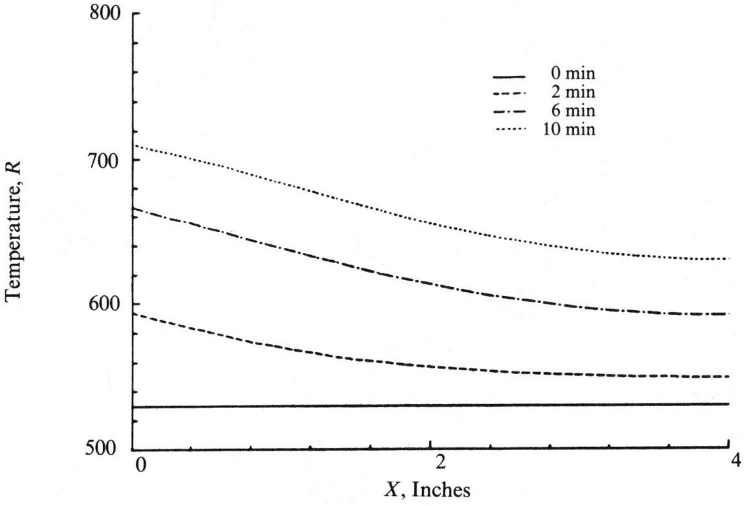

FIGURE 13.7. Flow/panel interface temperature distributions.

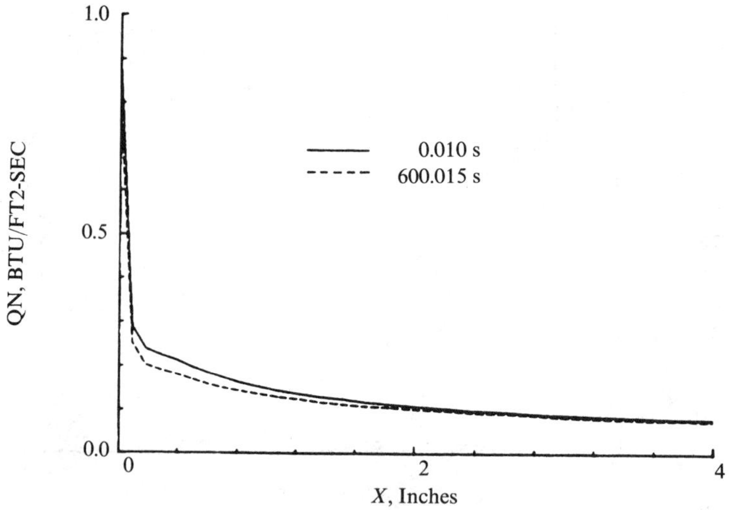

FIGURE 13.8. Panel heating rate distributions at two times.

interaction effects are not important to the panel structural response since the pressure loading is insignificant in comparison to the thermal loading on the panel. The structural analysis shows that the panel bowed up into the flow only a very small amount. Maximum upward panel deformations are less than 0.001 inch. Panel thermal stresses (Figure 13.10) are relatively high because the panel is constrained at its ends to restrict lengthwise thermal expansion.

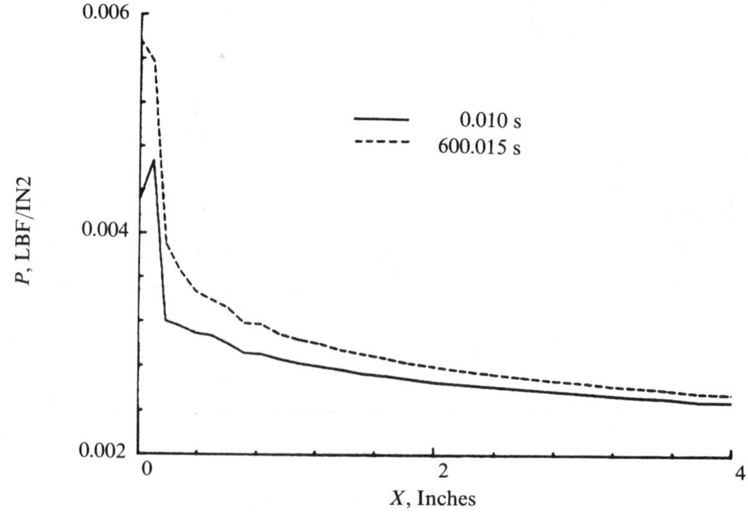

FIGURE 13.9. Panel pressure distributions at two times.

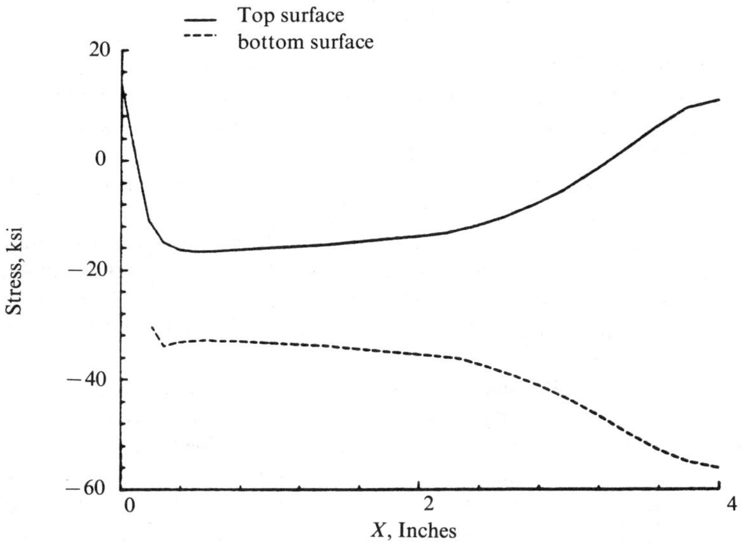

FIGURE 13.10. Panel thermal stress distributions.

For this example, the coupled analyses show that thermal interaction effects between the flow and panel are relatively modest, and deformation effects are negligible. Tests conducted at NASA Langley for Mach 7 flows show for the heating rates encountered at higher Mach numbers that thermal and deformation interactions between the flow and panel are appreciable. This example has demonstrated initial efforts of development of a coupled analysis approach for future applications to the more realistic problems encountered at high Mach numbers.

13.5. Concluding Remarks

Recent progress in the development of finite element methodology for integrating flow, thermal, and structural analysis is described. The finite element solution algorithm is highlighted, and the approach for coupling the analyses is described. Initial results for a Mach 3 flow over a thin panel are presented. For this example, flow/panel thermal interactions are modest, and structural deformation effects on the flow field are negligible. The research continues, and the approach is currently being applied to higher Mach number flows over panels where coupling effects are known to be more significant.

Acknowledgment. We are pleased to acknowledge the continuing support of the NASA Langley Aerothermal Loads Branch and the advice the encouragement of our technical monitor, Allan R. Wieting.

References

1. G.C. Olsen and R.E. Smith, "Aerothermal loads analysis for high speed flow over a quilted surface configuration," presented at *The AIAA 17th Fluid Dynamics, Plasma Dynamics and Lasers Conference* (*Snowmass, CO, June* 25–27, 1984), AIAA-84-1630, AIAA, New York, 1984.
2. G.C. Olsen and R.E. Smith, "Analysis of aerothermal loads on spherical dome proturberances," *AIAA J.*, **23**, No. 5 (1985), 650–656.
3. K.S. Bey, E.A. Thornton, P. Dechaumphai, and R. Ramakrishnan, "A new finite element approach for prediction of aerothermal loads—progress in inviscid flow computations," presented at *The AIAA 7th Computational Fluids Dynamics Conference* (*Cincinnati, OH, July* 15–17, 1985), AIAA 85-1533, AIAA, New York. Also NASA TM 86434.
4. E.A. Thornton and R. Ramakrishnan, "A finite element approach for solution of the 3D Euler equations," presented at *The AIAA 24th Aerospace Sciences Meeting* (*Reno, NV, January* 6–8, 1986), AIAA-86-0106, AIAA, New York, 1986.
5. E.A. Thornton, P. Dechaumphai, and G. Vermaganti, "A finite element approach for prediction of aerothermal loads," presented at *The AIAA/ASME 4th Joint Fluid Mechanics, Plasma Dynamics and Lasers Conference* (*May* 12–14, 1986), AIAA-86-1050, AIAA, New York, 1986.
6. E.A. Thornton and P. Dechaumphai, "Taylor–Galerkin finite element algorithm for transient nonlinear thermal-structural analysis," presented at *The AIAA/ASME/ASCE/AHS 27th Structures, Structural Dynamics and Materials Conference* (*May* 19–21, 1986), AIAA 86-0911 CP, AIAA, New York, 1986.
7. J.E. Carter, "Numerical solutions of the Navier–Stokes equations for the supersonic laminar flow over a two-dimensional compression corner," NASA TR R-385, July, 1972.